THE AGE OF SMOKE

History of the Urban Environment

Martin V. Melosi and Joel A. Tarr, Editors

The Age of Smoke

ENVIRONMENTAL POLICY IN GERMANY
AND THE UNITED STATES, 1880-1970

Frank Uekoetter

University of Pittsburgh Press

Published by the University of Pittsburgh Press, Pittsburgh, Pa., 15260
Copyright © 2009, University of Pittsburgh Press
All rights reserved
Manufactured in the United States of America
Printed on acid-free paper
10 9 8 7 6 5 4 3 2 1
ISBN 13: 978-0-8229-4364-8
ISBN 10: 0-8229-4364-6

Library of Congress Cataloging-in-Publication Data

Uekötter, Frank, 1970-
  [Von der Rauchplage zur ökologischen Revolution. English]
  The age of smoke : environmental policy in Germany and the United States, 1880-1970 / Frank Uekoetter.
      p. ; cm. — (History of the urban environment)
  Includes bibliographical references and index.
  ISBN-13: 978-0-8229-4364-8 (cloth : alk. paper)
  ISBN-10: 0-8229-4364-6 (cloth : alk. paper)
  ISBN-13: 978-0-8229-6012-6 (pbk. : alk. paper)
  ISBN-10: 0-8229-6012-5 (pbk. : alk. paper)
  1. Air—Pollution—Germany—History. 2. Air—Pollution—United States—History. 3. Air quality management—Germany—History. 4. Air quality management—United States—History. 5. Smoke prevention—Germany—History. 6. Smoke prevention—United States—History. 7. Environmental policy—Germany—History. 8. Environmental policy—United States—History. I. Title. II. Series.
  [DNLM: 1. Air Pollution—prevention & control—Germany. 2. Air Pollution—prevention & control—United States. 3. Air Pollutants—adverse effects—Germany. 4. Air Pollutants—adverse effects—United States. 5. Coal—adverse effects—Germany. 6. Coal—adverse effects—United States. 7. Environment—Germany. 8. Environment—United States. 9. History, 19th Century—Germany. 10. History, 19th Century—United States. 11. History, 20th Century—Germany. 12. History, 20th Century—United States. 13. Public Health—Germany. 14. Public Health—United States. 15. Smoke—adverse effects—Germany. 16. Smoke—adverse effects—United States. WA 754 U22v 2009a]
  RA576.7.G3U3413 2009
  363.738'70943—dc22   2008037355

CONTENTS

List of Illustrations	*vi*
Acknowledgments	*vii*
1. The Age of Smoke	*1*
2. Modern Times, Modern Problems: Controlling Smoke, 1880–1914	*20*
3. Pollutants and Politics: Air Pollution Control between the Wars	*67*
4. Beyond the Pall of Smoke	*113*
5. Going Local, Going National: The Postwar Divergence of Environmental Policy	*149*
6. Forerunners and Pioneers	*187*
7. Environmental Revolutions and Evolutions	*208*
8. Conclusion: Was the Environmental Revolution Necessary?	*260*
List of Abbreviations	*269*
Notes	*273*
Index	*341*

ILLUSTRATIONS

Fig. 1. Burbach steel mill, Germany, 1876   *9*

Fig. 2. Smoking enterprise in Munich, ca. 1906   *46*

Fig. 3. Efficient coal use, as pictured in a 1926 engineering periodical   *71*

Fig. 4. Gloomy morning in downtown Pittsburgh, 1940s   *84*

Fig. 5. Fly-ash control through electrostatic precipitators   *97*

Fig. 6. Walt Disney cartoon, 1955.   *117*

Fig. 7. Smokestack of the Völklingen steel mill, Saar region, 1962   *141*

Fig. 8. Cartoon from the annual report of the Allegheny County Bureau of Smoke Control, 1956   *165*

Fig. 9. Housing in front of a coke-oven battery near Bottrop, northern Ruhr Basin, 1976   *184*

Fig. 10. Cartoon from the *Chicago Sun-Times,* republished by the Public Health Service in preparation for the 1966 National Conference on Air Pollution   *222*

Fig. 11. The complex machinery of air pollution control, 1960s   *232*

Fig. 12. Scene from the film *Smog,* 1973   *254*

Table 1. An Overview of Important Smoke Abatement Associations   *22*

ACKNOWLEDGMENTS

This book is based on a German monograph that Klartext Verlag published in 2003. After numerous favorable reviews and a prize from the Society of Alumni of Bielefeld University, it was tempting to leave the manuscript unchanged for this American edition. In retrospect, however, thoroughly revising the manuscript and rewriting some parts entirely, including the introduction and conclusion, allowed a more careful consideration of the fact that urban environmental history in both Germany and the United States has changed in recent years, as has the general context. Ignoring these trends would have meant a missed opportunity.

While coming back to a completed project after half a dozen years can be tricky, it proved a rewarding experience, with inspiration and help coming from several sides. My first and most important thanks go to Joel Tarr, who hosted me during a very productive stay in Pittsburgh in the spring of 1999 and inspired this American edition in the first place. I am also grateful to him and to Martin Melosi for accepting this book into their urban environmental history series. Two anonymous reviewers gave crucial feedback on the project at an early stage. It was a pleasure to work with Cynthia Miller, of the University of Pittsburgh Press. The Hagley Museum and Library in Wilmington, Delaware, provided an ideal environment for revising the manuscript. Thomas Dunlap did an outstanding job translating major sections from the German original. Nonetheless, the result still strikes me as a bit more German than intended, and my only hope is that the reader will see this in a positive light.

I laid out my further debts, intellectual and other, in the foreword to the original German edition, but it is a pleasure to repeat them here: to my academic mentors, Peter Lundgreen, Joachim Radkau, and Klaus Tenfelde;

to my fellow graduate students in Bielefeld, Baltimore, and Pittsburgh; to the staff of more than one hundred archives and libraries in Germany and the United States; to Louis Galambos, who oversaw my first forays into the topic some fifteen years ago; to the friends who housed and entertained me during long research trips; to my parents and my wife, Simona. Funding came from the Studienstiftung des deutschen Volkes, the Deutsche Forschungsgemeinschaft, and the Deutscher Akademischer Austauschdienst. All these institutions and individuals helped me, I would say, to see more clearly through the age of smoke.

THE AGE OF SMOKE

1   The Age of Smoke

Smoke was the most severe air pollution problem of the late nineteenth and early twentieth centuries. Wherever coal was used in major quantities, smoke and soot, the typical by-products of incomplete combustion, infested the local atmosphere, provoking countless complaints and attempts at abatement. But this fact alone conveys an inadequate sense of what it meant to live in the age of smoke. The key characteristic of smoke was its pervasiveness: one did not simply live with the problem of smoke, but literally in it. In urban areas, smoke was everywhere: in cities large and small, in industrial and residential areas, in rich and poor neighborhoods. Almost all urban agglomerations were struggling with the smoke nuisance, from Berlin to Chicago, from scenic Heidelberg to industrial Pittsburgh. Smoke was a constant companion of urban life, a pollutant that every city dweller was inevitably breathing on a daily basis. Smoke stuck to facades and monuments, making for the pervasive gray so typical of late nineteenth-century cities. It entered homes, besmearing rugs, curtains, and anything else that was not safely tucked away. The faces and clothes of urbanites carried smoke's hallmark, and in some of the worst cities it was customary for white-collar workers to bring a second shirt to work, since the first was usually soiled by midday. In fact, many saw smoke as more than a material problem: smoke was the modern city's halo, a darkish cloud that was often the first thing visitors saw. Smoke was the symbol of urban gloom, a word that rhymed with *doom*, and not only for prophets of cultural despair.

    Of course, urban pollution problems were not in themselves new: they are likely as old as cities themselves. But two factors made the smoke nuisance particularly awkward. The first was the stupendous growth of cities during the nineteenth century. Both Germany and the United States saw the rise of vast urban agglomerations, many of which comprised more than a million people by 1900, along with a concentration of energy-intensive industries that had no precedent in the history of either country. The second was the sheer pervasiveness of coal use, specifically the use of soft coal, which was prone to creating smoke during combustion. With coal replacing wood as the dominant fuel during the nineteenth century, and with the per capita use of energy on a sharp upswing, coal was practically everywhere in late nineteenth- and early twentieth-century cities. It was used in homes and in industry, in transportation and power generation, for the production of electricity and as the basis for a burgeoning branch of industrial chemistry. Regions rich in coal deposits were thriving economically; regions distant from such deposits complained bitterly about their misery and often made frantic efforts to secure a reliable supply. Some cities were fortunate: for instance, New York City lay close to the only large anthracite deposits in the United States and thus had easy access to a type of fuel that was much easier to burn without smoke than soft coal. But most cities were not so lucky, with some suffering further if valley locations made them prone to inversions and poor ventilation. Smoke problems had thus become the rule in most German and U.S. cities by 1900, a constant reminder that modern society's dependence on coal came at a price.

The smoke nuisance challenged authorities nationwide, in both Germany and the United States. Many people agreed that fighting smoke had to be the key goal of contemporary air pollution control, the problem that regulatory agencies simply had to solve if they wanted to make any legitimate claim at environmental protection. But at the same time, existing laws and procedures quickly proved inadequate for an effective antismoke drive. American nuisance laws, as well as traditional German regulations, were cumbersome and complicated instruments that saw pollution as an isolated incident, not as a universal feature of modern life. As a result, discussions arose in both countries around 1880 over what to do about smoke, with input from industrialists, engineers, physicians, public health officials, and an enraged public. By the early 1900s, the U.S. air pollution debate of the Progressive Era—and its equivalent in Germany—had reached a degree of intensity that would remain unmatched until the 1960s.

Coal smoke, however, was not the first industrial air pollution problem. In Germany, sulfur emissions from copper smelters had been a political issue since the mid-nineteenth century; in the Western United States, conflicts between metal smelters and their neighbors over sulfur and arsenic pollution had escalated to what one observer called a "life or death struggle."[1] But the general public gave the lion's share of its attention to coal smoke, and urban smoke became a defining political issue. As a result, laws and institutional reforms were focused on coal smoke, far more than on any other air pollution problem. It is no exaggeration to say that coal smoke was the crucial institution-builder in the field of air pollution control, with countless smoke ordinances, rules, and regulations instituted and special "smoke inspectors" waging a desperate fight against this urban plague. Especially in the United States, smoke abatement became almost synonymous with air pollution control, until the mid-1940s. Urban smoke defined the development of regulatory bodies, and its effect remained significant even after the smoke nuisance itself was gone.

In retrospect, the fight against smoke looks like a protracted and mostly ineffectual battle—had it been otherwise, the age of smoke would have been a mere episode. But such a reading underestimates the huge extent of the problem and the obstacles to addressing it. Except in West Coast cities, coal was the dominant fuel of the early 1900s, and every smoke abatement strategy had to cope with a host of outlets: industrial enterprises, railroads, commercial buildings, apartment houses, and—not by any means least—domestic furnaces. As a result, launching a successful attack on smoke demanded clever strategies, significant resources, and a good deal of patience. To be sure, Progressive Era antismoke activists did hope for a quick solution; Pittsburgh's Smoke and Dust Abatement League even held a competition for a new municipal nickname in 1916, believing that the city's classic epithet—"smoky city"—would soon become obsolete.[2] But such hopes were quickly dashed, activists realizing that the smoke nuisance allowed at best "steady, though perhaps slow progress," as the Women's Club of Cleveland noted in 1923.[3] For an industrial society with almost universal coal combustion, the smoke nuisance presented a gigantic challenge.

Although smoke was a pervasive problem, it was not always perceived that way. It is impossible to discuss smoke without addressing class, race, and gender: pollution loads differed greatly according to one's place in society, and so did perceptions of the smoke nuisance. For owners of property in downtown areas, smoke was first and foremost a financial problem, and men in real estate thus became some of the most dedicated proponents of smoke abatement. The same holds for cleanliness, a key concern for antismoke activists, but also one with pronounced class implications. Notably, complaints did come from all segments of society, belying bourgeois stereotypes that cleanliness was of no concern to the lower classes.[4] But a bourgeois bias is still evident in the politics of smoke abatement, although American environmental historians have been somewhat reluctant to look into this issue. The American drive against smoke mirrored a socially exclusive understanding of urban air pollution—a limitation as much as an opportunity. As Mancur Olson points out in *The Logic of Collective Action*, citizens are usually disinclined to voice concerns that they share with a large number of people. After all, their personal gain is greater when they focus on issues that affect only them or a small group of people who will profit from their activism as "free riders." Olson notes that collective interests, then, often do not emerge as powerful motivators until they have gained support from "selective incentives"—for example, a class-specific definition of an environmental problem.[5] While the present work steers clear of a dogmatic interpretation of Olson's argument, a history of air pollution control cannot help but give it some credit: as a rule of thumb, protests were strong when rhetoric emphasized the norms and values of a distinct group and weak when air pollution was seen as a concern of basically every human being. Changing this logic, at least to a certain extent, may have been the greatest single achievement of the modern environmental movement.

Overall, compared with concerns about cleanliness and property, the "health argument" usually took a backseat in air pollution rhetoric during the age of smoke. As Adam Rome notes, "The medical argument against air pollution always was a hard sell."[6] Some contemporary observers already found this an odd situation. "Strange as it may seem, the housewife is far less concerned with health than she is with the fact that her draperies are soiled or that her neighbor will soil her hands if she touches her furniture," a public health officer declared in 1928.[7] But since the demise of the miasma theory, which implied a broad environmental understanding of the causes of disease, the general public had come to think of health hazards mostly in terms of bacteria and viruses. As smoke seemed unsuspicious from a contagionist standpoint, the health argument did not emerge as a major concern until smoke was in fact disappearing from America's urban areas.[8] Nevertheless, the health hazards of urban smoke were substantial, even though it would take several decades of epidemiological research to discover their true extent. The greatest hazard stemmed from particles up to ten micrometers in size, for such extremely fine particles penetrate the thoracic region of the lung. Researchers today are unanimous about the enormous health hazards of small dust emissions. In 1992, for example, a World Bank study put the human toll from particulate matter in the developing world alone at between 300,000 and 700,000 premature deaths per year. While this figure remains open to dispute, it gives an impression of the implications of the age of smoke from a health perspective.[9] Unbeknownst to contemporary discussants, smoke abatement was literally a question of life or death for thousands of people.[10]

The age of smoke lasted longer than the reign of coal, beginning with debates about new laws and procedures against urban smoke around 1880 and finally tapering off around 1970, when environmentalists started voicing concerns about other, often invisible pollutants. Thus, the chronology of the age of smoke is almost identical with what Charles Maier calls the "age of territoriality": a time span of roughly a century, when the regulatory abilities of nation-states were at their peak. Beginning in about 1860, centralized nation-states assumed a new kind of control over their territory. To some extent, this was because of advances in technology, like railways and telegraph networks, which allowed goods and information to travel to a country's periphery and back with unprecedented speed. At the same time, political reforms, like the Unification of Germany and Japan's Meiji Revolution, laid the foundations of powerful government institutions. Beginning in the 1960s, however, according to Maier, the technological, cultural, and sociopolitical scaffolding of the nation-state began to erode and fall apart, in part due to the rise of strong supranational agreements and institutions like the World Trade Organization and the European Union, but most prominently through the erosive force of economic globalization. The age of territoriality slowly continued to fade throughout the last third of the twentieth century.[11]

From an environmental history perspective, Maier's argument might appear somewhat counterintuitive. What about the plethora of environmental laws since the 1970s that demonstrate the enduring strength of the nation-state? How do the creation of the Environmental Protection Agency in 1970 and the passage of powerful Clean Air Acts in 1970, 1977, and 1990 fit Maier's argument? Maier, however, does not describe the age of territoriality ending abruptly, its collapse resembling that of communism in 1989. Rather, the slow trend of global competition gradually undermined the powers of nation-states—and certainly, few environmentalists would doubt that economic globalization has indeed hampered many policies over the last forty years. Maier's argument thus implies a deep irony for the history of the environmental revolution, one that few environmental historians have taken note of: at the very time when ecological concerns were gaining importance politically, the nation-state's power base was beginning to erode. Starting in the 1960s, environmentalists were struggling not only with the usual obstacles to air pollution control—wayward industrialists, deficient laws, lazy officials, and so on—but also with a long-term decline in the nation-state's regulatory potential. Against this background, the age of smoke emerges as even more crucial: never before or since was the nation-state so well suited to defining and enforcing codes of acceptable conduct and creating institutions to that effect. The age of smoke presented a historic chance to create a lasting regulatory tradition, one that Germany seemingly used far more effectively than the United States: the German regulatory system evolved from nineteenth-century traditions over several decades, whereas the American environmental revolution ultimately led to a break with preexisting traditions. The fact that the United States, unlike Germany, emerged from the age of territoriality without a firmly entrenched regulatory tradition may explain a great deal about the repercussions of U.S. environmental policy and its constantly shifting character.[12]

The moniker "age of smoke" carries a double meaning that is entirely intentional. Smoke, indeed, obstructed the vision of urban reformers in more than one sense. By focusing on smoke and other visible pollutants, they usually overlooked a wide range of other issues: lead and carbon monoxide from automobile exhaust, sulfur dioxide from coal combustion and metal refineries, cancerous pollutants from chemical factories and refineries, and so on. With visible damage so apparent, it was tempting to ignore questions about health hazards and focus instead on cleanliness and property issues. In hindsight, perhaps smoke abatement advocates might well have been more open-minded, seeing smoke as merely the most easily detectable among a host of air pollution problems, many more dangerous from a health and environmental standpoint than smoke and soot. But it is certainly easier to make this observation at a time when no employee still wonders whether his or her shirt will last beyond noon.

The priorities of activists during the age of smoke are not the only things that seem questionable in retrospect. Nothing would be more misleading than seeing the story that follows as one detailing merely a clash between polluters and their opponents: the history of air pollution control is full of hidden agendas that influenced the course of environmental policies, and a transatlantic comparison is a good way to identify these. Why did American engineers stress a professional duty to fight the smoke nuisance, whereas German engineers showed lukewarm interest, at best? Why did American industrialists, who originally fought smoke abatement tooth and nail, turn into defenders of air pollution control, while German industrialists were largely silent in public? Why did the famous Prussian bureaucracy fail so miserably in its drive against smoke, surpassed in efficiency by, of all things, the governments of large American cities, famously described by James Bryce as "the one conspicuous failure of the United States"?[13] And why did popular antismoke sentiment, a constant in both countries since the late nineteenth century, mirror wide fluctuations in civic activism? The age of smoke involved a host of separate interests, and this book makes a point of highlighting them—not because I see them as illegitimate or disturbing but because they are a part of the story that conventional narratives tend to ignore. The age of smoke was about far more than the pros and cons of air pollution; indeed, most of the time, it was not really about the legitimacy of air pollution control at all. By the eve of World War I, the group of industrialists who opposed smoke abatement had shrunk to a small minority in the United States and was almost nonexistent in Germany, yet control efforts remained inadequate for many years. It clearly took more than good intentions to fight the smoke nuisance; good intentions, in fact, were little more than a minor beginning.

The coal smoke nuisance is gone from the Western world, the battle over coal combustion shifting to other issues like sulfur emissions and, most recently, their contribution to global warming. But while few environmentalists are aware of this, the age of smoke continues to influence ongoing debates about environmental problems. Most people—and indeed, many scholarly works—see air pollution control as a recent invention, the foundations of environmental policy laid down around 1970, when the first Earth Day celebration drew an estimated twenty million participants on April 22.[14] But 1970 was an ending as much as it was a beginning: this outburst of environmental activism spelled the end of a regulatory tradition that had grown out of the smoke debates of the early 1900s. The rise of environmental sentiments, in other words, coincided with a crisis of the existing regulatory system. The agencies born during the age of smoke were no longer able to sponsor an energetic drive against air pollution, but they were strong enough to hinder the rise of a new regulatory system—and, perhaps even more important, strong enough to leave many industrialists and experts with the honest impression that they had done their homework. This book, then, offers a new perspective on the environmental revolution, a perspective that should interest every environmentalist who wants to move beyond a Manichean worldview: the environmental revolution was more than the overdue outcry of a suffering population—it was also a classic case of miscommunication, and it was much better at demolishing regulatory traditions than at creating them. It is impossible to understand modern environmentalism without the age of smoke.

A book on the history of air pollution control can probably no longer claim to investigate a neglected topic. Research into urban environmental history has been under way for more than a quarter-century now, and it shows little sign of diminishing. Some early publications on the subject stand out for their staunch attacks on industry and their presumed political allies. Their argument in essence is that the history of air pollution control was basically a history of willful negligence, the debate having been captured by industrial interests until the environmental movement entered the scene. However, recent publications have taken a more balanced approach, painting a more nuanced picture of the business community, paying more attention to the different parties involved, and examining more closely the cultural construction of pollution. The present book pushes this trend further, stressing the compromise inherent in environmental regulation: for all the political clout industrialists could muster, they rarely came out of air pollution debates with their original demands fulfilled. And on the one occasion when they did—namely, air pollution control in the United States after World War II—they would later pay dearly for their hegemonic ambitions.

This argument is prone to several kinds of misunderstanding, and it may be wise to confront these early on. First, speaking of the compromise nature of air pollution regulation does not mean that all parties benefited in equal measure. There can be little doubt that during the age of smoke, damage to property owners and the plight of housewives were far more significant than the losses industry incurred from fines, administrative proceedings, and inefficient fuel use. But environmental regulation is not a zero-sum game, where gains for one party inevitably imply losses for another. The smoke inspection approach, institutionalized first in Chicago in 1907 and subsequently copied all over the United States, provides a prime example here: it offered gains for city governments, engineers, industrialists, and antismoke activists alike. Conversely, the German approach to smoke abatement around 1900 did not satisfy the demands of either the public or the business community. The biggest winner in the German smoke debate was probably the bureaucracy, which successfully sustained a policy of processing incoming complaints in the least stressful manner. The German story thus also provides a reminder that it is insufficient to examine only the interests directly involved. While bureaucratic incompetence and failure of communication are not in anybody's interest, they clearly play an important role in the story that follows.

Second, an emphasis on cooperation and compromise in no way ignores or diminishes the fact that air pollution gave rise to vigorous complaints in both countries. Pollution caused enormous damage in Germany and the United States, and those who associate compromise with soothing industrial remarks of the "we have a common interest" variety are clearly off the mark. In fact, this book makes a point of putting public protests and campaigns front and center, for it is quite plain that in the field of air pollution control, little if anything gets done without pressure from an enraged

Figure 1. Many nineteenth-century pictures, like this 1876 depiction of the Burbach steel mill near Saarbrücken, Germany, celebrated emissions as modern industry's halo. However, when smoke and soot were a material reality rather than a symbol, industrialists were much more on the defensive. Image courtesy of the Deutsches Museum.

citizenry. Yet a book that spans almost a century cannot help but adopt a long-term perspective, and from this point of view, the limits of protests and campaigns become only too apparent. If it was difficult to start a powerful campaign against atmospheric filth, it was infinitely more difficult to sustain this campaign over the long term. As it emerges in this study, civic protest followed a typical pattern: some two or three years of intensive activism, followed by more lukewarm activities or even a total abandonment of the issue. Indeed, not many associations could claim more than a decade of sustained activism against the problems of air pollution. To be sure, there are a few examples of truly admirable endurance, like Cincinnati's Smoke Abatement League, founded during the Progressive Era, which continued to operate as the Air Pollution Control League of Greater Cincinnati until far into the environmental era. But impressive as these examples may be, it is painfully apparent that they are the exception to the rule. Before the rise of modern environmental organizations, civic protest was often haphazard and notoriously unstable, and regulatory styles inevitably reflected that fact.

In the end, regulatory compromise usually came about for lack of alternatives. Enraged citizens may be able to sustain a dedicated crusade for a while, but at some point, activism almost inevitably lags, and interest in compromise grows. Likewise, regulators and industrialists have an interest in some kind of a "gentlemen's agreement": after all, they are bound to meet time and again, and renegotiating the general terms of cooperation each and every time is, if anything, time-consuming and unnerving. In short, some kind of compromise was usually the path of least resistance: in the end, most people find it more advantageous to talk than to yell at each other. From an economic perspective, this might be seen as a quest to minimize transaction costs: a cooperative agreement is usually far less expensive for all parties involved than an all-out campaign. The question is not so much whether industrialists or protesters have to make concessions as it is what these concessions ultimately mean.

Thus, compromise is neither a cause for celebration nor a scandal—it is simply a fact of life that needs careful scrutiny. If this book demonstrates anything, it is that wholesale condemnation of cooperation is as shortsighted as general praise. Throughout history, "cooperation" has meant many different things. It was crucial for the development of the smoke inspection approach in the early 1900s, a strategy that was arguably the best approach possible under Progressive Era conditions. However, it was basically a smokescreen for industry's dominance during the 1950s and 1960s, when air pollution control progressed only slowly and with a rather narrow agenda. In Germany, while cooperation was originally a bureaucratic strategy focused on processing incoming complaints in the least stressful manner, it took on a more positive note in the postwar years, and the cooperative development of rules and regulations remains a prominent feature of German environmental policy to the present day. In short, no general alternative between strong state and weak cooperation exists. Cooperation can be tough, both on the problem and on the cooperators, especially if it is backed up in a sophisticated fashion by the powers of the state.

As a third caveat, I should emphasize that stressing compromise in environmental regulation is not meant to exonerate those industrialists who were fighting environmental regulation as a matter of principle. After all, the following material offers plenty of evidence that the evil industrialist is not simply a Hollywood invention, Erin Brockovich–style. Over the course of almost a century, a wide spectrum of attitudes must be acknowledged: from public-spirited corporate captains who felt an ethical obligation to curb their emissions, to narrow-minded industrialists who thought that air pollution control was nonsense. However, the vast number of businesspeople fell somewhere between these extremes. Had they not, one would be hard-pressed to explain why radical crusaders against pollution control drew criticism not only from civic activists but also from the industrial community. Even a representative of an oil and refining company, usually a less cooperative sector, admitted in the early 1960s that "a small percentage of industry is irresponsible."[15] Nevertheless, searching for "black sheep" and "smoking guns" misses the point: while it is clear in some cases who was wearing the white hat and who the black, most look pretty grayish in retrospect. Even when an industry obstructed solutions generally, like auto manufacturers did in the 1950s and 1960s, a closer look seems advisable. Environmentalists, in this particular case, were right to sense that the auto industry was dragging its feet on the smog problem, but they were wrong to suspect a conspiracy at the root of this reluctance. As shown later, it seems quite likely that the auto industry's behavior was much more incompetent. Environmentalists were also wrong to take Detroit's behavior as emblematic of all industrial attitudes, and they probably did themselves little good by generalizing in this way. After all, criticism of uncooperative industries becomes much more forceful if one doesn't simply see such behavior as "business as usual." The notion of the evil industrialist, so prominent around 1970, was thus the classic case of a self-fulfilling prophecy.

Therefore, if this study emphasizes the compromise nature of environmental regulation, this does not mean that cooperation is either good or bad; too much depends on the specific context. The situation is clearer, however, for the opposite of cooperation—namely, antagonistic modes of regulation. In fact, what emerges here is that antagonistic approaches are rarely ever the result of a conscious choice: usually, they are the result of accumulated anger and an urgent desire to "see things happen." A desperate fight against vested interests thus looks less like a clear strategy than a makeshift, an approach protesters use until they realize that talking with the other side may have some merit as well. Revealingly, the Progressive Era's antismoke movement did not make significant headway until it scaled down its aggressive rhetoric, turning from strict prosecution to education and consultation with smoke-generating businesses. Indeed, has antagonism, by and large, really produced impressive results since the environmental revolution of 1970? Significantly, the age of smoke involved a rather small amount of litigation—important because history shows that the side effects of court proceedings are huge. Time and again, litigation has produced bitterness on both sides, an escalation of demands and rhetoric, and the postponement of a solution. For nineteenth-century jurists, court battles over pollution problems were veritable nightmares: complex, time-consuming, and difficult to control once they had begun. Even organizations with a clear antismoke record, like the American Civic Association, were skeptical of lawsuits, noting that "people have a very healthy and proper disinclination to involve themselves in litigation over nuisances."[16] During the age of smoke, investments in control equipment were always far greater than investments in lawyers' fees—something that is no longer certain with modern environmental litigation.[17] In fact, few approaches have a worse record historically than litigation, and environmental historians are well advised to emphasize this point. In the twenty-first century, perhaps the best way to convert industry to an environmental agenda is to demonstrate that antagonism is bad business.

One of the most surprising trends in recent environmental history research has been an ongoing attempt to rewrite the history of air pollution control from an antiregulatory perspective. When the original version of this book was researched and written in the late 1990s, the dominant impression was that air pollution control before 1970 was mostly ineffective, if not nonexistent. As a result, I devoted a good part of my intellectual energies to proving that pre-1970 air pollution control, while not perfect by any means, does deserve some credit for making inroads against at least some pollutants. But recent publications, most prominently by Indur Goklany, in his book *Clearing the Air*, paint an altogether different picture. Promising no less than "the real story of the war on air pollution" in his subtitle, Goklany vigorously attacks the rationale for forceful federal policy by stressing the accomplishments of pre-1970 air pollution control. He asserts that pollution loads had been falling long before federalization and that, more generally, the United States was already on a path toward better environmental conditions in 1970. Goklany thus challenges the rationale for federalization: the state of the environment, he argues, would have improved just as well, or even better, had the states remained in control. Goklany outlines his "environmental transition hypothesis" as follows: "As a country becomes more economically and technologically developed, in order to improve its quality of life it first addresses immediate needs such as food, running water, basic medical services, electricity, and education. Once those needs are met satisfactorily, the country turns its attention to the other determinants of its quality of life, such as air pollution and other environmental matters."[18] And this shift, Goklany asserts, happened long before 1970.

Environmental historians have been reluctant to discuss this interpretation, in part because of the historian's natural distrust of books that promise "the real story" on anything but also because they see it as a mere "smokescreen" for Goklany's political agenda. To be sure, Goklany's bias is obvious: it would not have taken his membership in the libertarian Cato Institute, or his appointment as assistant director in the U.S. Department of the Interior's Office of Policy Analysis under the George W. Bush administration, to foster suspicions that his real goal was to discredit federal environmental policy.[19] And yet it seems too easy to simply dismiss his argument as a political ploy, for at least one reason: one of Goklany's key assertions—that air pollution loads had declined before 1970—does contain some truth. In fact, Goklany could have made his case even more persuasive had he done his research more thoroughly. In the early 1950s, industry's investments in air pollution control technology already lay in the range of 100 to 400 million dollars per year, and there are strong indications that these sums increased massively thereafter; a 1966 estimate cites an annual amount of 850 million dollars.[20] But do these figures support Goklany's argument?

Air pollution measurements usually go along with an active regulatory program, and the postwar years were no exception. Regulatory agencies existed mostly at the municipal level, having grown out of the smoke abatement tradition of the Progressive Era. Staffed with mechanical engineers, many of them hired to fight the smoke nuisance and little else, they commonly saw smoke and other particulate emissions as their primary issue, often achieving notable reductions in this regard. But at the same time, these agencies were less interested in pollutants that were not visible or otherwise available to the senses—and this was not their only limitation. The powers of municipal authorities inevitably ended at the city limits, and attempts to reach beyond these boundaries produced meager results. County or state agencies might have done a better job, but their development was hindered by the advocates of the smoke abatement tradition. As a result, solid accomplishments within cities went along with weak regulation in suburban areas, small towns, and the countryside, but nobody was taking measurements there.

Thus, the decline of pollution loads during the 1950s and 1960s does not signal a healthy regulatory system. Quite the contrary: the decline of particulate emissions was no more than the last meager accomplishment of a regulatory system that had long passed its prime, having grown from the ingenious solution of the Progressive Era into an enduring burden hindering reforms. The burgeoning environmental sentiment made it crucial to move beyond smoke and soot, but the regulatory establishment was dragging its feet in this respect, not for lack of awareness but simply because its own distinct interests stressed other priorities. This makes the story of postwar air pollution control ill suited for an attack on federal prerogatives. The merits of federal involvement are really quite plain in retrospect, and it is misleading to see the work of federal officials from a post-1970 standpoint.

During the 1960s, most of their work was directed toward strengthening local-control programs, rather than harassing them with federal intrusions. Two years after the first major piece of federal regulation, the Clean Air Act of 1963, funds for local and state programs against air pollution had roughly doubled, thanks to a generous program of federal grants. Later on, federal authorities were working toward a strong turnout in hearings pursuant to the Air Quality Act of 1967, offering training and other kinds of support to civic groups in the fight against pollution. But in the end, these hearings did not help regional programs as much as federal officials had wished. By providing a forum for the outburst of environmental activism around 1970, they ultimately discredited regional cooperation, nourishing the call for federal supervision. Federal involvement was thus the result of a highly complex chain of events—so complicated, in fact, that businesspeople were among the prime advocates of federal air pollution control in 1970.

Interestingly, Goklany does not discuss any of these factors or even indicate that he sees them as relevant. He does not trace the sequence of events leading to the 1970 Clean Air Act, or discuss the potential and limits of state regulation, or assess the cooperative spirit of postwar air pollution control; in his narrative, change takes place mostly "through some agency or the other."[21] His "environmental transition" is a strangely anonymous process, an almost magical trend that seems to have had neither advocates nor adversaries. There is no room in Goklany's narrative for the different parties involved in air pollution control: experts, officials, businesspeople, and active citizens. But environmental policy does not automatically follow a preconceived course, let alone a predetermined course for the better—it is made, and remade, on a daily basis, subject to twists and turns that no model can truly anticipate. In other words, what Goklany ignores is the fact that environmental policy, like every policy, is made by people, not by anonymous "processes" or "transitions." It is ironic, if not revealing, that a libertarian has forgotten this.

The present work was finalized as the antienvironmental revolution accompanying the presidency of George W. Bush seemed to be in its last gasp. I shall refrain from commenting on this antiregulatory onslaught here, since others have already done so, with much greater authority. However, it seems worthwhile to point out that Bush's antienvironmental revolution, or his attempt at one, underscores one of this study's recurring arguments—namely, the persistence of specific national paths of political development. After all, this full-scale assault on the accomplishments of the environmental era was only the latest sharp turn in a turbulent history: since 1970, radical shifts between aggressive policies and antiregulatory backlashes have been a hallmark of American environmental policy, so much so that William Ruckelshaus, the first director of the Environmental Protection Agency, spoke of a "pendulum" of environmental policy that swings constantly back and forth.[22]

European nations have followed the swings of this pendulum with a mixture of bewilderment and worry, pointing with pride to the solid and far less contested environmental regulations of European nations and, more recently, the European Union. Germans in particular have gotten used to looking down on the environmental record of their transatlantic partners, although this demonstrates little more than their short memory. As this book shows, the overall balance between the two countries since the late nineteenth century is really quite even. Around 1914, the best municipal smoke abatement program was in Chicago; but at the same time, no boilers were inspected more thoroughly for smoke emissions than those controlled by the Hamburg Society for Fuel Economy and Smoke Abatement. There was no German equivalent to the spectacular campaigns against domestic smoke waged in St. Louis and Pittsburgh around 1940. But then, the postwar reforms in the Ruhr area ultimately proved more sustainable than the U.S. smoke abatement tradition. To be sure, there are some exceptions where one country was generally ahead of the other one: due to the peculiar situation of Los Angeles, for example, the United States was quicker in controlling automobile exhaust, whereas German air pollution control was generally stronger in rural areas, the neglected side of the urban-centered American debate. But beyond this, one should be careful making generalizations about overall efficiency.

It is important to stress these diverse accomplishments, as they provide an antidote to the bias that frequently characterizes comparative studies. Fittingly, Samuel Hays notes in *Beauty, Health, and Permanence* that comparative studies often display "skepticism, even hostility, toward the more open and participatory political system in the United States and a greater admiration for the more closed and 'efficient' modes of decision making in Europe."[23] This book takes a different, more balanced approach: it focuses on the specific national conditions that shaped debates and decisions, assuming that a comparative perspective is better suited to highlighting these conditions than a focus on either country would be. In this approach, the purported "superiority" of European policies quickly looks dubious, at best. The German state may have been stronger and more autonomous than its American counterpart, but this was a burden as well as an asset, as the narrative will show, allowing officials to squelch a nascent debate about administrative reforms in the early 1900s, with severe consequences for the efficiency of control. Conversely, the more open and participatory character of the American political system has had mixed merits, as well. The civic leagues of the Progressive Era virtually created municipal smoke abatement out of nothing, and they were crucial in making it effective. But in the postwar years, the more open U.S. system allowed industrialists to influence policy to an extent that was probably unthinkable in Germany. Clearly, the issue is not whether one country enjoyed superior conditions, but rather what each country made of them.

National paths of political development, or "national styles of regulation," as they have come to be called, are one of those things that are more often assumed than actually studied. As a result, a tacit notion has slipped into discussions of "national styles of regulation" that deserves careful scrutiny: the notion of a "natural" path of political development, a "one best way" of regulation that suits the conditions of a particular country in the best way possible. For example, David Vogel argues, in his study of environmental policy in Great Britain and the United States, that "each nation does exhibit a distinctive regulatory style, one that transcends any given policy area," and that, as a result, "the characteristics of a political regime" define policy processes across the board.[24] In contrast, the present book argues for a more differentiated approach to national styles of regulation. On the one hand, it takes note of a host of national characteristics that are well known to students of either Germany or the United States: the important role of civic associations in American municipal politics and the weakness of Germany's civil society; the pride of German officials and the antistatist sentiment among German engineers; the declining legitimacy of the American business community during the 1960s; the campaign style of American politics and the greater amount of continuity allowed by the momentum of the German bureaucracy. On the other hand, this book also gives good reason to doubt naive essentializations about "national styles." For example, many studies of American environmental policy, including Vogel's, emphasize the antagonistic relationship between industry and regulatory agencies. However, the following narrative shows that for some sixty years, cooperation between businesspeople and regulators was in fact a hallmark of the American approach; indeed, a comparison between Germany and the United States in the 1950s would have found that America, not Germany, was the global custodian of cooperative air pollution control. As for Sturm Kegel's concept of "air pollution cooperatives," there is little doubt that they would have been declared a typical feature of the German approach if they had been successful. In reality, however, Kegel's idea never had a chance. Obviously, national styles of regulation do not determine a certain "best path" of development;

at best, they narrow down a choice between different approaches. Every national style of regulation is really a pluralism of possibilities.

It is important to emphasize this multitude of possible "national paths," as it aids understanding of the surprising efficiency of cooperative arrangements during the age of smoke. Throughout history, cooperative modes of regulation were usually at their best when they competed with a different, more confrontational approach; conversely, they were in danger of sclerosis when no alternative was in sight. For example, the rise of smoke inspection depended strongly on previous experiments with strict prosecution and fines. After prosecution had produced bitterness all around and little in the way of abatement, all sides embraced the less confrontational smoke inspection approach and sought to make it effective in order to forestall a return to fines and court proceedings. Likewise, the success of cooperative rule-setting within the VDI Clean Air Commission depended strongly on the threat of a state-centered alternative and on the fact that the federal government had reserved the right to ignore inadequate rules. Further, the lack of alternatives was clearly important for the decline of cooperative air pollution control in postwar America: why should industrialists strive to advance forcefully on the clean air front if cooperation was really the only realistic approach? Successful cooperation, then, seems to depend strongly on realistic alternatives, and it is important to see the cozy rhetoric that usually accompanied cooperative air pollution control in this light: if cooperation worked, it did so not because industrialists were nice people, but rather because they wanted to prevent a shift to less attractive policies.

The phrase "national styles of regulation" evokes notions of a holistic perfection, but environmental policy, and probably every policy, was far messier. For example, the reform of German air pollution control during the 1950s did not follow a grand design; it was a series of small steps, each contested, which ultimately produced a surprisingly enduring regulatory style. Likewise, the invention of the smoke inspector was originally a local solution for Chicago, propagated by the City Club of Chicago after consultation with Lester Breckenridge, a professor of engineering at the University of Illinois. It took numerous adaptations of the Chicago approach in other cities, and the failure of antismoke activists to develop a better strategy, for smoke inspection to become America's national style of environmental policy. At their inception, then, regulatory strategies are often improvised approaches, based on impromptu decisions made with nothing even vaguely resembling a blueprint for reform; they become national styles only after being replicated over time and in several places. National styles of regulation, as described in this book, are simply sets of practices that have gathered momentum.

Of course, this definition is to some extent based on the peculiarities of the topic. Decisions about air pollution control were mostly local and regional in both the United States and Germany before 1970; national legislation on these issues was weak in Germany and virtually nonexistent in the United States until far into the postwar years. And yet the importance of administrative routines should not be underestimated. In fact, the strength of bureaucratic routines emerges as one of the hidden forces of environmental history in this study—and one with ambivalent merits. It took some thirty years for the smoke inspection tradition to make possible the spectacular campaigns of St. Louis and Pittsburgh; but after that, it took another thirty years before the tradition was abandoned in favor of an approach that met the demands of the environmental era. "It is the unrelenting, day-by-day attention to detail that brings success," Raymond Tucker, the key figure of the St. Louis campaign and perhaps the best-known smoke inspector of all time, declared in 1941.[25] The age of smoke thus illustrates a classic dilemma of environmental policy: a good control program has momentum, but momentum makes it difficult to adjust to new challenges.[26]

I argue for a more nuanced and open understanding of national styles of regulation, analyzing the set of conditions that influenced policies in both countries, but emphasizing that these conditions in no way predetermined the course of events. While national styles of regulation have enormous momentum, they are not impervious to change. National styles of regulation are collective by nature, depending on the behavior of countless officials, experts, businesspeople, and citizens; and yet they leave room for courageous initiatives that open up new perspectives. With the words Ernest Renan famously used to describe French patriotism, national styles of regulation might be termed a "a daily plebiscite": a national routine that depends on certain political, social, and economic conditions and that may change, sometimes within a matter of days, as society itself changes.[27]

The conclusion of this book discusses whether the environmental revolution was really necessary in order to make air pollution control effective. While many environmentalists may find that question dubious, if not heretical, it seems unavoidable if the achievements of pre-1970 air pollution control are no longer ignored. Was it really necessary to abandon the smoke abatement tradition entirely in the United States and to begin the environmental era with a new set of policies and practices? Was there a chance for a more evolutionary process like that in Germany, where the more successful features of nineteenth-century regulatory traditions survived the outburst of environmental sentiments? Again, the goal here is not to depict one country as generally superior, but rather to learn about both by comparing their divergent responses to similar challenges. National styles of regulation depend on a host of preconditions, many of which are semiconscious, at best. Making these preconditions clear, and subjecting them to intensive scrutiny, is instrumental for making wise environmental decisions.

2   Modern Times, Modern Problems
CONTROLLING SMOKE, 1880-1914

The United States, 1880–1917

When American city dwellers of the Progressive Era talked about smoke, anger was usually their dominant state of mind. Since smoke had grown into a chronic problem of American cities in the late nineteenth century, complaints were as numerous as they were vigorous, often expressing indignation about the extent of the nuisance. Many urbanites thought that smoke was "a shame and an outrage on common decency," and the reasons are not difficult to understand in retrospect.[1] After all, coal smoke had numerous consequences, and they were all unpleasant. Perhaps most prominently, smoke was dirty and thus incompatible with bourgeois values of cleanliness.[2] "A city should be made attractive as a center for homes, as well as industrial establishments, and to this end, must be clean and bright," Pittsburgh's Chamber of Commerce declared.[3] In Chicago, American flags gave "the appearance of dishrags" after flying under the open sky for two or three weeks; in Baltimore, antismoke activists presented "what was formerly a lace curtain, but which had been so acted upon by sm[o]ke as to resemble the clothes in which mummies are packed."[4] For wealthy city dwellers, smoke implied enormous costs: one report in Boston declared that the smoke plague was "the cause of serious pecuniary loss." A home owner living next to a factory in the southern part of Buffalo calculated that the smoke nuisance had reduced the value of his property by four thousand dollars, and in Baltimore a Protestant clergyman complained that the "property values of a desirable residential section" were being driven down.[5] One advocate of the fight against smoke in Chicago estimated that the loss to his company from smoke amounted to about fifteen thousand dollars, while the head of the St. Louis Public Library put the annual damage to his books at ten thousand dollars.⁶ David Stewart, the head of the Anti-Smoke League of Baltimore, even declared quite openly that many members of the league's board were "personally interested on account of great injury done to their individual holdings of property."⁷ Other complaints called smoke "an aesthetic loss through discouragement of the ornamentation of buildings, homes and grounds" or attributed to smoke "a most depressing effect on the mind and spirits."⁸

In retrospect, it might seem that the situation was crying out for redress, but such deductions must be made carefully. "There is no such thing as absolute dirt," Ted Steinberg noted in a recent review essay, emphasizing that air pollution, like every environmental problem, is a social construct.⁹ Furthermore, no civic movement starts simply because a problem exists, and that is all the more true of late nineteenth-century cities, with their multitude of environmental, social, and ethnic problems. Eventually, a civic smoke abatement movement arose, with reform leagues, chambers of commerce, and even special smoke abatement leagues leading the fight, but a quick look at Germany, where no similar leagues existed, reveals that the formation of such a movement was no foregone conclusion.¹⁰ In fact, the movement probably would not have emerged at all if activists had not offered what Mancur Olson has called "selective incentives": a class-specific definition that stressed cleanliness and property values, thus transforming a general problem into the specific concern of a small, socially exclusive group of people. The elite background of many antismoke activists was thus not a minor blemish on an otherwise remarkable movement, as some environmental historians have thought, but a key requirement for its very existence.¹¹

This specific connotation of the smoke problem was never explicit. On the surface, declarations routinely stated that "the masses of the people are the chief beneficiaries."¹² However, it is usually not difficult to see through such rhetoric. The Smoke Abatement League of Cincinnati, for example, once stressed that it comprised "all classes of our people," only to single out among its members "bishops, clergymen, judges, lawyers, physicians, teachers, manufacturers, banks and bankers, merchants, hotel-keepers, railroad and coal men, together with many leading women"—certainly not a representative cross-section of the urban population.¹³ Moreover, complaints about damage to personal libraries and works of art also reveal that this movement gave voice to a fairly narrow segment of the urban population.¹⁴ Emphasis on the financial loss from the smoke nuisance pointed in the same direction: home owners lost substantial sums from frequent renovations and businesspeople from soiled goods, and members of the urban upper class were inevitably more concerned than anyone else. Revealingly, ownership

An Overview of Important Smoke Abatement Associations

City	Association (predominant gender)	Years of activity	Strategy
Baltimore	Anti-Smoke League (men)	1905–10	Prosecution through Health Department
	Women's Civic League (women)	1911–	Smoke inspection; engineer cofinanced with league funds
Chicago	Citizens' Association (men)	1880–95, 1905	Prosecution through Health Department
	Society for the Prevention of Smoke (men)	1892–93	Prosecution
	City Club (men)	1906–	Smoke inspection
	Anti-Smoke League (women)	1908–	Electrification of railroads
	Outdoor Art League (women)	1909	Planned cooperation with Department of Smoke Inspection
	Chicago Woman's Club (women)	1909	Planned cooperation with Department of Smoke Inspection
	Woman's City Club (women)	Before 1916	Cooperation with Department of Smoke Inspection
Cincinnati	Smoke Abatement League (mixed gender)	1906	Initially surveillance and prosecution with own force of inspectors; later smoke inspection
Pittsburgh	Women's Health Protective Association of Allegheny County (later merged with Civic Club) (women)	ca. 1890–95	Prosecution
	Engineers' Society of Western Pennsylvania (men)	1892	Prosecution
	Civic Club of Allegheny County (mixed gender)	1897–1919	Initially prosecution; later smoke inspection
	Chamber of Commerce (men)	1899–	Initially prosecution; later smoke inspection
	Smoke and Dust Abatement League (federation of multiple associations)	1912–	Smoke inspection

continued

An Overview of Important Smoke Abatement Associations, *continued*

City	Association (predominant gender)	Years of activity	Strategy
St. Louis	Engineers' Club of St. Louis (men)	Sporadic (1876, 1888, 1903, 1906)	Prosecution
	Citizens' Smoke Abatement Association (later merged with Civic League) (men)	1892–1904	Prosecution
	Wednesday Club (women)	1893, 1910	Prosecution
	Civic (Improvement) League of St. Louis (men until 1916)	1904–	Initially prosecution; later smoke inspection
	Women's Organization for Smoke Abatement (women)	1910–	Initially prosecution; later smoke inspection
	Million Population Club (men)	1911–12	Unclear

Source: Uekötter, *Rauchplage*, 521–53.

of land and real estate was pivotal in this debate. In large American cities, where neighborhoods often changed their character with amazing speed, the smoke plague played an important role, and real estate interests were therefore a prime mover behind public protests.[15] The weight of such interests is evident, for example, in the taxes assessed on the board members of the Anti-Smoke League of Baltimore: fifteen of the nineteen board members listed in the Baltimore Property Tax Book paid an average of $18,207 in taxes on land and real estate; five paid more than $50,000; and the maximum was $165,585, paid, significantly, by the league's chair.[16] Obviously, the league's campaign against smoke was not driven simply by "public interest," but by a very clear financial interest in the stabilization of real estate values.

The fact that many civic reformers were women has led some scholars to speculate about a peculiar "female approach" to the smoke issue.[17] Of course, gender roles had an impact on the perception of the problem, but attitudes were not rigidly gender-specific. On the contrary, many women addressed technological issues, a theme usually attributed to the male sphere. For example, Sarah Tunnicliff, the chair of the Clean Air Committee of the Woman's City Club of Chicago, demanded that "equipment . . . be installed which will insure complete combustion and at the same time prevent smoke."[18] The Woman's Club of Minneapolis took a similar view: "In many of our large buildings, . . . a wrong furnace construction is the reason for the smoking chimney," declared a report from the relevant committee, and the chair of the Smoke Committee of the Women's Civic League of Baltimore gave a report on work to date the programmatic title "Black Smoke Means Waste of Fuel and Loss of Boiler Efficiency."[19] Conversely, supposedly "female" perspectives were also voiced by engineers. Engineers, too, invoked moral arguments to denounce smoke-generating enterprises: "I have no more right to deluge my neighbor's premises with soot than I have to empty my garbage can over the fence line," declared Charles Benjamin at a meeting of the American Society of Mechanical Engineers. This view was shared by Pittsburgh's smoke inspector Searle: "The common good must always rise superior to individual interests."[20] Nor did engineers ignore the other aspects of the smoke nuisance: as late as 1926, a technical journal asserted that "gloomy weather breeds gloomy dispositions."[21]

In the light of this congruence, it was only consequential that intensive cooperation characterized the relationship between women's clubs and male reformers. Most striking was probably the connection between the Women's Civic League of Baltimore and the local smoke abatement engineer, the league paying half of his salary.[22] The situation was similar in Chicago, where the Outdoor Art League and the Chicago Woman's Club offered to help the engineers of the Department of Smoke Inspection keep tabs on smoking chimneys.[23] The Clean Air Committee of the Woman's City Club of Chicago published a poster with instructions on the efficient operation of furnaces, clearly created in consultation with official experts. The Woman's Club of Minneapolis campaigned for years to get city hall to hire an engineer; in the meantime, it distributed a list of experts who could provide reliable information on technical issues related to heating.[24] In general, women's organizations had few problems interacting with male smoke fighters: the Women's Health Protective Association of Allegheny County maintained a lively cooperation with the Engineers' Society of Western Pennsylvania; the smoke fighters of the Chicago Woman's Club conferred with smoke inspector Paul Bird even before they requested, at the plenary meeting of their own organization, the formal establishment of a committee; and before the women of the Wednesday Club, in St. Louis, set up a separate organization, they asked the Civic League whether this was acceptable.[25] It is not even possible to discern a typical sequence of male and female reform efforts. In Pittsburgh, the smoke question was first raised by the Women's Health Protective Association of Allegheny County, while in Boston, Rochester, and Cleveland, the Chamber of Commerce took the lead.[26] A male-dominated Anti-Smoke League was founded in Baltimore; St. Louis saw cooperation between the male-dominated Citizens' Smoke Abatement Association and the female Wednesday Club; and in Chicago, the fight against smoke was spearheaded in the first ten years by the Citizens' Association, an influential group of "law-and-order businessmen" in which Marshall Fields was the leading voice.[27] In short, male and female smoke fighters were in agreement on all essential questions of smoke abatement.[28]

It seems that this surprisingly harmonious relationship was the result of a simple insight: in light of the strong opposition to smoke abatement, the movement needed not only every man but also every woman. And indeed, a host of statements indicate that women were all but pushed into taking action: "Women should . . . , individually and through their organizations, protest vigorously against the present conditions," declared a 1907 essay, while the *Journal of the American Medical Association* recognized in women "a new and powerful source of aid in the antismoke campaign."[29] "I feel that very little can be accomplished . . . unless we are able to enlist the intelligent sympathy and the enthusiastic interest of the organized women of the community," explained a researcher at the Mellon Institute. In 1913, the International Association for the Prevention of Smoke was even discussing setting up a "ladies auxiliary."[30] When male and female activists joined a common front, the result was more than a simple addition of forces. Such cooperation made life much more difficult for the opposition within business circles: a male-dominated movement could potentially be weakened by exerting informal influence, and a purely female movement could be held up for ridicule—but an alliance of men and women was much harder to defeat. When the fight against smoke in Cincinnati got bogged down for a time, the Smoke Abatement League pointed to its women's wing as a kind of trump card: "Shall we have to appeal to Womankind again to arise in their might and beckon the men to follow them?"[31]

It is tempting to speculate whether this protest movement would have developed if the fight against smoke had entailed considerable costs. But the question of whether a choice had to be made between the smoke nuisance and prosperity was soon resolved: any technological objections to the fight against smoke were off the table after Philadelphia's Franklin Institute declared in 1897 that it was possible to combat the smoke nuisance and that it was worthwhile doing so.[32] In 1915, Charles Benjamin went so far as to state emphatically that the fight against smoke was "one of the simplest things in the world"—but this was true only from a narrow technical perspective.[33] After all, a promising campaign against smoke required not only technological means but also a strategy to promote their application, and finding such a suitable strategy would keep antismoke activists busy for decades.

In Search of a Strategy

The fight against the smoke nuisance usually began in American cities with the passage of a smoke ordinance. Many of these early ordinances come across as somewhat simplistic in retrospect, and for a reason. Apparently, the people behind these ordinances did not give all that much thought to their content, let alone consider alternatives. Rarely did the first generation of local laws contain more than a general prohibition of smoke. The typical ordinance declared smoke a public nuisance, stipulated the punishable quantity of smoke with more or less precision, and set the level of fines. Implementation was always entrusted to an already existing agency: in Chicago, Boston, and Baltimore it was the Health Department; in Pittsburgh the municipal works; in Philadelphia the steam boiler inspectors; in Cleveland the Division of Public Health within the local Police Department; in Milwaukee the building inspector, together with the Police and Health Departments; in Rochester the commissioner of public safety; in Cincinnati the fire-escape inspector.[34] If these institutions and agencies were adequately supported in their work, the smoke nuisance would soon vanish into thin air—or such was the hope of antismoke activists: "If the campaign against violators is carried on as vigorously as it should be St. Louis can be made practically a smokeless city in the next six months."[35]

All too often, however, such optimism gave way to growing disenchantment. This was especially evident in the case of Baltimore, where an Anti-Smoke League had won an ordinance in 1906 after a protracted struggle. "The league would hear complaints, investigate cases and collect evidence with a view to seeing that all preventable smoke emission is done away with," David Stewart, chair of the Anti-Smoke League, had declared in the euphoric mood prevalent on the eve of the signing of the ordinance into law.[36] His enthusiasm quickly evaporated, however, as case after case brought by the league lost in court.[37] Afterward, things quickly quieted down for the Anti-Smoke League: in the first year of its existence, it had sent out six circulars to its members; the seventh, following only three years later, already spoke of "exasperating conditions" in its title. The reason was obvious: according to Stewart's account, the jurors in the cases he had brought had not even considered finding the defendants guilty. "They considered it against public policy to enforce the law, or even to discuss its enforcement," a former juror had told him "in a great state of indignation."[38] The message was obvious: treating a respected businessman like a criminal simply because his factory was giving off some smoke—that was clearly going too far.[39]

The defeat of the antismoke lobby was not everywhere as complete as it was in Baltimore. But even in Chicago, where one of the most powerful organizations—the Citizens' Association—stood behind the fight against smoke, the result was less than impressive. Substantial numerical fluctuations in the annual reports of the Health Department indicate that the fight against smoke stood on shaky ground: in 1885, for example, only 58 lawsuits were filed; the following year, it was 210; by 1887, the number had dropped again, to 117.[40] Moreover, even if a case ended in victory, there was reason to doubt the permanence of the remedy: "In most cases there has been, as the result of a suit, an amendment for a few days in the annoying practice, followed by a relapse into the old ways," the Citizens' Association noted with regret. Even the Health Department, which otherwise made a point of radiating optimism in its reports, declared on one occasion that companies often relapsed into their usual routine, trusting that officials could not afford to engage in extensive surveillance, for lack of personnel.[41] "It is obvious that a force of four men is inadequate to cope with this nuisance systematically," the annual report of the Health Department finally noted in 1891.[42] By this time, resignation had long since taken hold of the Citizens' Association. After ten years of work, it noted with chagrin, "It will take time to make the movement effective, but it is the proper one and should be sustained."[43] Soon thereafter, the Citizens' Association withdrew from the battle against smoke.

These events unfolded even though the Citizens' Association had intensively supported the work of official agencies. Time and again, association members had appeared in court to help secure a verdict against a business.[44] The Chicago Society for the Prevention of Smoke, founded in 1892 with a view toward the following year's Columbian Exposition, a group whose antismoke activism picked up on the work of the Citizens' Association, even hired its own attorney, who brought a number of suits. In St. Louis, a "smoke philanthropist" by the name of James Gay Butler gave the Women's Organization for Smoke Abatement funds to pay for a lawyer, as well as six inspectors to monitor chimneys.[45] Understandably enough, such enforcement of municipal laws by private organizations drew intense criticism: "There is a man with more money than sense behind these prosecutions," someone griped about Butler's private police.[46] But the problem lay much deeper: in the course of such campaigns, prosecution of violations of the law regularly became an end in itself. Editorials were strident: "We will not exclude anyone from prosecution in our hunt for the lawbreakers"; "All offenders should be made to feel the power of the law."[47] "The department is trying to bring as many offenders to court before the spring months are apprehended," New York's health commissioner proclaimed resolutely, and Cincinnati's Smoke Abatement League reported to its members that it had retained one of the city's best lawyers "to 'fight the case to the bitter end.'"[48] Even Pittsburgh's Chamber of Commerce dismissed critics with a brusque comment: "Such protestations have to be ignored."[49]

Of course, the zeal with which officials and private organizations pursued these suits merely glossed over the internal contradiction of this strategy: the same entrepreneurs who were to be "enlightened" and "converted" (to use the language of the smoke fighters) were time and again put off by this approach. A person who was immediately threatened with legal action if he generated smoke would obviously find it difficult to develop a positive relationship with antismoke efforts. Moreover, the prosecution of smoke-belching enterprises was often laced with rhetoric that aroused little sympathy among entrepreneurs and industrialists. While the Anti-Smoke League of Baltimore assured the mayor in a letter that the organization felt no hostility toward industry, it did not hold back, making direct reproaches: "There is a very serious question as to whether a city really profits by a factory which is so run as to damage the neighborhood in which it is."[50] The Smoke Abatement League in Cincinnati reacted similarly: a notorious polluter, the league declared, "must be looked upon as one who openly defies the law, the rights of the people and the demands of municipal decency and should be treated as an enemy of the public welfare."[51] Although the league emphasized in the same passage that it had no sympathy for any kind of anti-industrialism, and that avoiding smoke was moreover "a source of economy to manufacturers who do it," such protestations naturally rang a bit hollow when these same manufacturers were arrested one by one, at the league's instigation.[52] When unhappy smoke fighters went so far as to call for one-thousand-dollar fines and jail terms and published the names of accused and convicted enterprises, the *Chicago Record-Herald* understandably declared, in 1902, that the result of the fight against smoke to date was merely "a spirit of bitterness between the health department and the men it seeks to reach."[53] Some smoke fighters began to realize that they had maneuvered themselves into a no-win situation: "Smoke inspectors are invariably damned if they do prosecute offenders, and damned if they don't," an official remarked at the second conference of the International Association for the Prevention of Smoke in 1907.[54]

The manufacturers' protests could be summed up in one sentence: "Smoke means prosperity." This slogan, tirelessly repeated by irate industrialists, attests to a deep displeasure over the approach of the smoke fighters, essentially casting them as opponents of progress.[55] The Illinois Manufacturers' Association, for example, spoke of "one of the many foolish agitations now being directed against the manufacturers of the state"; elsewhere, factory owners emphatically invoked the "smoking chimneys of prosperity."[56]

In several cities, the opposition was even formally organized. In 1897, for example, a Manufacturers' and Citizens' Protective Association of Chicago was set up, which unabashedly advocated "relief from the injustices of the smoke ordinance"; St. Louis and Baltimore had similar organizations.[57] This opposition was not necessarily narrow-minded: in some cases, protesting manufacturers might very well have a clean conscience. As a case in point, Chicago's Health Department declared that many factory owners were investing substantial sums in good furnaces but that the latter were often rendered defective or unusable through negligent operation. That the owners were in turn held criminally liable for this situation did seem, as officials themselves conceded, "almost like persecution, after they have devoted time and money in their efforts to comply with the ordinance"—but this consideration did not deter officials from taking action.[58] Similarly insensitive behavior was displayed by Cincinnati's Smoke Abatement League: while it likewise recognized that many of the manufacturers being sued had certainly not been inactive, it solved this problem by dragging owners, managers, and furnace operators into court to publicly clarify the question of responsibility.[59] That such an approach met with an unfriendly reception from manufacturers should come as no surprise. The manager of one factory in Rochester declared tersely, "We do not cherish the idea of being summoned before Court because our fireman has been careless."[60]

But what angered manufacturers more than anything else was the authorities' strict refusal to assist in solving the technical problems. "Manufacturers would cheerfully assist if the Department could recommend a sure cure for the evil," declared Chicago's Health Department, and the boiler inspector in St. Louis expressed a similar sentiment.[61] Both agencies, however, refused to take up this recommendation: "It is too dangerous a practice for me to make suggestions," was the official explanation in St. Louis.[62] As a result, furnace owners, with inadequate technical information at their disposal, were left to their own devices in the search for remedies. And when the Chicago Health Department, in its annual report, laid out detailed information on various methods for avoiding smoke but in the end presented the report itself as merely a nonbinding account of practical experiences that was by no means intended to promote recommendations, this must have struck prosecuted manufacturers like a provocation. Here was an agency that knew a great deal more about the techniques of smoke abatement than they did but that shared this information with no one, apparently dead set instead on simply persecuting smoke-generating enterprises.[63] Many manufacturers therefore were reluctant to deal with smoke abatement, opted for stopgap countermeasures to outwardly comply with the wishes of the agency, or behaved like the entrepreneur who told the Civic Improvement League in St. Louis that he would rather pay a monthly fine of fifty dollars than grapple with smoke-reducing equipment.[64] "He needs enlightenment," the league noted in response, but this merely reveals the ineffectiveness of a strategy that relied on prosecution in the face of reluctant manufacturers.[65]

Few people initially realized that manufacturers' wish for technical advice could also provide an opportunity for smoke abatement efforts. After all, the focus on criminal prosecution had never been a consciously chosen strategy; rather, it was the legal manifestation of the cumulative anger that had built up among antismoke activists over the continuing deterioration of the air quality in large cities. The notion of helping smoke-belching companies solve their technical problems therefore struck antismoke activists as rather absurd at first: "It cannot be expected, nor can the city be expected, to furnish manufacturers with consulting engineers," explained the president of the Smoke Abatement League in Cincinnati.[66] The Civic League of St. Louis initially would not hear of a consulting function either: in one report it demanded that polluters simply be hit with penalties "until it becomes cheaper to prevent smoke than to defy the ordinance."[67] Still, in the realm of practical smoke abatement measures, some took a conciliatory approach that took manufacturers' technical problems into consideration. In several cities, certain modifications, inconspicuous at first sight, signaled a turning away from rigid adherence to the strategy of prosecution. One popular conciliatory method was to postpone legal proceedings if a manufacturer was trying to address his problems.[68] Flexibility also existed when it came to setting fines, which could be used to make subtle differentiations: "I'm not going to impose the maximum fine upon any corporation or individual that gives the satisfactory proof of having taken steps to comply with the law," a judge in Chicago openly declared.[69] Interestingly enough, a similar approach developed—independently—in Chicago, St. Louis, and New York, where a suit was filed only after several written warnings had failed to produce any results.[70] When the relevant agency in St. Louis was accused of being too lax in its efforts, it responded by saying, "While in dealing with violators of the smoke ordinance, we were dealing with men who might be considered, technically, 'criminal,' we at the same time had to do with the best citizens of St. Louis," defending its decision "to conciliate [entrepreneurs] instead of flourishing a big stick."[71] Even the dogma against rendering technical advice began to crumble from time to time. For example, the relevant official in the Chicago Health Department—the very same person who a year earlier had proudly declared that his agency always responded to such inquiries evasively—noted that in response to complaints, he had on numerous occasions used "suggestions" to avoid initiating an official suit.[72] In 1892, a commission appointed by the mayor of St. Louis even issued the direct recommendation that the city, as a supplement to the usual local ordinance, establish a separate committee for experiments with smoke-reducing heating, so as to create an unbiased source of technical information for affected manufacturers.[73] Initially, these were merely attempts to keep the antagonism between entrepreneurs and agencies within certain limits and to make it easier to pursue prosecution in the day-to-day work of the bureaucracy. Yet the mere fact that such approaches took shape, tacitly accepted as unavoidable concessions, indicates that the prosecutorial strategy was in crisis. In retrospect, these behaviors thus point to the transition to a different approach, one that envisaged, alongside punitive measures, positive incentives for manufacturers. These steps condensed into a comprehensive strategy after 1905, when the American movement against the smoke nuisance entered a new phase.

The Invention of the Smoke Inspector

In 1906, the City Club of Chicago took up the issue of urban smoke. At first, it looked like the usual campaign against smoking manufacturers and their accomplices: "Smoke is about as necessary to industry as a dirty factory is to good business," the chair of the "smoke committee" declared emphatically, and the club filed a suit for negligence against the official in charge with the Civil Service Commission.[74] It was not long, however, before a more moderate tone crept into the City Club's statements. In part, this was because the members of the City Club were businesspeople themselves and therefore had certain reservations about an overly rigid policy of prosecution; in part, it was because the organization from the outset drew on the expertise of engineers. From the very beginning of its work, the relevant committee had hired a mechanical engineer by the name of Robert Kuss as its secretary, and for a "Smokeless Chicago Meeting" in February 1907, the City Club recruited Lester Breckenridge of the University of Illinois as a speaker.[75] Not only did the City Club thus gain technical competency, but it also encountered ideas about departing from the paths of prosecution and punishment. "This problem can never be solved except by hearty co-operation," Lester Breckenridge declared in his speech, suggesting the establishment of "some sort of a bureau" as an information clearinghouse and advisory agency.[76] Such talk obviously impressed the members of the Committee on Smoke Abatement: when they presented the first annual report about the committee's work in April 1907, they made it clear that merely firing the sitting officials would not be sufficient. Rather, the City Club would now opt for a cooperative approach, and it already had concrete ideas about what this might mean: "The engineers and assistants employed would inspect the plans for proposed installations; inspect completed work, determine by observation the effectiveness of the installed plant and aid in every way possible to adjust the conditions to conserve the interests of the plant owners as well as the general public."[77]

At first glance, advice and education might seem weak substitutes for prosecution and fines, but it is important to recognize the technological peculiarities of smoke abatement. From a chemical standpoint, smoke and soot were the typical by-products of imperfect combustion. Thus, smoke abatement was largely synonymous with optimizing firing conditions, and this usually went along with gains in the efficiency of coal use—obviously an attractive prospect for consumers of fossil fuels. But at the same time, optimizing firing conditions required attention to a wide range of factors: type of fuel, furnace construction, operation, boiler type and size, and chimney draft had to be aligned to achieve combustion without smoke and soot. Engineers were confident that they could find a solution for each decent boiler that minimized smoke, but this required attention to the specifics of each individual furnace; no panacea would generally do away with smoke. Looking at these specifics presented a quintessential engineering challenge, one that few nineteenth-century entrepreneurs bothered to take up by themselves. As a result, surveys of firing conditions usually painted a dire picture. One engineer, who in 1906 inspected the furnaces in central St. Louis at the behest of the Civic League of St. Louis, found only 15 percent of the installations in good condition; 55 percent had deficiencies, and 30 percent he labeled simply as unacceptable: "hot, poorly ventilated, dark and cramped for room." The engineer's notes about individual setups were harsher still, as when he quoted the foreman of one installation with twelve boilers as saying, "Only negroes can be retained as firemen because of the close quarters and intense heat."[78] In short, an opportunity existed to apply engineering knowledge in such a way as to minimize smoke and save coal, and it was this combination that the City Club had in mind when stressing advice and cooperation.

Such considerations evidently met with a sympathetic reception from Mayor Fred Busse (in office since April 1907), a coal merchant with German ancestors who owed his election largely to the support of probusiness circles. At any rate, the program that was institutionalized in Chicago in the summer of 1907 bore a striking resemblance to the ideas of the City Club: henceforth, the city's smoke inspector was to function not only as an enforcement agent but also as an advisor to the city's business community. "The fundamental idea underlying all of the work of the department has been not only to complain to a citizen about the smoke and insist that it be stopped, but to show him how to stop it and to give him assistance in every possible way toward that end," declared a report from the Department of Smoke Inspection.[79] In this spirit, smoke inspectors first carried out a detailed inspection of the technical installations in smoke-generating enterprises, at the end of which the entrepreneur was presented with concrete recommendations for improvements. Punishment resulted only when the business was still producing smoke after an adequate deadline for implementing the recommendations had passed and the entrepreneur had shown himself to be insufficiently cooperative.[80] A clever system of monitoring chimneys made it possible to control with great precision whether regulations were being followed.[81] Eventually, too, new construction or refitting of furnaces had to be supervised and approved by the authorities to prevent unsuitable installations from being set up in the first place.[82] This, in a nutshell, was a comprehensive program to systematically improve the status quo of furnace technology.

This new approach quickly met with enthusiastic approval. The *Chicago Record-Herald*, at the time a kind of mouthpiece for Chicago's battle against smoke, was soon full of praise: "There is no warrant whatever for questioning the good faith and the steadfastness of the commission and the inspector," one commentary declared after two years.[83] Others could only agree. A Chicago engineer who had complained bitterly about local conditions in a book published in 1906 proclaimed in the revised 1915 edition that the city had assumed the leadership in the battle against the smoke nuisance and now possessed "probably the best organized smoke inspection department in the world."[84] Mrs. Culver, chair of the "smoke committee" of the Outdoor Art League, commented with respect on the approach of the new chief engineer, Paul Bird: "He is doing the best he can under the circumstances."[85] And the City Club gave itself a big pat on the back: "We believe that it is a matter for congratulation on the part of the City Club that such results have accrued from its original undertaking," a committee report noted in 1908.[86]

The new approach began to receive acclaim from beyond the city as well. Within a matter of months, Chicago's strategy became the shining example that large American cities sought to emulate. "A dawn of better methods is at hand," announced one article as early as 1908, while the professional journal *Power* wrote five years later that Chicago's department was "probably more widely and favorably known than any other in the United States."[87] "The country has come to look upon the Chicago Department as the ideal one," noted John O'Connor of Pittsburgh's Mellon Institute. The

U.S. Bureau of Mines recommended a local ordinance that corresponded to the Chicago model in all but minor details, and the first official act by Baltimore's smoke inspector was a trip to Chicago to gather information about the new approach.[88] In other cities a similar strategy carried the day, without direct reference to Chicago. The Smoke Abatement League in Cincinnati, for example, embraced the new strategy in several steps. At first, it was interested—in the words of its president—only in the chimney's upper end: "What goes on at the lower end of the stack is the exclusive business of . . . the proprietor, the engineer and the fireman."[89] Three years later, however, the league organized a meeting on the premises of the Business Men's Club at which entrepreneurs were brought together with technical experts.[90] And two years on, the league's fundamental approach was reconsidered: "There was a time when harsh, severe measures were the best, the only ones to employ; . . . but that time should have passed," declared the annual report for 1911. A year later, a municipal committee, under the chairmanship of the league's vice president, drafted a new local ordinance that was largely in line with the Chicago approach.[91] In Pittsburgh, the Smoke and Dust Abatement League was behind the passage of two local ordinances that had been drawn up by a committee of technical experts appointed for this very purpose; in St. Louis, meanwhile, the Civic League, the Business Men's League, and the Women's Organization for Smoke Abatement joined forces to come up with a concept that reformed the local fight against the smoke nuisance.[92] Boston, too, saw the establishment around 1910 of a combination of advice and punishment. Here the model was not Chicago, though, but the Hamburg Society for Fuel Economy and Smoke Abatement. While the assertion that the inspiration for this new approach came from a German seaport was not correct in terms of its actual content, this attribution probably had a better ring in Boston than invoking a railroad hub somewhere out West.[93]

Nevertheless, environmental historians have often taken a critical view of this change from prosecution to advice, labeling it, for example, a "go-slow, educational approach."[94] But this perspective overlooks a crucial point: smoke-generating enterprises continued to be prosecuted. What really changed was not punishment as such, but its function: no longer the only instrument of control, punishment now had as its sole purpose to compel unwilling entrepreneurs to cooperate. "The Department has continued to prosecute violators, when all other methods have failed to accomplish the desired results," reported the responsible official in St. Louis.[95] Raymond Benner, the head of Pittsburgh's Smoke Investigation, also favored this stance: "The legal appeal should be the last appeal. If, however, such an appeal is made it should be made effectively."[96] Chicago's smoke inspector

Paul Bird was even more blunt: "If we do our part to assist the smoke violator to correct his habits we can prosecute with a better grace."[97] These were by no means empty words: between October 1, 1909, and September 30, 1910, Bird filed no fewer than 1,040 suits against smoke-belching businesses.[98] The level of punishment also rose dramatically: since fines were now no longer the rule but the ultima ratio in cases of stubborn recalcitrance, fines of five hundred dollars and more were repeatedly levied in Chicago—a sum unheard of during the first phase of the city's antismoke movement.[99]

From this perspective, the smoke inspectors' recipe for success was that they knew how to combine the demands for effective antismoke action, including credible enforcement, with the concerns of entrepreneurs. The result was a broad coalition of winners. With smoke inspection, entrepreneurs gained technical advice on matters of furnace technology, which they had been asking for from the outset. In addition, they profited from the fact that measures for smoke abatement frequently also improved the efficient use of coal—a fact that smoke fighters never tired of pointing out.[100] For instance, the Civic Improvement League of St. Louis noted that "in every case the use of the better apparatus is accompanied by a saving in fuel amply large to pay a good return on the necessary investment"; others promised the entrepreneurs "marvelous reductions in fuel cost."[101] Arthur Hamerschlag, both the founding president of the Carnegie Institute of Technology and chair of Pittsburgh's Smoke and Dust Abatement League, went even further in 1915: "The dividends earned by coal and iron users installing efficient apparatus for eliminating waste are greater than from any other similar investment in the plant."[102] Further, the emphasis on coal savings established a connection to the theme of efficiency and thus to a key contemporary buzzword: smoke inspection was thus "in harmony with the best spirits of the times."[103] Whoever was active in antismoke efforts, therefore, could feel himself at the forefront of progress: "Every part of modern manufacturing is tending toward the elimination of waste, and the saving on the smoke is directly in the line of progress."[104]

This kind of promotion was not without effects in business circles. Among industrialists, at any rate, it soon became common knowledge that avoiding smoke and economizing furnace operation pointed in the same direction, and the fight against smoke gained enormous popularity among entrepreneurs.[105] A poll by the Boston Chamber of Commerce revealed that 1,148 of its members supported the new approach, 60 were opposed, and the number of invalid votes was 12.[106] The City Club of Chicago uncovered a similar result in a 1910 survey focusing exclusively on manufacturers who had been sued over smoke emissions the previous year. Most entrepreneurs readily admitted the facts, and in other respects, as well, the defendants found surprisingly little to complain about in the conduct of the authorities.[107] In Pittsburgh, too, industry opposition had evaporated: "The time has arrived when the city's right to regulate its smoking stacks and chimneys is no longer disputed," explained the head of the city's fight against smoke in 1914.[108] In fact, the number of unhappy voices from business circles declined noticeably after 1910; even in the industrial city of Lowell, Massachusetts, the official in charge found only three or four companies in 1912 that were fundamentally opposed to antismoke efforts.[109] This seems peculiar only at first glance. During the Progressive Era, any protest against an improvement in "efficiency" required courage, in addition to a good dose of obliviousness to the Zeitgeist. And in any case, the chances of prevailing against the elaborate set of punitive instruments in the hands of the smoke inspectors were not very good.

This change of heart among entrepreneurs pleased the members of the civic antismoke movement. After all, from the beginning, it had been their goal to push these men toward cooperation, and for all their zeal in prosecuting punishable emissions, this approach had never become unquestioned dogma. In fact, on occasion, activists had even voiced some discomfort over whether the smoke issue truly justified treating respected citizens like criminals—which comes as little surprise in a movement that was in large part recruited from these very respected circles.[110] At any rate, the antismoke movement transitioned to smoke inspection with remarkable unanimity. Some organizations—for example, the Smoke Abatement League in Cincinnati, the Chamber of Commerce and the Civic Club in Pittsburgh, and the Civic League along with the Women's Organization for Smoke Abatement in St. Louis—even followed the trajectory from prosecution to consultation backed by sanctions over the years. To those involved, the change in strategy seemed almost like an inevitable process to which there was no alternative: "It was generally felt that . . . the Department should be so organized that the citizens could receive more instructions as to what to do to prevent smoke and have sufficient time to bring their plants up to the standard," explained Sarah Tunnicliff, chair of the Clean Air Committee in the Woman's City Club of Chicago.[111] And the Women's Civic League of Baltimore called the new approach simply "the only logical way."[112]

At the same time, the rise of the smoke inspector in no way meant that civic organizations lost importance. On the contrary, they continued to enhance the work of the smoke inspectors in a constructively critical way. The work of these groups thus typically had two sides: on the one hand, civic organizations functioned as political lobbies of official organs; on the other hand, they tried to control administrative work as thoroughly as possible, in a variety of ways. In the case of Chicago, this control was assured by the fact that Thomas Donnelly—the head of the City Club's campaign—had been appointed chair of the Smoke Abatement Commission that the city had set up to provide oversight.[113] In addition, in 1908 and 1909, the City Club carried out detailed investigations into the conduct of city officials, bringing about a number of modifications aimed essentially at stiffer punishments for smoke-generating businesses. Yet another survey examined whether bureaucratic implementation of the regulations was indeed free of corruption.[114] The Chicago Woman's Club's activities were similar: on the one hand, it strove to prod the officials in charge to work in the most effective way possible; on the other hand, it did everything in its power to procure the necessary political backing for this approach.[115] In St. Louis, the Civic League and the Women's Organization for Smoke Abatement jointly complained to the mayor when they saw the independence of municipal officials threatened. Pittsburgh's Smoke and Dust Abatement League discovered, through a review of an official report, that the measures taken by smoke inspector J. M. Searle had been inadequate in three of four cases; after his dismissal, the league set up its own committee "to keep in touch with the work of the Bureau of Smoke Regulation."[116] When the Women's Organization for Smoke Abatement in St. Louis was unhappy with the work of city officials, it temporarily hired its own inspectors to keep tabs on chimneys and had a lawyer bring observed violations to court as a way of putting pressure on the smoke inspector—who, in response, initiated many more criminal proceedings than he had before.[117] And Pittsburgh's smoke inspector J. W. Henderson (Searles's successor) came under intense pressure from civic reformers when the Smoke and Dust Abatement League began to push for criminal proceedings around 1915.[118] Revealingly, Henderson's chief problem was a dearth of suitable candidates to take to court: no entrepreneur would rashly pick a fight with an alliance that comprised nearly all important circles of the urban middle class, from the Chamber of Commerce to the Allegheny County Medical Society, to the two large local universities and the Congress of Women's Clubs.[119]

These episodes, in particular, reveal that cooperation with bourgeois reformers could very well turn uncomfortable for city officials. Still, these officials were generally convinced that such cooperation was indispensable: "It is absolutely essential that a smoke inspector have the backing of a well-formed and strong public opinion," explained Paul Bird.[120] These were not just empty words. After all, the city engineers owed their jobs to such reform associations, which also offered effective political flank protection. In at least two cases, officials themselves came out of the antismoke movement: E. P. Roberts, chief smoke inspector in Cleveland from February 1912, had been for three years the chair of the committee to combat smoke in the local chamber of commerce; Robert Kuss, after working for the City Club of Chicago, had similarly joined the local Department of Smoke Inspection.[121] "Unless [the city smoke inspector] can get the cooperation of the public, of the citizens, he has an up hill job," explained Cleveland's smoke inspector.[122] It would therefore be wrong to speak of a change of leadership in the antismoke movement: it was in fact dominated by intensive, although not always conflict-free, cooperation between engineers and technical laypeople, and a preponderance of the engineers did not develop until the period after World War I.

While the change in strategy thus also had advantages in the eyes of the civic antismoke movement, engineers clearly derived the greatest benefit from it. After all, it was now naturally presumed that the fight against smoke was to be undertaken by engineers with expert knowledge of furnace technology: only they possessed the technical know-how to provide entrepreneurs with concrete and reliable recommendations. "Practical results cannot be expected unless the department is filled with men who are qualified, by training or experience in the field of engineering, to fill the position," proclaimed the Civic League of St. Louis as early as 1906. As Chicago's Thomas Donnelly declared apodictically, "Smoke Abatement is not primarily a matter of legislation, but of engineering and administration."[123] This was not something furnace technicians needed to be told twice: within a few years, smoke abatement had become a solidly established area of work for engineers. And engineers repeatedly noted proudly that their work made possible a solution that was agreeable to all sides: "Smoke abatement is one civic activity that results, generally, in profit to the abater as well as in good to the community."[124]

Of course, such statements sought to cultivate the engineers' self-image as problem solvers par excellence, but they did aptly reflect a general sentiment. For example, the Rochester Chamber of Commerce opened a 1911 brochure on smoke abatement with a pronouncement: "Economy, civic pride and common sense all call for the abatement of the smoke nuisance."[125] The observation that there was in the end no reason for the smoke nuisance to continue to exist goes straight to the heart of the new approach. It was no mere coincidence that smoke inspection often emerged from the interaction among businesspeople, reform associations, and representatives of official agencies: advice backed by sanctions had advantages for all actively involved groups, without entailing direct disadvantages for other groups. A result of imperfect coal combustion, the smoke nuisance allowed a solution that put no one at a disadvantage—and at the same time, no one had come up with a better strategy. Chances are that smoke inspection was the best possible approach under Progressive Era conditions, and it would take no less than a quarter-century before an even better strategy against smoke was put to the test in St. Louis. The stagnation of the smoke abatement movement during most of the interwar years was indeed a direct result of the dramatic advances made during the Progressive Era.

Increasing Momentum

After several years of smoke inspection, antismoke activists were generally confident. "All agree that there has been a great improvement in smoke conditions . . . within the last year," the Chicago Woman's Club claimed in 1910. The journal *Power* described the change in Pittsburgh as "almost beyond belief."[126] The Smoke Abatement League in Cincinnati claimed in 1917 that emissions from industrial plants, office buildings, and trains had dropped by 75 percent.[127] Optimistic remarks of this kind were indeed not mistaken, as measurements generally indicated a decline in smoke emissions. The Pittsburgh office of the U.S. Weather Bureau recorded a continuous drop in the number of smoky days since 1912, and measurements made by the Mellon Institute in 1923 showed that the precipitation of smoke and soot particles had gone down by a full 70 percent, compared to identical measurements made ten years earlier.[128] In Chicago, the Woman's City Club calculated that the number of observed violations per hour and inspector had declined by 24 percent between 1910 and 1923, even though the agency's budget had been slashed nearly in half over the same period.[129] Finally, a comparative study at the end of the 1920s confirmed the general impression: Baltimore, which lagged behind the general trend and did not pass a useful local ordinance until 1931, had suddenly become the dirtiest city in the United States, even though the situation there had been assessed as "comparatively easy of solution" as late as 1912.[130]

With the merits of smoke inspection mostly undisputed, the smoke abatement drive gained momentum, and the sense of crisis among antismoke activists around 1900 became a fading memory. Perhaps the best indicator of the new climate of vigorous optimism was the establishment of a professional association of smoke abatement officials in 1906.[131] The International Association for the Prevention of Smoke, which changed its name to the Smoke Prevention Association in 1915 (the international character of the association stemming merely from the membership of a handful of Canadians), quickly acquired crucial importance within the antismoke movement. At the association's annual conferences, members discussed technical and legal problems in the fight against smoke and worked toward standardizing and consolidating their efforts. The resonance that these conferences achieved was certainly substantial: for example, the eighth conference, held in 1913 in Pittsburgh, drew about 250 attendees.[132] But there was even more going on at these meetings. They allowed an esprit de corps to be forged, not least by the sharing of war stories. The story of a Memphis inspector chasing a smoke-belching locomotive all over the city on his motorcycle, while taking incriminating pictures, makes it clear that the pronounced self-confidence of smoke inspectors also had its roots in these meetings.[133] In any case, there was little room for self-doubt at the conferences: "I believe that smoke elimination is the greatest single movement that the world has undertaken up to this time for the advancement of human happiness and health," declared the association's president at the 1915 conference.[134]

Needless to say, the American antismoke movement was not able to achieve a comprehensive victory over the smoke nuisance. To begin with, there were physical and practical limits to smoke abatement: even the most agile smoke inspector could not prevent the production of light smoke below the punishable threshold. In addition, the American fight against smoke almost entirely ignored a substantial group of emitters, domestic furnaces, not least out of strategic calculation: it was easier for activist citizens to take up the fight against industrial furnaces than to set their own houses in order. Furthermore, the meager personnel resources of municipal offices imposed limits on what was possible; even the Chicago Department of Smoke Inspection, comparatively well funded at $48,000, considered its personnel level inadequate and demanded that its budget be increased to between $125,000 and $150,000.[135] Above all, however, the fight against smoke in the United States was hindered by mobile sources of smoke, among them especially locomotives; the antismoke movement only recorded partial successes in this regard. Locomotives in fact contributed considerably to the urban smoke nuisance: the Chicago Department of Smoke Inspection blamed them for 43 percent of the city's total emissions, and even the pro-railway study conducted by the Chicago Association of Commerce put their share at precisely 22.06 percent.[136] Moreover, the technical possibilities were significantly more limited for locomotives than for conventional installations. Locomotive furnaces had a substantially higher power density, given the same grate area, than stationary furnaces, and under certain operating conditions—for example, a heavy tow weight or an inclined track—it was hardly possible to avoid a stepped operation and thus the generation of a lot of smoke.[137] Still, railway companies took pains to limit avoidable emissions after the antismoke movement had attained a certain weight of its own—

even critical observers granted as much.¹³⁸ It was not rare for these companies to hire personnel at their own expense to bring their smoke problem under control, and in some cities—among them Chicago and Cincinnati—railroad smoke inspectors merged into distinct supervisory organs.¹³⁹ While the work they did should not be underestimated—indeed, the Joint Smoke Inspection Bureau of the Railroads of Chicago had more employees than the much-vaunted municipal Department of Smoke Inspection—their efforts were never able to truly satisfy the champions of the American fight against smoke.¹⁴⁰ And so people continued to look for other solutions—and this search, ironically, held the seed of a tremendous defeat for the antismoke movement.

When a state law compelled the railway lines into Manhattan to electrify, reformers saw this move as the precedent they longed for, even though the New York law was only indirectly concerned with the smoke nuisance: it was a response to a horrific accident in the smoke-filled railway tunnel under Park Avenue, caused by an express train that had gone through several signals for lack of visibility.¹⁴¹ Moreover, American smoke fighters willfully ignored the fact that the conversion concerned exclusively passenger traffic.¹⁴² Electrification was now everywhere seen as the dictate of the hour: "The word 'electrification' seems to be attractive to everyone, save those charged with the responsibility of obtaining the money required to meet the enormous outlay of capital," one representative of the Pennsylvania Railroad aptly noted.¹⁴³ In retrospect, the antismoke movement's electrification dreams seem overly ambitious: their goal became comprehensive electrification, rather than using electricity first where its strengths could be fully utilized—for example, in suburban traffic with its frequent braking and acceleration or on tunnel routes.¹⁴⁴ In Chicago, there was even serious discussion about electrifying the world's largest railroad junction, even though there was not a single electrified freight terminal at the time—a project that took the antismoke movement to the very threshold of megalomania.¹⁴⁵ Still, the smoke fighters pursued this goal with a zeal that can probably be fully understood only if one considers that railway companies were at the time among the most unpopular large enterprises.¹⁴⁶ Needless to say, activists were not successful: protest against the smoke nuisance had grown into a broad movement, to be sure, but forcing companies that were intent on minimizing their fixed costs to invest millions of dollars was obviously beyond the movement's power.

It is important to acknowledge this defeat, as it shows that the smoke inspection approach had its limits. It brought only gradual change to the problem of railroad smoke and did virtually nothing to combat domestic smoke emissions. But with respect to emissions from industrial and commercial boilers, it is hard to deny the merits of smoke inspection: it clearly encouraged the use of more sophisticated firing equipment and led to significant reductions in urban smoke emissions. Of course, it did not rid cities of smoke within months, as early activists had hoped; but given the pervasive emissions, these expectations had been unrealistic from the onset. Even more important, smoke inspection was not only an effective but also an efficient strategy. In view of the limited resources of the antismoke movement, the division of labor between municipal and civic forces was a clever move: while smoke inspectors were doing the time-consuming and technologically demanding job of inspecting, civic leagues focused on monitoring inspectors' performance and fending off political interference. Furthermore, the combination of advice and control meant that enforcement targeted specifically those industrialists who thought that smoke abatement was nonsense, and the group of obstructionists shrunk rapidly as a result. Because every industrialist who honestly tried to do something about his smoke problem was safe from prosecution, the business community had effectively made its peace with smoke abatement by the early 1910s. The bitterness of the early smoke abatement drives thus quickly dissipated, and smoke abatement became uncontested to an extent rarely found in environmental regulation.

Notably, the debate over strategies died down within the smoke abatement movement: except in the case of railroad electrification, nobody felt that a different approach, even if superior, had any chance under the prevailing conditions. To be sure, not all cities adopted the smoke inspection approach during the Progressive Era. West Coast cities favored petroleum as their fuel, thus making efforts at coal-smoke abatement pointless.[147] New York City drew much of its coal from anthracite deposits near Scranton, Pennsylvania, and since anthracite burned smokelessly even under adverse firing conditions, the city pursued something of a unique path in smoke abatement until far into the twentieth century.[148] From the viewpoint of the smoke abatement community, perhaps the greatest disappointment was that smoke inspection initially failed in the South. There were no significant activities in New Orleans before World War I, Baltimore saw the collapse of smoke abatement work after the local smoke inspector was fired under somewhat obscure circumstances in May 1915, and the state of Alabama even passed a law in 1915 that specifically prohibited municipal smoke abatement.[149] But these cities would eventually catch up with the national trend, thus demonstrating that Chicago's 1907 approach had gathered an almost irresistible momentum. For decades, smoke inspection would be the American way of fighting air pollution.

Germany, 1880–1914

By the eve of World War I, the United States had a multitude of civic leagues supporting smoke abatement, smoke ordinances in most major cities and cadres of engineers to enforce them, and a powerful professional association. Comparatively, the balance sheet of German smoke abatement looked bleak. There were few associations, civic or otherwise, that worked on the smoke nuisance, and no national organization. While the American drive against smoke was gaining momentum, the German fight seems dispirited in retrospect: most urban areas entertained a lukewarm interest in the issue at best, with bureaucrats focusing mostly on incoming complaints while they worked with inadequate laws; for many officials, the favored remedy was raising the smokestack so that emissions would be carried beyond the polluter's immediate surroundings. Some cities had found promising solutions, Hamburg maintaining the most energetic fight with its Society for Fuel Economy and Smoke Abatement, but these accomplishments remained the exception. Germany did have an academic journal titled *Rauch und Staub* (Smoke and Dust) since 1910, something that the more practically oriented American smoke inspectors did without until after World War II. In all likelihood, this journal was the first of its kind worldwide, but its fate is revealing: after the first few issues, it shrank dramatically in size, narrowly survived World War I and its aftermath, and then largely abandoned pollution issues in favor of other themes. After several changes of publisher and vain attempts to make the journal more profitable, it finally ceased publication in 1933.

In its first issue, *Rauch und Staub* defined its mission as "providing a meeting point for the movement against smoke."[150] This was right on the mark, for the German smoke debate was much more fragmented than its American counterpart. While smoke was the subject of a lively and controversial exchange among businesspeople, officials, engineers, and civic activists in the United States, these groups remained mostly isolated in Germany, rarely trading views and opinions with the same candor that characterized debates in Progressive Era America. It is rewarding to examine these groups separately, analyze their interests and perceptions in both countries, and trace how they contributed to each country's specific path. Such an approach makes it clear that the American crusade, successful as it was, relied on far more than an energetic citizenry—and that the German debate originally started off much more promisingly. Germany, in fact, had a regulatory tradition on smoke that went back at least to the mid-nineteenth century, an efficient and mostly corruption-free administration on the state and municipal levels, and a business community that had learned to live with bureaucratic interventions. But building on these positive elements, and using these resources, turned out to be much more difficult than expected.

Citizens Filing Complaints

The most striking difference between the German and American smoke debates was the degree to which affected city-dwellers participated. While dozens of associations took up the smoke nuisance in the United States, the pressure of an organized public was almost completely absent in Germany. No countrywide organization was able to get off the ground: the Austrian Society for Combating the Smoke and Dust Nuisance had no German counterpart.[151] This was true in spite of the fact that citizens' discontent was clearly as pronounced in Germany as in the United Sates. In Dresden, for example, a member of the city council declared that the smoke issue was a "question of culture, a battle between dirt and cleanliness."[152] A citizen of Bremen lodged a formal complaint "that the maids had to clean the porches and sidewalks three and four times this morning, since the flakes were falling as thick as snow in winter."[153] As in the United States, cleanliness and property damage were the crucial concerns, health only a distant third: indeed, a member of Stuttgart's city council noted that smoke was "more a matter of comfort than a matter of health," even while arguing for a smoke ordinance.[154] At a session of the Munich city council, no one objected to the statement that "every reasonable person who has the general welfare in mind will have the desire to eliminate the very troublesome smoke and soot nuisance."[155] Counterarguments along the lines of the American saying "Smoke means prosperity" were hardly ever heard in Germany. For Germans, smoke abatement was simply a matter of common sense.

In Germany, however, this view was not put on the political agenda of an organized reform movement. Organizations and individuals only rarely pushed for a general reduction of smoke emissions, and these occasional initiatives were generally short-lived. The petition that the East Suburban District Association submitted to the Municipal Council of Leipzig in 1880 remained that organization's only statement on the smoke issue; the same is true for the letters written by the Association for the Promotion of Dresden and Local Tourism in 1899 and by the Hanover Association for the Improvement of the Eastern Section of the City that same year.[156] These petitions had as little appreciable effect as the effort of a Frankfurt citizen who in 1905 called in vain for much tougher measures to fight the smoke nuisance.[157] Still, in a few cases, civic initiatives did lead to certain reforms. For example, a local police ordinance was passed in Stuttgart—in the final analysis the result of the efforts of the lawyer and member of the state assembly August

Becher, who had sent a petition to the Württemberg Ministry of the Interior in 1881, in his capacity as chair of the Downtown Civic Association.[158] And in Chemnitz, a commission of engineers and business owners for several years advised the city council on technical questions.[159] Few organizations, however, were able to marshal the energy for sustained activity. The most active were probably the Association for Public Health in Hanover, which monitored local efforts for more than a decade, and the Polytechnic Association in Munich, which appointed two commissions on the issue.[160] The civic activities of American women are entirely without parallel in Germany, even though smoke must have also been a source of constant frustration for German housewives. According to one complaint, some women in Frankfurt did demand "that the entire neighborhood undertake a protest march to the municipal authorities to show them the cleaning rags." But this intention was never put into practice, in Frankfurt or elsewhere.[161]

The political value of these occasional initiatives was further diminished by the fact that they remained strikingly vague in their demands. While American antismoke activists had no inhibitions about drafting laws and accusing officials of negligence, German citizens spoke emphatically in generalities. August Becher left his demand at a simple "request for measures that would reduce chimney smoke as much as possible." The Association for the Promotion of Dresden and Local Tourism was barely more precise in expressing the desire "that the most complete possible burning of smoke become mandatory in all larger factory installations in and around the city." The commission in Chemnitz did not even take up the administrative aspects of the smoke issue.[162] This vagueness arose less from political incompetence than from the understanding that harsh attacks on the officials in charge would probably be counterproductive. "I must object to the way in which judgment has been rendered on the enforcement of the smoke ordinances by the police," Berlin's factory inspector Tschorn responded, hearing criticism of official actions at a meeting of the Association for the Promotion of Commerce and Industry (Verein zur Beförderung des Gewerbfleißes).[163] The distance that separated city residents from the administration, a classic theme of German history, here placed severe limitations on an open exchange of ideas.

Of course, there were thousands of complaints in Germany about individual smoke-belching chimneys, and officials—as we shall see—dutifully processed them. But it was hardly ever possible to move from complaints about individual emitters to a general debate about the poor quality of urban air: the myth of the German bureaucracy and the statist sentiment among the general public were simply too strong. Perhaps not everyone went so far

Figure 2. Smoking enterprise in Munich, ca. 1906. The German administration was quite effective when it came to complaints about individual chimneys, but finding a strategy to combat the urban smoke nuisance in general was a different matter. Image courtesy of the Stadtarchiv München.

as the author of a letter to a Stuttgart newspaper, who solemnly declared, "I have confidence that our authorities will, before further palpable harm is done, offer our good city and its residents the necessary protection by taking moderate steps against the transgressors."[164] However, the firm belief that the strong German state would take on the matter with its usual thoroughness is unmistakably present in many complaints: fighting smoke was a matter for the authorities, not, as in the United States, an issue that called for the involvement of all parties concerned.[165] On one occasion, a high school principal withdrew his petition with the explanation that "after the information given to me, [I assume] that the relevant authorities will see to it that potential damage to the neighborhood will be prevented."[166]

Those who lodged complaints thus placed their hopes in the strong state. But the image of the state was not without ambivalence: the power of the state in authoritarian Germany was both welcome and sinister— a behemoth that was better kept on a short leash. A considerable number of statements thus reveal that people were wavering between the desire to fight smoke and the fear that regulations would throw the door wide open to bureaucratic despotism. One member of Munich's municipal government voted for the local police ordinance "only reluctantly," demanding that the officials in charge of enforcement "be reminded ... always that they should not act in a harassing, but in a reasonable manner." A council member saw in the same regulation "the dubious advantage that in the hands of overzealous officials it can become another tool to harass homeowners."[167] It was not understood that one could address such concerns with a carefully devised strategy, so deep was the fear of being "left completely at the mercy of the building police": "Once the latter is headed by a zealous technician, the matter would be very serious."[168] The ideal of the engineer who was vigorous but fair in his actions, which characterized the American antismoke movement—"the smoke problem calls for broad men, men who have a sound fundamental training, men of imagination and men of force"—was always foreign to the German smoke debate.[169]

Industrialists without a Voice

Since the inception of the American antismoke crusade, businesspeople had played a prominent role in the ongoing debate. Some supported the smoke abatement drive from the beginning; others maintained that "smoke means prosperity." The German business community, in contrast, remained notably silent on the issue. Few industrial organizations bothered to talk about smoke, and when they did, they often produced lackluster statements: smoke abatement, a memorandum commissioned by the German Association of Industrialists admitted, was "a perfectly proper demand."[170] Accordingly, authorities noted an overwhelmingly cooperative attitude on the part of entrepreneurs. "In most cases the owners of furnaces readily admit that a smoke and soot nuisance from their installation exists and promise to remedy it as best they can," reported the Stuttgart machine engineer Emil Kerschbaum, and other officials reported similarly.[171] Braunschweig's factory inspector went so far as to regard the occasional punishment of industrialists as unnecessary, "because successes can be achieved also without preceding fines."[172] Having lived with a strong and self-confident bureaucracy for decades, industrialists usually refrained from fighting demands for smoke abatement tooth and nail. Of course, they did not like to be summoned for smoke violations, but they were realistic enough to know that, living in Germany, they would somehow have to make their peace with bureaucracy.

Added to this was the fact that the smoke issue touched on an area of entrepreneurial activity few industrialists wished to address. As a result, entrepreneurs largely exhibited a profound lack of interest: technical firing questions simply did not interest them. "There is still a whole host of companies where the boiler room is literally closed to the factory owner. He walks through all facilities, but almost regularly avoids the boiler room, where the sign 'No entry' is prominently displayed," a Berlin engineer wrote.[173] Even the chair of the Central League of Prussian Steam Boiler Supervisory Associations, who as the representative of an entrepreneurial organization had reason to show restraint, complained about the "laziness and indifference of owners of furnaces."[174] In short, from the entrepreneur's perspective, the furnace man's workplace was literally a "black box."

As a result, ignorance and helplessness were often typical reactions of industrialists when officials cited them for smoke violations. "The relevant inventions, and improvements accumulate . . . more and more, to the point where it is nearly impossible for those industrialists who are not specifically technicians to pick from the large number of smoke-burning apparatuses and smoke-consuming furnace systems those of good and best quality, and among those the ones best suited for the specific purpose," a technical journal noted toward the end of the nineteenth century.[175] Many entrepreneurs therefore cultivated a pronounced conservatism on matters of furnace technology: as long as technical changes were not absolutely unavoidable, things were left as they were. "Industry . . . is not very receptive to innovations in this area," the Düsseldorf machine engineer Pasinski noted with a tone of regret.[176] Entrepreneurs resorted with remarkable frequency to low-smoke types of coal or coke to eliminate smoke nuisances, since this—though costly—posed the fewest problems in terms of operational safety.[177] As the director of the Magdeburg Association for Steam Boiler Operation aptly noted, "The furnace is the *nervus rerum* of a factory, and every owner is right to shy away from experiments with the same."[178] Though antismoke activists might complain bitterly about entrepreneurial lethargy, the owners of commercial furnaces, if in doubt, preferred to follow a lethargic motto: No experiments!

Moreover, in the few cases in which organizations of entrepreneurs were able to muster the energy to send a letter of protest, their demands remained strikingly moderate—a far cry from the loud complaints with which American industrialists confronted antismoke activists. Following passage of a local statute, for example, the bakers' guild in Braunschweig merely asked "for an interpretation of the regulation that is favorable to the undersigned guild." The Bremen Chamber of Commerce, at the very beginning of an official statement, took "the general position of recognizing it as the duty of the authorities to pass regulations in caring for public health and cleanliness so as to diminish, or, to the extent possible, entirely eliminate the nuisance from soot and smoke associated with larger industrial enterprises, harmful factory waste water, and so forth."[179] Even the petition from the Association of Berlin Metal Industrialists—perhaps the strongest protest letter from a German entrepreneurial association—by no means advocated abandoning the fight against smoke altogether.[180] All the appellants were hoping for was a less bureaucratic handling of the regulations, one that acknowledged the peculiarities and technical complexities of this particular issue. "The industrialist engaged in a tough struggle for survival would—of this I am convinced—happily do everything in his power to spare his fellow citizens from any nuisance, only he wants to have as little contact as possible with the Berlin policeman," declared one building inspector at a meeting of the Association for the Promotion of Commerce and Industry.[181] German entrepreneurs did not dare to dream of anything more.

That German entrepreneurs were right to set themselves a limited goal from the outset is already evident from the fact that officials rigidly rebuffed even moderate demands. The bakers' guild in Braunschweig was coolly informed by the municipal police that "the actions taken must adhere to the statute in question, and that it is in the interest of the local master bakers to devise ways and means of eliminating the smoke and soot nuisance possibly caused by baking activities." The Berlin factory inspector Adalbert von Stülpnagel—incidentally a retired major whose military duty had been done with an artillery unit—dismissed critics with the remark that his office had "the duty to provide remedies for justified complaints about smoke nuisance using the legal means at its disposal, and it cannot accept the demands of the industrialists for the sole consideration of their interests as justified."[182] Even the Prussian minister for trade and commerce responded to petitions from Berlin interest groups with a clear warning that industry "should not expect that the generation of excessive smoke will meet with leniency in the future."[183] Evidently, German officials would not discuss the modalities of enforcement with those subject to their oversight.

Ultimately, German businesspeople behaved cooperatively for lack of a choice. At first glance, this was an asset for the German fight against smoke, for it spared German officials the bitter battles that American antismoke activists were waging with reluctant industrialists. But in the end, the apathetic stance of the German business community was more of a burden to smoke abatement than contemporary observers thought; after all, it meant that only few attempts were made to bring businesspeople on board, as American reformers did by stressing that "smoke means waste" and by hiring professional smoke inspectors to consult with violators. While the idea of fighting inefficiency and waste through wise governmental management was the key to American conservation, similar lines of reasoning remained foreign to most German businesspeople, and the same held true for the German engineering community.

The Reluctant Engineer

In order to understand the behavior of German engineers in the debates about smoke, it is crucial to see that they played a dual role. On the one hand, they were crucial experts on the technology of smoke abatement: mechanical engineers knew how to minimize smoke, what this cost, and what it meant in terms of fuel savings. On the other hand, they had a professional stake in the burgeoning debate. The fight against the smoke nuisance offered an opportunity to boost the professional standing of engineers, and even more important, it offered the prospect of jobs. This rendered mechanical engineers both recognized experts and an interest group, and while both roles generally pointed in the same direction, they were by no means synonymous. The key difference was evident in how engineers as a group felt about smoke abatement positions in the two countries. American engineers were keenly interested in such jobs—not surprising for a group feeling a strong sense of their professional obligation toward society.[184] "It is absurd to talk of putting this matter into the hands of the police or of the health officer. The official having charge of this work should be a trained engineer," went a typical statement.[185] In contrast, German engineers showed only lukewarm interest in these job prospects and, more generally, in the political aspects of smoke abatement. German engineering had a long tradition of opposing bureaucratic regimentation, the most vigorous criticism directed at the subordinate position of engineers within the German administration, where they routinely had to work under the auspices of legal experts.[186] However, the fight against smoke offered nothing more than such subordinate jobs, and this dampened interest in smoke abatement jobs within the German engineering community. While American engineers always kept an eye on their professional interests in debates over smoke, technical precision ruled supreme for their German colleagues. As a rule of thumb, German engineers' statements were more sophisticated technologically but less helpful in political terms.[187]

Statements about coal savings, usually a by-product of smoke abatement measures, are a case in point. American engineers were quick to realize that the combination of antismoke measures and more economical operation of furnaces could constitute the basis for a promising strategy. As a result, they never tired of propagating the principle that "smoke means waste"—and they did not shy away from vivid formulations: "A streamer of black smoke is the black flag of a pirate confiscating a part of the nation's resources," proclaimed an engineer with the U.S. Bureau of Mines.[188] German engineers were much more restrained in this regard. They emphasized the complexity of the matter and argued that there was "a more complicated connection between economical firing and smoke abatement than one might expect."[189] A simple equation on the American model was, "expressed as a generality, false," declared a presentation made to the Hanover branch of the Association of German Engineers (Verein Deutscher Ingenieure, or VDI). Consequently, "a smoking chimney" must "not be simply presented as proof of a defective furnace."[190] German engineers also freely stated that the economic gain from low-smoke burning was generally not very high. While American engineers for the most part did not quantify the financial losses suffered by entrepreneurs so as to avoid undermining the political effectiveness of the slogan "smoke means waste," fastidious German engineers emphasized "that the generation of smoke mostly does not entail the great loss of fuel that those not familiar with the matter generally assume."[191] While American engineers, in other words, took great pains not to weaken the link between the fight against smoke and economical firing, their German counterparts killed a strong argument for combating smoke by qualifying it to death.

The different maxims underlying their actions become clearer through an examination of how engineers in both countries dealt with the problems of domestic coal. No serious argument was made that a substantial portion of the smoke nuisance in large cities stemmed from domestic furnaces. Of course, for the practical fight against smoke, this fact was rather awkward: domestic coal fires posed particular technical problems, and their sheer numbers alone made it virtually impossible to subject them to regulatory oversight. In the United States, the situation was exacerbated by the fact that members of civic antismoke leagues were among those creating the domestic coal problem. Consequently, American engineers were eager to downplay the importance of the domestic coal problem: "The smoke from these domestic fires is not believed to-day to form an important part of the problem," said one report, which put the share that domestic burning contributed to Chicago's smoke nuisance at a whopping 2 percent.[192] Estimates by German engineers were substantially higher and—one may presume—much closer to the truth.[193] Pasinski reckoned the share of domestic fires at between 40 and 50 percent; other experts, too, believed that "it is the immense number of small domestic fires to which city air 'owes' the majority of its soot."[194] Apparently, German engineers couldn't care less that such statements earned the practical fight against smoke the stigma of being incomplete and half-hearted.

As a result, German and American engineers held different views on the issue of regulation. American engineers never disputed the need for smoke ordinances to back up the educational work of smoke inspectors. In contrast, German engineers took a much more ambivalent stance, their antibureaucratic feelings resulting in a constant vacillation between two positions: on the one hand, there was no question that the fight against smoke was in principle a legitimate bureaucratic activity; on the other hand, it was felt that technical progress would solve the problem in any case and that "the question in general must be left to its natural development and promotion by Germany's engineers."[195] It was best, then, not to say anything at all, and German engineers usually refrained altogether from making statements on regulation until they were approached by some bureaucrat seeking advice.[196] And if taking a position became unavoidable, its details were often contradictory. In a commission of the Prussian Ministry of Trade, for example, the VDI director Theodor Peters declared that "the improvement cannot be forced upon an industrial enterprise" and that smoke-reducing measures must instead "emerge out of the matter itself, out of a better understanding and advantage"—at the same time as he came out in support of a police ordinance.[197] Revealingly, the chair of the Württemberg Association for Steam Boiler Supervision—a professor at the polytechnic school in Stuttgart—declared at the organization's annual meeting that an expert opinion from the association had "prevented unenforceable police regulations from being enacted."[198] To put it pointedly, in debates over strategy, German engineers' concern was not to ensure, on the basis of their expertise and experience, the most effective action possible; rather, their chief goal was to prevent bureaucrats from generating too much nonsense on an issue of technical regulation.

This also made it nearly impossible for a German engineer to contemplate a connection between the fight against smoke and economical burning. To be sure, it was known that, from a technical point of view, the two issues overlapped, but while the fight against smoke was seen as a matter for the authorities, improvement in the utilization of coal was regarded as the natural path of technological progress.[199] Consequently, it was unthinkable for a German engineer to propagate the slogan "smoke means waste" as a guideline supporting bureaucratic measures against smoke: the notion that technological progress could achieve a breakthrough only with police intervention was all but anathema to a German engineer.[200] Yet precisely this fusion of economical burning and the fight against smoke was the credo of American engineers: after all, improving the use of resources through regulatory intervention was exactly in line with the contemporary conservation movement.²⁰¹ Probably the clearest manifestation of this way of thinking was the proposal that a "National Smoke Abatement Conference" be held under the aegis of the American Society of Mechanical Engineers (ASME), an idea that went back to the engineer Morris Cooke. Not only would engineers at this conference explain, in language understandable to laypersons, the basis of the fight against smoke, but doctors, chemists, architects, physicists, and other experts would simultaneously examine the smoke nuisance from the perspective of their particular disciplines. According to Cooke, the goal of the conference—ostensibly aimed at achieving the maximum public impact—was "to educate the public to a better understanding of the problem and of the lines along which relief should be sought."²⁰² Even if this conference was eventually cancelled, because the New York power-plant lobby (the "utility crowd," in Cooke's derogatory words) dug in its heels, the declared intention hit a nerve, and the petition's signers included, in addition to several board members of the ASME, prominent antismoke activists like Lester Breckenridge, Charles Benjamin, William Goss, Paul Bird, and Frederick Winslow Taylor.²⁰³ Characteristically, the St. Louis–based engineer Victor Azbe came up with a nearly identical proposal some twenty years later.²⁰⁴

The VDI, the German counterpart of the ASME, acted very differently: it targeted not the broad public of laypersons, but the community of professional engineers, commissioning a detailed monograph on the technical aspects of combating smoke.²⁰⁵ This study, very eagerly received among German engineers, was considered the standard work on the subject, but needless to say, it held little appeal to the broader public. This had not been the VDI's intention, in any case: the unequivocal attitude of German engineers was that they had done their duty by clarifying the technical issues.

Enter the Bureaucracy

The German bureaucracy was ultimately the key policy broker in the German debate over the smoke nuisance. Their role came about mostly by default: since industrialists, engineers, and the general public all kept a low profile when it came to policy, bureaucrats had enormous leeway in decision making. And civil servants had no problem with this: a wealth of quotes illustrates that the German bureaucracy from the very beginning looked upon the fight against smoke, quite matter-of-factly, as their exclusive field of activity. "Not being a vigorous proponent of smoke burning [is] a legitimate reproach," wrote Bremen's factory inspector Hermann Wegener, while his Berlin colleague Tschorn, at a meeting of the Association for the Promotion of Commerce and Industry, noted boldly that "the state authorities have always taken a lively interest in the matter."²⁰⁶ A Dresden official declared to the city council: "I do not believe it is necessary for me to say that the municipal administration for its part, as well, will work hard to preserve the reputation that our city enjoys as a clean, salubrious, and neat city." The reaction was "multiple shouts of bravo."[207]

A series of legal regulations did in fact place German officials in a comparatively comfortable situation. The Regulation on the Installation of Steam Boilers, of September 6, 1848, already contained a relevant clause: ever since the passage of this ordinance, Prussian authorities, "in all appropriate cases, especially in the operation of boiler systems in proximity to human dwellings," could wield a stipulation that made low-smoke operation mandatory for owners.[208] This requirement, referred to by contemporaries as the "smoke clause," was introduced in other states after the Prussian Industrial Code (Gewerbeordnung) was incorporated into Reich law, and it was generally seen as a useful regulation. In the case of businesses that did not require a permit, the Allgemeines Landrecht (General Prussian Code of Law) offered other possibilities for sanctioning them, since the code allowed for police intervention to ward off health threats.[209] And building codes often also contained paragraphs that made it possible to take steps against the emission of excessive amounts of smoke.[210] A few cities additionally enacted smoke ordinances, though the rigid manner in which the Prussian Supreme Administrative Court administered the law made the creation of such local laws impossible in most of Germany's large cities.[211] Finally, civil law also contained some relevant provisions, but they declined in importance vis-à-vis public law over the course of the nineteenth century.[212]

The value of this foundation becomes clear when the legal situation in Germany is compared with that in the United States. There, smoke ordinances, enacted after long and protracted struggles by the civic antismoke movement, were repeatedly declared void in court.[213] Indeed, not until 1916 did a Supreme Court decision clarify once and for all the legality of municipal antismoke efforts.[214] Beyond were only the vague stipulations of the "nuisance law," which, according to environmental historian Christine Meisner Rosen, were the province of a highly unstable and incoherent jurisdiction.[215] German authorities were thus already, in purely legal terms, in a much stronger position than their American counterparts. While German legal provisions were somewhat dispersed and sometimes awkward to wield, officials generally had the requisite legal means at their disposal when they wanted to intervene. As a result, German officials sometimes flexed their muscles: "Let no one believe . . . that the success of every individual factory, the greatest possible expansion of its operation and the distribution of the highest possible dividend, is an end in itself, the achievement of which must not be impeded by the authorities under any circumstances," the district administration (Regierungspräsident) of Hanover informed representatives of various associations.[216]

However, it is not difficult to see through such staunch rhetoric. The trouble began with the simple question of responsibility: who was really in charge of smoke abatement? When smoke abatement became a pressing problem, a large number of institutions already existed to deal with related problems: the building police, the fire police, the general police, factory inspectors, the technical personnel of municipal enterprises, medical experts, and steam boiler inspectors. It therefore seemed logical not to create any new institutions to combat smoke, but to obligate existing ones to fight against smoke as a secondary activity. This approach typically resulted in arrangements that placed smoke-related work on a number of different shoulders. In Braunschweig, for example, complaints were handled by the police directorate once the factory inspector had given his expert opinion.[217] In Frankfurt, the building police were in charge, with recourse, if necessary, to the expertise of the Association for the Supervision of Steam Boilers. In Munich, for a time, both the city building office and the Polytechnic Association delivered expert opinions. And in Stuttgart, the responsibility for prosecuting smoke violators lay with the city police office, which used the city's machine building engineer to inspect steam boiler installations and central heaters and the local fire inspector—an official who was actually charged with eliminating fire hazards—for all other firing installations.[218] At times, even officials themselves could not keep the situation straight. When the first larger debate about the smoke issue got under way in Hanover, there was initial "confusion among the most involved agencies . . . over whether the state or the municipal police would be responsible for measures against the smoke from factory chimneys."[219]

For the most part, little thought was given to the efficiency of these arrangements. The crucial point was the desire to handle the new tasks with existing administrative resources. In extreme cases, this could give rise to regulations that, de facto, put the fox in charge of the henhouse. For example, the Heidelberg city administration decided that the inspection of smoke-generating enterprises should be carried out by—of all people—the director of the municipal gas, water, and electric works, even though the enterprises he oversaw were no doubt among the biggest polluters in an industry-poor city like Heidelberg. The inevitable result was an exceedingly conciliatory inspection practice, since the city's expert—as he himself put it in a report—"believes that you should not throw stones if you are sitting in a glass house."[220] Still, the Heidelberg approach remained something of an exception, and to their credit, most German administrations secured an impartial expert witness on pollution cases. However, the more pertinent problem was that such wide distribution of responsibility constrained bureaucratic initiative, as did the fact that smoke inevitably ranked low on the bureaucrats' list of priorities. And why should they engage in a vigorous fight against smoke if their core task was in a different area? The involvement of existing organs led almost invariably to lackluster enforcement that was essentially limited to the processing of incoming complaints. "Where there is no plaintiff—there is also no judge," the Munich City Building Office opined quite nonchalantly in a report to the government, going on to say, "If ... the office continues the current practice of letting complaints come to it, it will surely have fulfilled its legal obligation."[221] It stands to reason that such considerations were typical of most bureaucrats.

Most officials dealing with the smoke nuisance in a secondary capacity were thus convinced that they were doing what was necessary, and subjectively, these feelings are perfectly understandable. Nobody likes to admit that he has not satisfactorily accomplished a task, and this applies especially to members of the German bureaucracy, with its almost legendary self-confidence. It was therefore inevitable that many officials whose fight against smoke was in fact only half-hearted gave unrealistically rosy accounts of their work in reports. This is especially obvious in a Prussian government survey conducted shortly before World War I, in which officials and authorities signaled, one after another, their belief that they had the problem under control.[222] The situation in Düsseldorf was also typical: in January 1909, the Health Commission, at the instigation of the city physician, called for "restrictive measures." This request was unanimously rebuffed by the officials at whom this demand was directed. "I would be greatly interested to find out on what facts the Health Commission's assumption of a general, strong increase in the smoke nuisance is based," responded Düsseldorf's factory inspector in a huff.[223] Machine engineer Pasinski also rejected the charge of negligence: "All factories are aware of the emphatic efforts of the Düsseldorf city government, and with few exceptions everyone is striving to take them into account when setting up new installations"—self-praise quickly unmasked by the fact that Pasinski himself, a short time later, tried to instigate a reform debate.[224]

Particularly serious were the consequences that flowed from the tense relationship between officials and engineers, hinted at earlier. The prerogative of the administrative jurist and the subordinate position of the engineer within the bureaucracy constituted an imbalance that, at the least, placed a strain on constructive cooperation. A smoke ordinance for Berlin sought by the Prussian Ministry for Trade and Commerce—the centerpiece of what was probably the most ambitious reform project in the German fight against smoke—was undone at the last moment by a misunderstanding between the ministry and its expert commission.[225] Of course, there was no dearth of passionate criticism charging that technical experts were being marginalized by officials trained in the law. "Such questions cannot be decided in the same old way at the conference table, but require real knowledge," explained the hygienist Theodor Weyl.[226] The Munich smoke fighter Karl Hauser expressed a similar sentiment in 1913: "Unfortunately there is still far too much writing, only, and the question is dealt with far too much around the conference table."[227] Needless to say, this made little impression on administrators: as they saw it, these were merely the protests of a correctly marginalized interest group. The de facto dependence of the bureaucrats on technical experts is evident in the fact that they repeatedly copied page after page of the experts' reports, presenting them as their own positions, but the aura of having the final say was important to the typical German government bureaucrat, cultivated with diligence.

The communication problem between bureaucrats and engineers found its most striking expression in the futile search for a universal panacea.[228] A technical remedy that would be able to eliminate the production of smoke in all cases was, in a sense, the administrators' dream: such an apparatus could be made mandatory without officials needing to understand how it actually worked. And so the Leipzig city council declared in 1881 that it had so far dispensed with a police ordinance because, among other reasons, "there do not exist, according to the current expert opinion, any smoke burning apparatuses that accomplish all their promises under all circumstances and with all boiler types."[229] In 1889, the Prussian minister of public works also rejected general regulations on smoke avoidance, "because so far not one of the installations carried out has sufficiently proved itself, to the point where it could be recommended for general introduction."[230] However, the minister believed it was "useful to await further experiences in this area"—evidently he had not given up hope for a one-stroke solution.[231] Even if engineering experts were tireless in reiterating that the wish for a universally applicable installation was a pipe dream—"there is no smoke-consuming installation and none will ever be invented that is usable for all cases," a typical report stated—this dream proved impossible to eradicate.[232] As late as 1913, the provincial governor (Oberpräsident) of Posen issued a call that "experiments in this area be promoted with state subsidies."[233]

German officials were similarly hesitant in their outreach toward civil society. In a way, this approach was completely opposite to the openness of the political process in American cities, where reform leagues could draft legislation, consult with officials, and sit on oversight commissions. What was routine in the United States, however, encountered stiff opposition in Germany, as the Düsseldorf engineer Pasinski discovered in 1914 when he made the only documented attempt to organize the German fight against smoke along the model of American smoke inspectors. Shortly before he launched his ill-fated effort, Pasinski had been visited by "higher officials" from Chicago, Pittsburgh, and New York, and he was inspired by their account of U.S. engineers' vigorous fight against smoke.[234] But he did not get very far: when Pasinski, in a memorandum, suggested forming a commission to "document the interest of the citizenry in improving conditions," his boss responded angrily, noting indignantly in the margin, "Unnecessary! A commission will at best be an impediment [to fighting smoke]." This was the end of Pasinski's suggestion.[235] The proud German bureaucrats had no interest in a public discussion of their own work, and they made this very clear.[236] The Braunschweig police authorities even succeeded in shifting responsibility for the shortcomings of their own work onto the public: "Punishment, however, would occur far more frequently, thus better furthering the purpose of the statute, if the public supported the efforts by ruthlessly reporting nuisances that have occurred, and not, as in so many instances, being reluctant to report offenders to the police."[237] Thus, bureaucrats always had a very clear conception of the role of the public: citizens could express their unhappiness by filing a police report—and that was about it.[238]

Further, the debate about smoke within the bureaucracy was proceeding at an exceedingly slow pace. The topic was consistently low on the agenda, and administrators preferred to postpone it if there was more urgent business. Ultimately, then, the smoke debate was administered rather than carried on—a further drag on officials' already less than exuberant attitude. It took more than three years to get the local ordinance in Stuttgart passed; in Braunschweig nearly four years elapsed between the initial declaration of intent and the enactment of the local statute; and the local law the Dresden city council passed on February 10, 1887, went back to a motion brought by the same body on October 1, 1879.[239] Other debates, too, were dominated by this leisurely pace: the Prussian trade minister's decree of February 5, 1901, was based on a vote taken by his expert commission two years earlier, and the Hamburg Committee of the Citizenry took more than eight years to deliver its report—during which time it met a grand total of nine times.[240] Is it surprising, then, that the results of such deliberations were often dispiriting, never demonstrating the kind of imagination and cleverness called for by such a tricky problem?

For all the shortcomings of the bureaucracy's approach, one thing must be said for the German officials: they processed incoming complaints thoroughly and without prejudging the outcome. Oftentimes, bureaucrats agreed with complainants that there was indeed a problem: in Stuttgart, where the work of officials can be traced in detail over more than two decades, well over 80 percent of complaints were judged substantively correct; in Nuremberg, three out of every five complaints filed between 1903 and 1914 were seen as justified.[241] Given that convoluted procedures for dealing with complaints did not exactly promote vigorous action, these figures are impressive—except that the mere processing of complaints did not in itself constitute an effective strategy against the smoke nuisance itself. After all, toward the end of the nineteenth century, the crucial problem was no longer the small number of smoke-belching chimneys but the daily emissions from hundreds, indeed thousands, of businesses—quite apart from the fact that there was, of course, no guarantee that people would lodge complaints against the worst offenders. In the final analysis, the result of the leisurely though fairly reliable processing of complaints—the specialty of German authorities—was not so much an effective way of keeping the air clean but a way to channel—and in part neutralize—public discontent. If authorities in Chemnitz reacted to a complaint about smoke in a particular neighborhood, signed by twelve petitioners, by singling out four offenders and taking action against them, this did not solve the problem, but it did take the edge off citizen protest.[242]

Clearly, then, when German authorities reviewed complaints, their approach was far from a policy of "zero tolerance." As a rule, they assessed smoking chimneys against the backdrop of a certain "normal level" of emissions that they did not consider punishable. For one thing, this seemed a simple matter of fairness. Given that there were thousands of smoking businesses, why should only those that someone complained about be punished? "During my inspections I was not able to notice any conspicuous smoke nuisance for the neighborhood; at any rate, it is not likely to be worse than the many other commercial establishments of the city," the Bielefeld district physician noted about a brewery.[243] A Cologne building inspector drew a revealing distinction between serious cases, where he sought "to establish tolerable conditions," and "usual nuisances," which were not the state's business.[244] A key factor throughout was the authorities' strong interest in resolving complaints with the least possible effort so they could return to their "real" tasks. A report from the Cologne Office of Factory Inspection demonstrates the absurd consequences that could flow from such bureaucratic logic: the smoke-belching steam boiler of the Stollwerk company was explicitly commended because it had not caused any work for officials, thanks to its tall chimneys! "Even though thick black clouds of smoke are almost continuously emitted day in and day out from the comparatively tall chimneys, clouds that drift far and are very conspicuous, so far I am not aware of any complaint about smoke nuisance," the official noted unabashedly. Obviously, it never occurred to him that he might consider steps against smoke clouds on his own initiative.[245] For most officials, everything was fine as long as no one was complaining.

The same motivation, to minimize their own workload, was probably also crucial for the lenient attitude that authorities took toward smoke-producing entrepreneurs. For the police president of Hanover, it was "imperative in every instance... to proceed first of all with moderation." In Cologne, "amicable encouragement" was used to "work toward the installation of relevant equipment." The Munich building inspector Karl Hauser advocated enforcement "in the most considerate manner while taking into account the specific, pertinent circumstances," and he set up an office hour for questions concerning the technology of smoke abatement, frequented every year by several hundred interested parties.[246] At the ministerial level, too, officials did not indulge in rigid legalism: in its decree on the Breslau Police Code, the Prussian Ministry for Trade and Commerce argued for "an especially careful handling," while the Saxon Ministry of the Interior noted in its decision on a contested case that "it might have been appropriate to attempt, with more effort than was done, to achieve the desired outcome through negotiations."[247] The smoke problem was tricky and technically demanding; moreover, in Germany, nobody was really opposed to fighting smoke. This being so, why should one not negotiate first, rather than resorting right away to an official order?

Still, this cooperative attitude should not be taken to mean that officials were soft on pollution, something authorities took great pains to emphasize. While officials were willing to negotiate informally with entrepreneurs, they always reserved the right to proceed strictly according to the letter of the law. In other words, they cooperated, but another approach was always an option: "The simple fact is that in order to achieve real results, official coercion must be used where instruction yields nothing," the Hanover district administration declared, while the city physician of Cologne argued that officials should resort to licensing requirements or the Allgemeines Landrecht "where amicable negotiations with the owners produced no result."[248] The degree to which the threat of punishment and gentle persuasion operated hand in hand is evident from a statement by the relevant office in Hamburg: "The only way to accomplish something is through a vigorous, resolute approach, namely through repeated visits and by convincing the owners themselves that the chimney is emitting too much smoke," declared the Steam Boiler Inspection of the Building Police, which proudly reported that only in two cases had it been "compelled to file criminal charges."[249] If an entrepreneur was being flagrantly obstructionist, however, the authorities' response was for the most part forceful: "The temporary shutdown of a business must be admissible as the means of last resort for cases of stubborn intractability, if the measure is to be securely and successfully enforced at all," opined the head of Dresden's Building Police Office—and in a few instances businesses were in fact shut down.[250] Not infrequently, the tone that crept into bureaucratic correspondence suggested that entrepreneurs were good-natured children who only occasionally got a little out of line and had to be brought back to their senses with measured doses of punishment: it would "require only the goodwill of the owners of said commercial installations to achieve the desired result," said one form letter sent by the administration of Chemnitz to smoke-generating businesses.[251] Needless to say, entrepreneurs did not appreciate this attitude, but in the end most of them relented: they knew full well that if push came to shove, the authorities had the upper hand.

Overall, the German bureaucracy presented an ambivalent picture. Few officials had any doubt that smoke abatement was a task for the German state, but few sought to tackle the problem aggressively. They readily accepted complaints from citizens, seeing many of them as justified, but their work ultimately touched only on a fraction of the problem. German officials enjoyed a degree of autonomy that was unknown to American civil servants, but this situation proved a burden when it came to communicating with engineering experts and the general public. Like American smoke inspectors, German officials were willing to cooperate with businesspeople on a somewhat informal basis, but they did so merely because this was the path of least resistance. In short, the smoke debate transformed the German system of air pollution control far less than it did the American one. By 1914, German environmental policy still saw pollution as an isolated incident that could be dealt with on an ad hoc basis, rather than as a common feature of modern society. This apparent inability to respond to the change in the nature of the pollution problem constituted the crucial failure of the German bureaucracy. The age of smoke had dawned, yet German bureaucrats adhered to a case-by-case strategy utterly inappropriate to a modern industrialized society.

Cities as Agents of Change?

It is worth noting in this context the peculiar situation of German cities. American city governments, as noted earlier, became the hub of air pollution control, mostly because no other governmental level was willing to take on the issue.[252] Late nineteenth-century German cities, however, were in a more advantageous situation, since state officials had been active in pollution control for decades. For example, the licensing procedure under German commercial regulations had first been enacted in Prussia in 1845, and even this law was merely a codification of previous practices. German cities, however, did not simply sit back and let state officials do their jobs: they had a special incentive pushing them toward activity, not only because urban areas bore the brunt of the smoke nuisance. After all, Imperial Germany saw the heyday of what contemporaries called "municipal socialism": powerful city governments eager to counter the social and environmental problems of modern society. Why should cities spend millions to provide clean water and sanitary living conditions and not strive to control smoke as well?[253]

German cities did in fact begin quite early to lay the legal basis for efforts to combat smoke. A police ordinance was passed in Breslau as early as 1874, followed by ordinances in Nuremberg (1876), Braunschweig (1883), Stuttgart (1884), Dresden (1887), Heidelberg (1890), and Munich and Freiburg (1891).[254] In the United States, on the other hand, the first "smoke ordinance" was passed in Chicago in 1881, and other cities were much slower to follow suit: Cincinnati in 1881, Cleveland in 1882, Pittsburgh in 1892, St. Louis in 1893, Minneapolis in 1894, Milwaukee in 1896, Boston in 1901, Providence in 1902, Baltimore and Rochester in 1906.[255] But in the final analysis, such ordinances were simply declarations of intent. Every attentive observer knew that what ultimately mattered was the enforcement of these regulations. For most German smoke abatement activists, legal provisions were the least of their problems.

In fact, German cities took a wealth of individual measures that emphatically demonstrated the goodwill of the municipal administrations. The Chemnitz city council introduced a special licensing requirement for the operation of baking ovens, which at the time had a bad reputation as exceedingly strong smokers.[256] In Karlsruhe, the price of the coke produced in the gasworks was deliberately set at such a level that using it was no more expensive than conventional coal burning; in addition, the gasworks offered the assistance of its employees in the case of technical problems.[257] In Aachen, the city administration and the office of the steam boiler inspector jointly financed the hiring of a furnace instructor, who took up the job of encouraging city firemen to operate their installations more appropriately.[258] Bielefeld hired a female furnace instructor "who goes from house to house and instructs women how to stoke ovens and stoves with coal and coke." Other cities sought to tackle the problem of domestic firing with special information leaflets.[259] But this effort, and others like it, always lagged well behind the requirements of an effective smoke abatement strategy: there was no system or comprehensive plan, and many of these measures relied on voluntary participation and compliance. Only three cities had the kind of comprehensive program to raise the status quo of firing technology that American smoke inspectors had undertaken: Hamburg, Munich, and Dresden.

In Hamburg, the fight against smoke was spearheaded by the Society for Fuel Economy and Smoke Abatement, founded in 1902. An association of industrialists, this society had been set up at the urging of the Industrial Commission of the local Chamber of Commerce when the government of the city-state was planning a police ordinance against the smoke nuisance. Its work centered around a regular inspection of its members' furnaces by engineers and furnace instructors hired by the society, with the emphasis always on achieving an operation that was both low in smoke and economical. Although membership was voluntary, registered members were obligated to follow the directives of the society's engineers.[260] While the Hamburg Society thus had clear corporatist elements, the strategies in Dresden and Munich were based on special local laws, which—unlike in other cities—were implemented autonomously and systematically. In Dresden, the position of a technical supervisory official for furnace installations was created in 1901, with the Fire Police; this official, and the furnace instructor hired the following year, made up the Office for Smoke and Soot Abatement.[261] The engineer who took this job preferred to offer advice and instruction, resorting to sanctions only when an entrepreneur proved recalcitrant.[262] In Munich, too, an engineer headed the office in charge of smoke abatement, and here, as well, preference was given to amicable influence over criminal prosecution. In addition, the Munich Office of Furnace Inspection, set up within the City Building Office, was active in experimentation.[263] Statistics document systematic action against the smoke nuisance in these cities: in 1910, 497 cases were processed in Munich and 760 in Dresden, and American smoke inspectors never came close to the frequency of inspections undertaken by the Hamburg Society for Fuel Economy and Smoke Abatement, which in 1908 had four engineers for 792 furnaces.[264] Promising solutions were thus entirely within the realm of possibility in Germany: the approaches pursued in Hamburg, Dresden, and Munich strikingly demonstrate the kind of effective institutions that were possible in Germany once the bureaucracy got its act together.

However, a comparison with American smoke inspectors also reveals the downsides to these stories of success. Officials in Munich and Dresden lacked the kind of supportive lobby among city residents that was the centerpiece of the American fight against smoke. "The population is not resistant to the efforts," was the assessment in Dresden of the public's support for the bureaucracy's actions, and Munich officials expressed a similar view.[265] Moreover, success stories were ultimately rooted in individuals' initiatives: from Hermann Blohm, as the chair of the Industrial Commission of the Chamber of Commerce in Hamburg; from Dresden's privy councilor Dr. Hempel, who vigorously pushed for a reform of smoke abatement measures during deliberations over local regulation of chimney sweepers; and from the Munich building inspector Karl Hauser, who skillfully expanded his heater-inspection mandate into a potent tool.[266] Finally, German strategies lacked the sort of visibility that, for example, characterized the Department of Smoke Inspection in Chicago. Hauser and his Dresden colleague Rebs did pen reports about the steps they were taking, but these attracted little attention, while attempts in Stuttgart and Düsseldorf to create comparable institutions were unable to point to a shining example of success. And while the work of the Hamburg Society for Fuel Economy and Smoke Abatement was widely and for the most part positively noted, no one was prompted to establish a similar organization; the society's only offshoot was founded in Helsinki, Finland.[267] As a result, the effective approaches in Hamburg, Dresden, and Munich remained largely isolated exceptions that did not function as any kind of model; in fact, no link can even be detected between the three cities' organizations.

How can this lack of energy within German municipal administrations be explained? In all likelihood, the fact that big-city bureaucracies were themselves the source of considerable smoke emissions did not exactly encourage municipal smoke fighters. This was, in a sense, the dark underside of the infrastructure of municipal services: slaughterhouses, baths, the workshops of the transportation services, pumping stations for the water supply, gas- and electricity works, and other installations repeatedly came in for stinging criticism for their emissions of smoke and soot. "The municipal enterprises smoke the most," a Munich councilman explained in a reproachful tone, while the Stuttgart professor of engineering Carl Julius von Bach complained that "the furnaces of the buildings and enterprises of the state and municipality . . . are quite frequently among the worst offenders."[268] Municipal electric works, in particular, were often the target of fierce public attacks. There was hardly another type of enterprise that was more apt to generate smoke: fluctuations had to be compensated for without delay; the rapidly rising consumption of electricity overtaxed the boiler plants time and again; and finally, the most smoke was usually generated at dusk and was therefore highly visible.[269] "If a private factory produces a little more smoke once in a while, you can be sure that someone will immediately show up and forbid it, but when a municipal enterprise endangers the health of the residents, no one gives a hoot," a Stuttgart resident complained about the local electricity works. The chimney of the electricity works in the Dresden neighborhood of Friedrichstadt even received an unflattering nickname.[270] Of necessity, then, city governments set a bad example in matters of smoke, which suggested a cautious approach when it came to practical smoke abatement measures. "Everyone against whom we wish to enforce the regulations will say, 'Magistrate, why don't you lead by good example and first bring your own furnaces in line with the rules,'" a Munich councilman observed during deliberations over a new smoke ordinance.[271] While this was no reason to dispense entirely with smoke abatement, it was certainly reason enough to exercise caution in matters of smoke and to put the problem off for another day.

Hubris

In the early 1910s, the Smoke Abatement League of Cincinnati sought to learn from other countries what to do about the smoke nuisance. It cast a wide and truly global net, requests going to numerous European countries, Canada, Cuba, Mexico, Egypt, South Africa, India, China, Japan, Australia, and several South American countries.[272] One recipient was the German consulate in Cincinnati, whose officials, however, were unaware of the requests sent to other countries. And so the consul passed the request on to the Reich government in Berlin, with the proud comment that this request was "renewed evidence of the happy fact, widely observable, . . . that the Americans regard Germany as a model country in matters of urban order and administration."[273] This letter is arguably a curious detail in the history of the transatlantic exchange about air pollution problems, but it is also a revealing one, demonstrating the differences between the German and American approaches to the smoke problem. While the Americans were self-critical and eager to learn, the German officials were convinced that they had everything under control. Even after the outbreak of World War I,

American reformers had no reservations about declaring that "we need to take a lesson from German organization."[274] All the while, German officials were convinced that they did not need to take lessons from anyone.

Thus, the pride of the German consul was not only mistaken but part of the problem. It is true that the smoke debate in Germany could begin from a point that American antismoke activists had to struggle to reach. Further, Germany had a strong and largely corruption-free administration, a solid legal basis for the fight against smoke, and many well-trained engineers knowledgeable about the technical means for reducing smoke. And while American activists had to contend with vigorous criticism of the "smoke means prosperity" variety, the struggle against smoke in Germany encountered hardly any meaningful criticism. In 1911, Gustav Lang, a professor at the Technical College in Hanover, took it as a generally accepted "fact" that a remedy was "urgently needed"—and that "opinions still diverge only over the 'How' of this remedy."[275] Yet the German bureaucracy was able only in a few exceptional cases to fashion a promising strategy out of these favorable conditions—while other relevant groups, especially engineers and the broader public, could not and would not help it out of its predicament. In other words, the German smoke abatement debate failed not in spite of the German bureaucracy's strength, but because of it.

The result was an outcome that left all parties losers. To be sure, the losses were by no means equally distributed: the toll for Germany's housewives was surely much larger than it was for German industrialists, who suffered losses from inefficient coal combustion. Nonetheless, all parties ultimately paid a price. Officials failed to live up to their pledge to stop the smoke nuisance, engineers failed to gain jurisdiction on smoke issues, businesses paid more for coal than they needed to, and the general public suffered from an excessive amount of airborne pollutants. In contrast, the American smoke debate ultimately sponsored a win-win scenario: businesses saved money through efficient coal use, engineers secured jobs for themselves as municipal smoke inspectors, and civic reformers felt jubilant about a strategy that was as efficient as it was effective. The contrast should have made proud German officials blush with shame, but this probably would have been asking too much of the German bureaucracy.

3   Pollutants and Politics

AIR POLLUTION CONTROL BETWEEN THE WARS

After the intensive debates in the early 1900s, the postwar discussions in both the United States and Germany seem like a lukewarm postscript, devoid of the sense of urgency and the reform spirit so prominent earlier on. The only exceptions are the spectacular campaigns in St. Louis and Pittsburgh around 1940. In focusing on the tricky issue of domestic smoke, these campaigns dealt with an issue that had given advocates of smoke abatement headaches for more than a decade. Much has been made of Raymond Tucker's pivotal role in the St. Louis campaign, but in the end, Tucker merely implemented an approach that antismoke activists had been debating since the mid-1920s. To be sure, it clearly took courage and integrity to spearhead the St. Louis cleanup, but the campaign itself had been "in the air" for some time. Tucker's success gave new legitimacy to the smoke abatement community, which had been somewhat lackluster for most of the interwar years, but the ensuing boost was a decidedly mixed blessing: that the smoke inspection approach continued to dominate air pollution control long after its prime was a result, not least, of the spectacular campaigns in St. Louis and Pittsburgh.

The American Smoke Abatement Movement, 1917–50

It comes as little surprise that America's entry into World War I in the spring of 1917 represented a turning point for the movement against the smoke nuisance. In theory, the fight against smoke, which was focused on boosting the efficient use of fuel, did not clash with the demands of the war economy, especially since the Bureau of Conservation, set up in October 1917 within the U.S. Fuel Administration, was pursuing quite similar goals.[1] But in many quarters, the struggle against smoke was considered out of place under wartime conditions, and the fact that companies were often stretched to the maximum confronted municipal officials with a problem against which the usual technical modifications of furnaces could do little.[2] Whatever leeway for action remained under these conditions shrank to a minimum when the transportation problems in supplying coal, already smoldering before the war, escalated into a massive crisis. Eventually, the railroads were so overburdened that in December 1917 mines could produce at only 60 percent capacity.[3] The federal government initially reacted by closing down some enterprises—one spectacular decree in January 1918 ordered that all factories east of the Mississippi close for several days—and then by establishing zones for the coal supply.[4] Beginning on April 1, 1918, coal was to be mined and used only within officially determined districts in order to avoid long transportation routes and to free urgently needed transport capacities.[5] This decree was all but disastrous for smoke abatement efforts: now furnaces could no longer be supplied with the type of coal for which they had been designed and approved, but only with the types that happened to be available in any given district.[6] Wartime also offered some businesspeople a chance to settle old scores. Samuel Insull, for example, in his capacity as chair of the State Council of Defense, called upon the mayor of Chicago to suspend the existing local laws, no doubt in part because his power plants had been repeatedly attacked for generating a lot of smoke.[7] When antismoke activists tried to sell the fight against smoke as patriotic or—like Pittsburgh smoke inspector Henderson—even advocated a countrywide fight against smoke by federal authorities, these efforts amounted to little more than whistling in the dark.[8]

Probably the only meaningful success of the smoke abatement movement during World War I was that it did not terminate altogether. The basic institutional framework was essentially preserved, so that only a few months after the end of the war, matters seemed to pick up seamlessly where the prewar period had left off. As early as February 1919, a large-scale campaign was launched in Salt Lake City, which received substantial support from the Bureau of Mines and was also watched with interest by other cities. In September 1919, a "Clean Air Week" was held in Chicago.[9] "There is reported renewed activism by civic authorities and organizations in the campaign to eliminate the smoke nuisance," *Power Plant Engineering* noted in 1923.[10] But under the surface, a certain disenchantment could be felt. In the prewar period, activists had frequently behaved as if the final solution of the smoke problem were imminent. "Our triumph is certain; for although in the past Smoke has reigned, the Future is ours," declared Cincinnati's Smoke Abatement League at the conclusion of an annual report.[11] After the war, however, a different tenor held sway: "Smoke abatement has progressed so rapidly in Pittsburgh from 1914 on that some people began to think there was no further need for a League to battle against the smoke nuisance," explained the secretary of the Smoke and Dust Abatement League in 1919. "And then came 1918! During 1918, Pittsburgh returned to the 'Dark Ages' of 1907–1912."[12] The painful reality of how quickly successes evaporated under the conditions of the wartime economy remained living memory long after 1918.

And the sobering experiences did not end with the war. More than once, antismoke activists now found themselves confronted with problems that strike American urban historians as all too familiar: strong fluctuations in funding and thus in the level of personnel, political pressure, and simple corruption. An especially flagrant display was offered by Chicago antismoke efforts in the wake of the widely praised reform of 1907. As early as 1912, a crisis in municipal financing necessitated a cutback in staffing levels, which must have struck antismoke activists, who had in fact been hoping for a considerable budget increase, as a major setback.[13] The following year, Osborn Monnett (Paul Bird's successor) uncovered a case of corruption in his own ranks, which led to the firing of nine coworkers.[14] And a little while later, the department's political independence came to an end when a dentist was appointed chief smoke inspector—to the dismay of the journal *Power*, which saw the crown jewel of the antismoke movement degenerating into the plaything of the usual political horse-trading.[15] In 1924, the department budget was a mere $25,000, which meant that the amount granted in 1910 had declined by about half; only five officials were available to keep watch on chimneys, and two engineers inspected and approved more than four thousand new furnaces each year.[16] By December 1927, the very survival of the antismoke effort hung in the balance when the city's finance commission decided that the department, whose staff had grown to thirty employees, would have to let go of twenty-nine.[17] An intervention by several organizations, however, resulted in the rehiring of nine engineers, and within a year activists had gotten a local law passed that reorganized the department along the principles of 1907.[18] Still, antismoke activists could not enjoy the $63,520 approved for 1929 for long: between 1929 and 1933, the staff of the department shrank from twenty-two to seven, presumably as a result of the Great Depression.[19]

This tumultuous sequence of events in the "city of the big shoulders" illustrates a more general point: during the interwar period, the municipal fight against smoke was often little more than a fight to preserve the status quo. For example, the budget of the Division of Smoke Inspection in Cleveland dropped from $11,500 in 1913 to $4,498 ten years later, a trend made even more serious by the fact that its officials had in the meantime also been put in charge of overseeing the city's ice machines.[20] The situation in Boston was little better: six inspectors were spread out over an area of about 470 square miles.[21] "None of the cities in the soft-coal districts have ever had sufficient appropriations really to cover the field," Osborn Monnett wrote in 1926, and the Great Depression exacerbated this situation: a 1936 poll by the Smoke Prevention Association showed that of thirty cities of two hundred thousand or more inhabitants, only seventeen had full-time smoke abatement officials.[22] Political pressure also impeded the work of smoke engineers: "All too frequently has politics made smoke-abatement campaigns at least partially ineffective," complained William Christy, head of the Department of Smoke Regulation in Hudson County and one of the most aggressive antismoke activists of the interwar period.[23] The simple reality was that the mantra of the professed political neutrality of smoke inspectors, touted by engineers and reformers, was not echoed by all municipal politicians.[24] At a meeting in St. Louis in 1924, attendees watched a film with a revealing title: *The Trials and Tribulations of a City Smoke Inspector*.[25]

Part of the problem was that civic activism regarding smoke had declined notably since the heyday of the Progressive Era. As a result, the once-typical situation of a smoke inspector's work being attended and protected by a civic association had ceased to be the rule by the 1930s: among large American cities, no more than a third reported an active smoke abatement movement in 1936.[26] Of course, this was due primarily to the repercussions of the Great Depression, but it also stemmed from the fact that the institutionalization of smoke inspectors had raised the level of knowledge necessary to participate in civic activities. Whereas before World War I, it had been sufficient to evoke the extent of the smoke nuisance and call in general terms for remedies, fine-tuning was now the order of the day. A citizens' movement that aimed not simply to participate in the antismoke movement, but lead it, would therefore need technical expertise—and that was generally in short supply among civic reformers. As a result, the Women's City Club of Cleveland retained its own engineers, "who will give engineering advice on the many problems which are constantly coming before us"; the Citizens' Smoke Abatement League in St. Louis even had its own "technical division," with a total of ten subcommittees.[27] "It is important that a smoke campaign be founded on a sound engineering basis," declared one member of the latter organization.[28] And this clearly raised the bar for civic engagement after 1918.

Just how much the climate had changed can be seen, for example, in

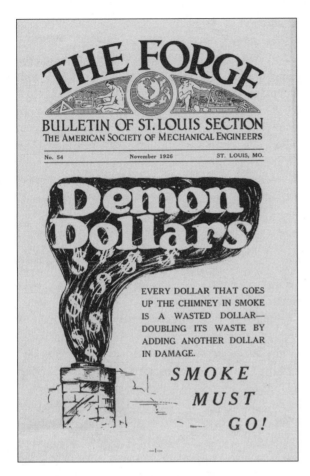

Figure 3. Efficient coal use was consistently the key selling point for smoke abatement in the United States, as shown in this picture from a 1926 engineering periodical.

the Women's Civic League of Baltimore's struggle over the hiring of a smoke inspector for Baltimore. The campaign began in 1924, but it was not until May 1933 that the position was filled permanently. So slow was progress on this issue that one citizen decided to wear only black clothing and thereby start a mass movement—"A black shirt brigade with smoke and not Mussolini as the inspiration," as the *Baltimore Sun* caustically remarked.[29] In Cleveland, at times up to twenty women from the Women's City Club acted as chimney watchers to record violations of the local ordinance, sometimes on a part-time basis, as they occasionally had their chauffeurs help out so they could take a break and do the shopping.[30] Soon, however, the smoke inspector's processing of these reports was seen as inadequate: meeting minutes record complaints about the official's "lamentable disinterestedness," and the group later successfully lobbied for a new, more energetic successor.[31] Afterward, however, interest declined notably within the Women's City Club, civic activism in Cleveland as a whole shrinking so much that the local smoke inspector complained at an ASME meeting in 1928, "There does not seem to be any loud demand for improvement nor any particular praise for work accomplished."[32] Only rarely was the situation as favorable as in Cincinnati, where the Smoke Abatement League (founded in 1906) remained active throughout the interwar years and, until the Great Depression, even maintained its own staff of inspectors, who checked on compliance with the stipulations of the law along with the municipal Smoke Department.[33] More typical were short-lived campaigns that were mostly a flash in the pan, although they do indicate a continuing interest in fighting smoke. For example, when the Chicago agency found its survival under threat, as described earlier, no fewer than three hundred people turned out for the crucial hearing.[34] In 1936, the establishment of a women's antismoke group in Salt Lake City, at the initiative of Cornelia Sorenson Lund, quickly attracted the support of around five thousand people.[35]

The most active civic campaign of the interwar period, however, took place in St. Louis. At its center stood the Citizens' Smoke Abatement League, founded in 1926, which represented the entire spectrum of groups involved in smoke abatement, from the Chamber of Commerce to the local association of engineers and the Women's Smoke Abatement League.[36] "The public are interested as never before, not only in St. Louis but in all parts of the United States," the league proclaimed in its founding charter.[37] That this impression was not entirely off the mark, at least in St. Louis, was demonstrated by one of the group's fundraising appeals, which elicited more than $190,000 for smoke-abatement efforts, a sum that dwarfed all previous efforts.[38] Pittsburgh's "Smoke Investigation," by comparison, the most ambitious research project of the prewar years, had cost a mere $40,000.[39] The Citizens' Smoke Abatement League used this money to finance an education campaign that was unequaled in its size and breadth: the league employed as many as thirty-two individuals as technical advisers on firing, ran its own school for furnacemen, placed a wealth of articles in the local press, and even managed to get the proper operation of domestic firing apparatuses taught in the city's schools.[40] However, the St. Louis campaign had a tragic flaw: it was set to run for only three years, which meant that the league ran out of money just as the Great Depression began. After years of energetic activities, then, the league slid into a grave financial crisis and was eventually dissolved in April 1935.[41]

Yet in spite of all the problems afflicting bureaucratic practices, one thing should not be overlooked: on the whole, the fight against smoke continued throughout the interwar period. No large American city could afford to abandon smoke abatement efforts altogether after 1918 without suffering a serious loss of face: a certain minimum level of bureaucratic activity was part of municipal-politics etiquette. Moreover, there was unanimity on the question of how this battle against smoke could best be organized: "Cooperation has always been found the best working tool where successful smoke abatement work has been done," the consensus ran—as straightforward as it was broad.[42] Inefficiency remained the key argument in efforts to battle smoke: "That smoke is waste, and absolutely unnecessary waste, is patent to all," declared a speaker at a 1922 meeting of the Smoke Prevention Association.[43] The central role of engineers also remained uncontested: "A Division of Smoke Inspection should be headed by a bureau chief of recognized engineering ability," the Women's City Club of Cleveland demanded, and no meaningful opposition could be found.[44] Mostly friendly approval also continued to be heard from business circles. "Plant operators are cooperating in a way that, a decade ago, would not have been expected," explained Harry B. Meller of Pittsburgh's Mellon Institute in 1931.[45]

As a result, cooperation remained the typical pattern of behavior and criminal prosecution the exception.[46] But this did not by any means amount to negligence or apathy, especially since some smoke inspectors began to adopt a somewhat sterner approach if they ran into opposition. For example, William Christy did not shy away from pushing for the dismissal of furnacemen and employees if they could not be prevailed upon to cooperate.[47] His colleagues in Salt Lake City lost no time when it came to checking on early-morning emissions of smoke in the wintertime: they procured a searchlight, speculating openly that its beam of light, visible for miles, made factory personnel feel subject to constant observation.[48] Chicago's antismoke activists were even given the right to compel the temporary shutdown of an enterprise—a power they used, as *Power Plant Engineering* reported with satisfaction: after several warnings to a bank had gone unheeded, one inspector went into its furnace room, grabbed a fire hose, and doused the smoking fire. The next day, according to the journal, there were no more complaints about the operation of this boiler.[49]

From the perspective of business owners, potential savings remained the central sales pitch: "Smoke means . . . a waste of money; and a waste of money is bad business management," wrote one entrepreneur in the journal *Factory*.[50] But this certainly did not mean that the fight against smoke had taken on a momentum of its own: the coercive power of local ordinances remained indispensable. No one knew this better than the residents of Salt

Lake City, where the local laws were abrogated in the mid-1920s after years of successful antismoke efforts. Within a few days, the city saw a relapse into conditions that were believed to be a thing of the past: one month after the prohibitions were lifted, measurements showed that the total duration of the generation of thick smoke had increased by a whopping 775 percent over the previous year.[51] Although Hudson County saw the establishment of an Industrial Smoke Abatement Association, in which businesspeople could exchange information about smoke abatement, this example did not catch on anywhere else; this group's existence was no doubt due entirely to the energetic William Christy.[52] Indeed, at a meeting of the Smoke Prevention Association, the group's founding president readily admitted that he had initially been anything but enthusiastic when he heard about the Hudson County organization: "My personal thought was that it was just one more political nuisance tacked onto industry."[53] Entrepreneurs tried to reduce smoke not out of inner conviction, but because laws were in place and because the local smoke inspector was usually a pleasant and accommodating man.

All in all, the interwar smoke abatement movement leaves a highly ambivalent impression. On the one hand, there was every indication that smoke abatement was here to stay. The Smoke Prevention Association held annual meetings; cities like Baltimore and New Orleans, without meaningful smoke abatement efforts before 1917, finally joined the movement; and addressing smoke had become an established field of research. By 1924, the Mellon Institute in Pittsburgh already listed eighteen hundred publications on the topic.[54] On the other hand, the movement exuded a strong feeling of stagnation, along with the air of having failed in its original pledge. "It is certainly a fact that the public is not satisfied with the results," one essay noted at the beginning of the 1930s, and indeed, in the interwar period no one was truly satisfied with the state of smoke-abatement efforts.[55] The proud smoke inspectors of the Progressive Era were now more ambivalent in their self-descriptions, sometimes even comparing their own status to that of a dog catcher—"A matter of a man or two, a job or two, a small appropriation of variable and uncertain amount, to be abolished in a fit of economy and reestablished under pressure of a vigorous minority."[56] It would take a new challenge to reenergize the movement, and from the mid-1920s, it became increasingly clear what this challenge would be: the issue of domestic smoke.

The Revenge of Success

From a regulatory point of view, domestic smoke posed an awkward problem. After some time, smoke inspectors usually achieved some control over a few thousand industrial firing installations, but from there it was an enormous step to the supervision of hundreds of thousands of domestic fires. To the vast number of emission sources was added the level of technology, noticeably lower in domestic stoves. As one technician wrote, "These stoves have changed somewhat in outward appearance, but have not been fundamentally changed in principle since the stove was invented by Franklin."[57] When antismoke activists nonetheless began to spend an increasing amount of time discussing domestic smoke during the interwar period, there was a simple reason: in the long run it was not possible to keep silent about it. As soon as the fight against smoke had been waged in a city for several years with some success, the problem could no longer be denied. In a sense, this was the revenge of success: the more the emissions from trade and industry declined, the more obvious the domestic smoke problem became. Measurements revealed that between one-half and two-thirds of urban smoke came from domestic sources.[58] "It will be admitted by even the casual observer that heating plants are responsible for a large part of the air pollution from smoke," the *Bulletin* of the Woman's City Club of Chicago declared, and Osborn Monnett noted in 1926, "The real problem today is the small heating plant."[59] During the Progressive Era, these small heating plants had been exempted from local ordinances.[60] Now there was rising pressure to include domestic firing installations in the fight against smoke. But what, exactly, was to be done?

Initially, the emphasis was on educating the population. In May 1926, Frank Chambers, in Chicago, organized a series of meetings with the local janitors' union that were described as "educational in character"; the Women's City Club of Cleveland published a brochure entitled "How to Fire a Furnace"; the St. Louis smoke commissioner and the local coal industry distributed 130,000 cards with similar information; and the Boy Scouts were deployed to educate the urban population in Pittsburgh, Salt Lake City, Oakland, and a few other cities.[61] But opinions rapidly diverged about the success of these activities. It was only natural that at first reactions were optimistic; Osborn Monnett, for example, claimed that a group of pupils in Memphis had achieved a 60 percent reduction in the smoke nuisance.[62] A few years later, however, he had few encouraging things to say about such activities: "Most of these campaigns have shown inappreciable results," he wrote in 1941.[63] Even an expert consultant who supported educating coal consumers was unable to detect any sweeping success in 1935 on the issue of domestic burning.[64] But what was the alternative to education? To be sure, by the mid-1930s, a number of workable stoves allowed for low-smoke operation even with coal that was high in volatile matter, but clearly, it would take quite a few years before these stoves were in general use.[65] And was it even possible to control the use of such stoves in households without maintaining a veritable army of inspectors? Beginning in the 1920s, a growing number of antismoke activists came to believe that in fighting the smoke from domestic stoves, little could be achieved with either technical specifications or brochures about proper operation. More and more people thus began to think that the only feasible approach would center on the choice of coal.

Indeed, it seems that over time the idea of restricting the choice of fuel virtually thrust itself upon antismoke activists. At least this is the impression left by the course of the debates within the Citizens' Smoke Abatement League in St. Louis. In principle, hardly any other organization had a greater reason to emphasize the value of educational work: after all, the league had invested a six-figure sum in such activities. Still, even here, the notion gradually prevailed that there was no way of getting around at least a partial ban on certain types of coal. Within less than two years of the league's foundation, its board was discussing coercive measures. "Some members of the Board are in favor of trying to pass a drastic ordinance requiring people to burn smokeless fuel unless they can burn soft coal smokelessly," the league president explained at a board meeting on January 17, 1928.[66] For the time being, the board decided not to go down this path, but a new statement of the organization's principles was soon drafted, giving the group's consultants the option, under certain circumstances, of recommending a change in fuel.[67] While this was an internal move, not yet made public, in deference to the league's ongoing educational campaign, a public step was taken a little over a year later. On May 28, 1929, the league adopted as official policy the demand of one of the league's most prominent members—that low-smoke fuel be made mandatory for all furnaces and stoves that could not be made smokeless in any other way.[68] Internally, the league hesitated to call for an ordinance to this effect, since it had originally decided not to take any position on local laws.[69] But even this inhibition was overcome in the end. "It is high time to adopt restrictive legislation against soft coal, unless it can be burned absolutely without smoke," a board meeting concluded in 1932.[70]

Elsewhere, too, the idea of regulating the coal market was gaining traction. "Cannot all agree that the time has come for . . . an ordinance . . . requiring everyone to burn smokeless fuel or to burn smoky fuel smokelessly," wrote one member of the Quorum of the Twelve Apostles of the Mormon Church in 1936 to Cornelia Sorenson Lund, head of the Women's Chamber of Commerce in Salt Lake City.[71] It's possible that the implementation of such a proposal might at least have been tried in the 1930s, had the global economic crisis not banished all such notions from the sphere of the politically feasible for several years.[72] And so when a new local ordinance was prepared in St. Louis in 1934, a corresponding clause was stricken during the draft stage: "This drastic provision was eliminated ... because of the certainty that the Ordinance would not be passed if it included this feature"—an assessment that was strikingly confirmed when the slimmed-down version of the bill also failed to pass the city council.[73] As it was, limiting the choice of coal was consistently controversial within the antismoke movement, concerns focused less and less on the utility of such an approach and increasingly on its political feasibility. "It is doubtful ... if any law could be enforced which absolutely prohibits the burning of soft coal," Henry Obermeyer wrote in 1933.[74] Still, the traditional dogma of the antismoke movement—the emphasis on pursuing smoke abatement with locally available types of coal ("we must burn today's coal in today's furnace")—was beginning to crumble.[75] The only questions were whether it would be possible to mobilize sufficient support for this approach and whether the antismoke movement would be strong enough to sustain this strategy over the long term. There was only one way to find out: someone, somewhere, would have to try.

Raymond Tucker's St. Louis Campaign

In the fall of 1937, Raymond Tucker was appointed commissioner of smoke regulation in St. Louis.[76] Tucker was new to the antismoke movement: "I come before you as a new-comer in this field of smoke abatement," he explained shortly before assuming his office, in a speech to the Engineers Club of St. Louis.[77] Tucker had been a professor of machine building at Washington University since 1920, before the Democratic mayor Bernard F. Dickmann named him his personal advisor.[78] Tucker therefore had a clear party affiliation when he assumed his office—rather unusual among smoke inspectors, who were generally careful to maintain strict political neutrality. When Dickmann's Republican successor kept him in his post, no one was more surprised than Tucker himself.[79]

It quickly became clear that Tucker would not be content with simply administering the status quo. Already the ordinance passed in February 1937, substantially influenced by Tucker, had caused a stir with a novel clause: for the first time, such a law set upper levels for the ash and sulfur content of coal. If fuel material intended for St. Louis had more than 12 percent ash or 2 percent sulfur, these pollutants had to be removed by washing the coal.[80] This provision, which contemporaries referred to as the "washing clause," was aimed at reducing air pollution from fly ash and sulfur dioxide, and it led to a decades-long enmity between Tucker and the coal interests in the neighboring state of Illinois. Almost immediately, a coal merchant tried, unsuccessfully, to procure a court order against the clause.[81] The coal lobby was not in principle opposed to smoke abatement efforts; in fact, it had initiated some praiseworthy programs. In 1938, for example, the publisher of the journal *Coal Heat* explained at an annual meeting of the Smoke Prevention Association that the coal industry was now employing around two hundred consulting engineers.[82] Since the 1920s, the demand for coal had been stagnating, while petroleum and gas were gaining more and more market share.[83] The smoke given off by coal was increasingly seen as a competitive disadvantage vis-à-vis the smokeless fuels petroleum and gas: "Smoke is probably the greatest factor in determining the future status of coal as a fuel in congested centers of population," *Heating and Ventilating Magazine* had noted as early as 1927.[84] In response, the coal industry had become more active in the fight against smoke, especially in the 1930s. In Kansas City, the consulting on firing technology was performed almost entirely by engineers from the Coal Dealers Association; in Atlanta, the Coal Merchants Credit Association helped the women of the Business and Professional Women's Club in a study of the state of technology in the city; and in Cincinnati, smoke inspector Charles Gruber kept the local coal merchants apprised of their customers' smoke problems so as to include them in the search for remedies.[85] At the same time, however, the coal industry, clearly, was resolutely opposed to any regulation of the fuel market.

The industry's vehement reaction to the 1937 St. Louis ordinance and its "washing clause" was therefore no coincidence. A key question loomed behind the conflict over fly ash and sulfur dioxide: Did the city have the right to demand that locally used coal meet certain quality standards? "The smoke ordinance . . . has established the fundamental principle that the City of St. Louis has a right to dictate the type of fuel which may be burned in the City of St. Louis," Tucker explained to the Engineers Club of St. Louis.[86] In other words, legally there was no longer an impediment to outlawing fuel that produced copious amounts of smoke. Other statements Tucker made between 1937 and 1940 clearly reveal that he was pursuing more ambitious goals. In June 1938, at a meeting of the ASME, he candidly declared that the existing ordinance was provisional: "At present, St. Louis has passed an ordinance that is not the ultimate solution of the problem." He was vague, however, when it came to the changes he had in mind: "Time and experience, no doubt, will point the way toward the ultimate solution."[87] Tucker was more concrete in his first annual report: "The solution to the smoke problem in St. Louis will not be had until there has been provided, a fuel for domestic users that will not smoke." And then he added, more obscurely,

"Some form of mass attack will have to be pursued."[88] The man, it seemed, was just waiting for his opportunity.

And so, when a serious smog episode descended upon the city in November 1939—a day deemed "Black Tuesday" in the city's history—Tucker knew what had to be done. Exactly four days later, he submitted to the mayor a detailed proposal that would place the local smoke abatement effort on a new foundation.[89] "We have arrived at a point where more drastic steps will have to be taken," Tucker proclaimed at a public meeting; the existing ordinance, Tucker declared, "has accomplished everything we desired it to accomplish."[90] Indeed, *drastic* was an entirely apt description of Tucker's proposal: henceforth, gas-rich coal was to be used only if the construction of a furnace could guarantee smokeless operation. "The trouble must be cured at the source and be as near fool-proof as possible," Tucker believed. The goal should not be to fight the problem after the fact, but to create conditions under which smoke could be generated only through deliberate manipulation.[91] This argument proved persuasive. On April 8, 1940, the St. Louis city council approved an ordinance that corresponded in all essential aspects to Tucker's proposals. For the first time in the history of the American antismoke movement, a local law was passed that included all sources of municipal smoke.[92]

Passage of this law was probably possible only because Tucker's proposal was exactly in sync with sentiment among the city's population: since the smoky November days, the residents of St. Louis had been calling for radical measures with grim determination. "The time has come for new and more drastic steps," proclaimed the local engineers' association.[93] James Ford, the chair of the Smoke Elimination Committee set up by the mayor, which transformed Tucker's ideas into law, uttered similar sentiments: "We feel that the public was never so united on anything in the history of the city as on this elimination of smoke."[94] Of course, emotions were running no less high among the representatives of the coal industry. "Coal will not suffer unwarranted accusation, and . . . it will assert itself against confiscation of its rights and property," thundered Richard Wood as the representative of the Coal Exchange of St. Louis in December 1939.[95] Of course, such threats had little effect on Tucker. On the contrary, he responded by going on the offensive. When he sensed that the coal industry was out to subvert his efforts, he reacted by further radicalizing his already ambitious program. With little hesitation, Tucker dropped his original plan of gradually phasing in the ordinance over a period of about two and a half years and staked everything on a bold move: as of May 1, 1940, coal could be sold to the owners of hand-fed stoves only if it had a gas content of less than 23 percent.[96] For an agency that had a total of thirteen inspectors, this was a daring undertaking.[97] Failure would have discredited the antismoke effort in St. Louis for years to come, not to speak of the consequences for the American antismoke movement as a whole. Tucker was therefore risking a great deal—and he won. Although his decision led to a massive rise in the price of coal in the winter of 1940–41, the regulations remained in force. Even during World War II, the ordinance was not suspended. The breakthrough had come.

Contemporaries viewed the victory over the smoke nuisance in St. Louis as the work of Raymond Tucker, and he did little to dispel this notion: "I look upon it as my plan," he wrote in a private letter in 1946.[98] Tucker had at most disparaging things to say about earlier efforts: "[The] history of smoke elimination in the last 73 years in St. Louis should conclusively prove to its citizens that legislation will not abate smoke," he declared.[99] If Tucker thus presented his campaign as a departure from decades of lethargy, he did so for a simple political reason: anyone propagating radical measures against the smoke plague had better keep silent about the fact that efforts to solve the problem had been under way for fifty years. In point of fact, Tucker's campaign was not a radical break with an earlier, unsuccessful tradition; rather, it represented the culmination of the smoke inspection tradition. It was based on the achievements of decades of antismoke efforts; even the strategy itself was anything but new. Tucker was simply putting into effect measures that the Citizens' Smoke Abatement League had called for some five years earlier.

Revealingly, Tucker never had to canvass for support for his approach, and alternative strategies never had a chance before the Smoke Elimination Committee.[100] "I think it is agreed now . . . by all of the editorial writers, that our problem is the use of this Illinois soft coal," a citizen explained at a public meeting on December 5, 1939—when Tucker's reform proposal was exactly three days old.[101] In addition, Tucker's campaign benefited from the fact that the industrial smoke problem had long been solved. As early as January 1929, Osborn Monnett, the leading engineer of the Citizens' Smoke Abatement League, had explained that industrial installations were now, with few exceptions, "in good shape."[102] "Industry does not present a problem," noted Tucker at a session of the Smoke Elimination Committee.[103] Industry, indeed, remained entirely untouched by the regulation of the fuel market; it continued to burn gas-rich coal from Illinois without creating any significant problems. "Not a single industrial power plant to my knowledge in the City burns low volatile coal," wrote James Hinman Carter, Tucker's successor as smoke commissioner, in May 1946.[104] According to Carter, only vociferous and persistent protests from the coal industry had forced St. Louis to continually point out how important the choice of fuel was: "We have never claimed it did the whole job."[105]

Beyond that, in retrospect, Tucker's extreme luck must be emphasized. From the perspective of antismoke activists in St. Louis, America's entry into World War II came at exactly the right moment: the agency had had a year and a half to settle on a work routine without impediments. Moreover, the mobilization of the war economy boosted the demand for coal, which temporarily mitigated the displeasure of Illinois mine owners.[106] Above all, Tucker benefited from the fact that fairly close to St. Louis was a source of coal that produced far less smoke than the types of coal mined in Illinois. It is likely that Tucker's campaign would not have survived its first year without the presence of sizeable coal deposits of useable quality in the neighboring state of Arkansas; in the winter of 1940–41, 1,115,000 tons of low-gas coal were burned in St. Louis, an amount that exceeded the consumption of the previous year by over threefold.[107] Tucker also had the Interstate Commerce Commission to thank for this, because it had approved favorable freight rates for the coal from Arkansas.[108] Finally, Tucker's campaign was also favored by the geographic situation: since St. Louis obtained its coal from outside Missouri, the local coal industry had considerable problems bringing its influence to bear. Moreover, it was relatively easy to control the transport routes for coal, since the Mississippi lay between St. Louis and the coalfields. At times, Tucker posted some officials at bridges to stop the uncontrolled trade in coal.[109]

Finally, it is crucial to see Tucker's campaign through the lenses of race and class. The fight against smoke continued to be a concern of the city's upper middle class. As had been the case since the Progressive Era, real estate interests played a central role: "The biggest boosters for [the campaign] were the St. Louis Real Estate men," an outside observer reported.[110] As a result, the Smoke Elimination Committee was dominated by the city's upper class: its members included a banker, a stockbroker, a real estate broker, the president of the St. Louis Medical Society, and the research director of a large chemical plant.[111] This dominance made it possible to systematically downplay and marginalize the consequences that regulation of the fuel market would have for the lower strata of the population. In fact, a certain chill cannot be avoided when one reads how coldly the reformers assessed the consequences of their approach. For example, the chair of the Smoke Elimination Committee made only this terse comment about the problems of the poorer segments of the population: "It would not paralyze them if they had to spend $20.00 more a year."[112] The Chamber of Commerce, as well, saw no reason to take up this issue in detail: "Let us cross those bridges as we come to them."[113] When the city's director of public welfare, in October 1940, warned against a rise in coal prices during the winter, invoking the specter of tens of thousands of people in underheated apartments, Tucker responded with the arrogant comment that the man would be well advised to look after the problems of his own agency with equal zeal, without so much as hinting at a solution: "Eventually a rational solution will be obtained for this particular phase of the smoke problem."[114] And when Tucker pointed out that the rise in the price of coal had to be set against the drop in cleaning costs, the chair of the Smoke Elimination Committee noted tersely, "You could not get the Negroes to figure that."[115]

All in all, there is good reason to be skeptical about an excessively personalized interpretation of the events of 1940. Still, the campaign would have been hardly conceivable without the determined leadership of Raymond Tucker. It was Tucker's personal integrity that repelled all attacks from lobbyists without effect, and it was Tucker's bold decision to enact the ordinance immediately that invested the campaign with authority and put a stop to attempts to sidetrack the drive against smoke. Tucker defended only one thing more fiercely than the provisions of the ordinance, and that was the professional standards of his profession. Indeed, of the eighteen positions authorized by the ordinance, he filled only thirteen, unable to find sufficiently qualified candidates for the rest.[116] "I believe that the engineering profession at the present time is on trial," Tucker told the Engineers Club of St. Louis in 1937. Given the events that followed, it is not difficult to imagine that Tucker saw himself as defense attorney and star witness rolled into one.[117] For all the support that he received from fellow activists and the broader circumstances, Tucker remained the key figure of a campaign whose success rested in no small measure on his personal ability to get things done. Never before had a single engineer possessed such power in the American movement against air pollution—and, as time would show, never thereafter.

A Movement's Apogee

From the beginning, the events in St. Louis had unfolded before the eyes of a national public. Respected magazines like *Life* and *Business Week* wrote repeatedly on the local efforts, and the report of the Smoke Elimination Committee drew wide attention; only four days after its publication, Tucker spoke of "innumerable requests for copies."[118] For all their obvious sympathy, however, observers were never quite sure whether these events would simply end up being an exotic exception. And their skepticism was not without reason: anyone who took a closer look at what was happening in St. Louis would certainly be justified in entertaining doubts about whether Tucker's approach could be implemented elsewhere. Thus, the new approach of the St. Louis campaign enjoyed an invaluable advantage when it attracted the interest of a city that was linked to the smoke nuisance like none other: Pittsburgh. It is, in fact, difficult to imagine another event that would have demonstrated more effectively the transferability of Tucker's approach than the passage of a similar ordinance in a city whose smoke problem had been virtually proverbial for as long as anyone could remember. One headline in *Business Week* put it in a nutshell: "Smoky City, Too."[119]

Since the campaign in Pittsburgh has been exhaustively described by Joel Tarr, a detailed account here can be dispensed with.[120] Briefly, in Pittsburgh, as in St. Louis, all owners of stoves and furnaces were given a simple alternative: either set up their installations in such a way as to guarantee smokeless burning of coal or purchase only specific kinds of fuel.[121] And here, too, coal merchants were compelled to support the effort by making a special permit mandatory for their trade—a permit that could be revoked if a merchant supplied coal that did not meet the specifications.[122] In Pittsburgh, the industrial smoke problem could be seen as having been solved long ago, as it had been in St. Louis: "We can more or less forget them and be grateful we don't have to fight big industry," noted a "smoke memo" in January 1941.[123] And as in St. Louis, the fight against smoke was chiefly a project of the city's upper class. A survey conducted in 1946—shortly before the ordinance went into effect for domestic burning—found an approval rate of only 36 percent among low-income individuals.[124] To put it in somewhat pointed terms, the people outside of the city's economic and political elite gave their approval to the campaign above all by not mounting a clear protest.[125] Finally, the antismoke effort in both cities profited in the immediate postwar period from the fact that natural gas replaced coal as the most important fuel in private households within a few years, thanks to newly constructed pipelines and lower costs.[126] Here, notably, the word *profited* reflects the perspective of antismoke activists, because in retrospect a different perspective might be taken: in both cities, reforms were implemented just in time for smoke inspectors to claim credit for the decline in the smoke nuisance that accompanied increasing use of natural gas.

The important thing, however, was that Pittsburgh was able to avoid a grueling conflict with the coal industry. While the relationship between city and industry was certainly not devoid of tensions—after World War II, the coal industry pushed insistently but unsuccessfully for an extension of the

Figure 4. A gloomy morning in downtown Pittsburgh. After the successful 1940s campaign, antismoke activists used such pictures to demonstrate the progress that had been made. Image courtesy of the Archives of Industrial Society, University of Pittsburgh.

deadlines for converting domestic stoves—the parties were always intent on deescalation in the face of divergent interests.[127] "I have yet to hear any coal men who said they wanted to fight smoke abatement in Pittsburgh," declared the managing director of the Retail Coal Merchants Association at a hearing in 1946.[128] The supporters of antismoke efforts, too, showed little interest in avoidable quarrels. "We want to make this city smokeless without hurting anybody," emphasized Abraham Wolk in his capacity as chair of the Pittsburgh Smoke Commission. I. Hope Alexander, the director of Pittsburgh's Department of Public Health, chimed in: "The coal people didn't want to co-operate," he reported, describing the situation in St. Louis, but "we don't have that situation here."[129] When Pittsburgh set up a smoke abatement program at the county level in 1947, the man appointed to head the Allegheny County Bureau of Smoke Control was Thomas Wurts, a longtime friend of the president of the Pittsburgh Consolidation Coal Company—a circumstance that would have been unthinkable in St. Louis.[130]

Thus, if St. Louis had demonstrated that the problem of domestic smoke could be addressed by regulating the fuel market, Pittsburgh showed that this was also possible without running battles with coal interests. In other words, what was possible in these two cities was, in principle, possible everywhere. And so Pittsburgh and St. Louis became for many years the shining examples that other cities sought to emulate. In October 1947, James Ford reported visitors from more than 150 cities, while, according to Tucker, the number of written inquiries stood at 221.[131] "Public interest in smoke abatement is now at a high level, since St. Louis has demonstrated that pollution of urban atmospheres really can be controlled," began a 1942 article in the *Cincinnati Journal of Medicine*. The president of the Woman's Civic Forum in Nashville saw St. Louis as "the model for the country," and a professor at the Carnegie Institute of Technology in Pittsburgh, who had studied the St. Louis approach on site together with Ely and Alexander, complimented Tucker by calling him one of "those relatively few men who have made outstanding contributions, not only to their home town, but for the health of the whole United States."[132] Osborn Monnett, a kind of "grand old man" of the American antismoke movement, declared, "We are approaching a new era in smoke abatement."[133]

Large American cities, then, were evidently able to deal with their emission problems on their own. "One needs only to consider such outstanding examples as St. Louis and Pittsburgh to realize that communities can be cleaned up," noted one author in the mid-1950s.[134] Only one group was unable to embrace Tucker's breakthrough: the coal industry. In part, its response was certainly constructive when it placed greater emphasis on technical advice and research—with results that were acknowledged even by Tucker's supporters. "It has made the coal people really smoke conscious," observed the head of Washington's antismoke efforts regarding a welcome side effect of the events in St. Louis.[135] But the coal industry's response also included a rather underhanded campaign, in which the Coal Producers Committee for Smoke Abatement played a prominent role.[136] In the end, however, few were taken in by their efforts; even Cincinnati—where the Coal Producers Committee was headquartered—passed a 1947 ordinance modeled after St. Louis's.[137] Thanks to St. Louis and Pittsburgh, municipal smoke abatement enjoyed a reputation like never before, and it lasted for a while—longer, perhaps, than was in the interest of the movement and its cause.

For until 1945, the fight against smoke had always been waged with the suggestion that the final victory over the smoke nuisance would resolve all the significant air pollution problems of America's cities. Of course, this narrow focus on the smoke problem had never held the status of unquestioned dogma: Pittsburgh's Smoke and Dust Abatement League had dealt with the problem of airborne dust as early as 1915, and toward the end of the 1920s, the Woman's City Club of Chicago had repeatedly discussed the dangers of automobile exhaust. These, however, were individual initiatives, sporadic in nature.[138] It was only after World War II that the totality of air pollutants moved into the center of public interest. "Smoke abatement is not the only objective; it is air purification," declared the 1951 *Fuels and Combustion Handbook*, under the heading "misconceptions and illusions in smoke-prevention measures."[139] This view was also now shared by William Christy: "Smoke abatement no longer is enough—today's programs cover smoke, dust, and fumes."[140] The importance of the smoke nuisance, in contrast, declined in public awareness, now widely regarded as a problem that was sufficiently under control. "The battle against smoke has been won in many communities, but the longer and more difficult task of combating other air pollution lies ahead for most industrial cities," said the managing director of the Air Pollution and Smoke Prevention Association of America at the beginning of the 1950s.[141] "Most people forget how bad the smoke used to be," James Ford declared in his capacity as the former chair of the Smoke Elimination Committee in St. Louis, and this touch of nostalgia was by no means unique.[142] Herbert Dyktor, the head of the Division of Air Pollution Control in Cleveland, wrote in a letter in 1950, "In general, it can be said that smoke ... is not the problem it used to be."[143]

In order to understand the shifts in American air pollution control in the postwar period, it is crucial to keep in mind that the St. Louis and Pittsburgh campaigns remained part of the collective memory for years to come. As a result, the municipal smoke abatement movement accumulated enormous prestige: smoke inspectors had succeeded in liberating two notoriously dirty cities from the smoke nuisance, a success that hardly anyone in the broader public had expected. There was thus no question that municipal authorities would play an important role in the fight against the remaining air pollution problems; in many places, the local agencies were simply too popular to refuse to *take on* the new challenges. However, it remained to be seen whether they were strong enough to *meet* these challenges.

German Air Pollution Control, 1914–45

As in the United States, World War I was a rupture in the German drive against smoke. However, while the American debate eventually rebounded, the German one never recovered from the war-induced setbacks. After all, the German smoke debate had already been lacking momentum before 1914, and the adverse conditions of the interwar years, with their rapid succession of economic and political crises, were anything but conducive to an effective smoke abatement drive. It therefore comes as no surprise that the institutional fight against smoke now dropped even below 1914 levels. The Hamburg Society for Fuel Economy and Smoke Abatement not only suffered from a decline in membership since the beginning of the 1920s; it also changed its character, increasingly becoming a conventional engineering office with the economics of heating as a focal point of its work, the drive against the smoke nuisance taking on secondary importance in its day-to-day activities. After World War II, the society was dissolved without much ado.[144] The Office for Smoke and Soot Abatement in Dresden was abolished in October 1920 so that the experts on firing technology employed there could be transferred to the Heating Advisory Office of the Coal Agency.[145] And the Heating Inspection Office of Munich's Municipal Building Inspection was able to resume its prewar activities to only a limited extent in the 1920s. As Munich's Administrative Report of 1927 laconically put it, "The monitoring of industrial and commercial firing installations on the basis of the local police regulations pertaining to smoke and soot nuisance were resumed on a limited scale."[146] What this meant was that there was no longer a single German city in the 1920s with a systematic antismoke program.

Consequently, the smoke debate, even where it still existed, conveyed a dismal picture. In Berlin, a 1925 attempt to pass an ordinance against smoke and soot was thwarted by "the impossibility of its implementation"—the third unsuccessful attempt in the history of Berlin's fight against smoke.[147] At the beginning of the 1930s, Braunschweig even saw a discussion about repealing the local 1883 ordinance against the smoke nuisance; in the "Police Ordinance on the Abatement of the Generation of Noxious Odor, Dust, and Smoke," which was to take its place, regulations concerning smoke stood alongside those for compost heaps, rug beating, and rubbish pits.[148] Even when the Prussian Landtag called upon the state government in 1929 "to take measures to eliminate the intolerable, noxious soot nuisance produced by factories for the surrounding residents," no official action was set in motion; the pursuant decree by the minister of trade amounted to nothing more than a lame appeal to the subordinate agencies "to continue to pay special attention" to the problems in question.[149] A survey conducted by the Association of Official Engineers of the Machine and Heating Sectors arrived at this sobering conclusion in the mid-1920s: "Up to now, a general, purposeful drive against the smoke and soot nuisance has not materialized in Germany."[150]

What remained were complaints. At first glance, this might seem rather pitiful next to the citizen participation of American campaigns. After all, not every citizen was ready to lodge a formal complaint with the authorities, since this could easily become a strain on neighborly relationships. Moreover, environmentalists often take a rather skeptical view of complaints by neighbors, and with good reason; Raymond Dominick, for example, distinguishes between NIMBY ("not in my backyard") protests and the actions of true conservation groups in his history of the German environmental movement.[151] And in any case, complaints could also be made to the smoke inspector, who had a much greater interest in achieving quick results, since his job depended in the end on the goodwill of the citizenry. But such an assessment ultimately, and inaccurately, perceives the German approach exclusively through the lens of the smoke inspection approach. As much as the lack of any system in the German fight against smoke deserves to be criticized, the German approach also had some significant advantages. The limited focus on excessive polluters, for example, offered important opportunities in rural areas, where individual, clearly identifiable perpetrators could be targeted. Although the problem of air pollution in the countryside was fairly marginal, the contrast with the smoke inspection approach, which remained a purely urban institution, is worth noting. Further, German efforts to keep the air clean always encompassed all pollutants that were in any way dangerous, not only smoke and soot—traditionally the province of the smoke inspector. How to evaluate these advantages against the meager German smoke abatement record is a matter of debate, but they do underscore one of the recurring themes of this study: comparisons are not simply a matter of "better than."

It is, of course, not easy to make general statements about how German authorities dealt with incoming complaints. The number of rejected complaints was substantial, but so was the number of cases in which the authorities took action—but even then, it is difficult to assess their measures. For example, an official's response to a complaint in Cologne—"the entrepreneurs are endeavoring in every way to live up to the demands to be made of them"—can be interpreted in various ways: as the result of earlier complaints, as a preemptive effort by an entrepreneur aware of the wishes of his neighbors and the officials, or as the pious wish of an official who was hoping to spare himself a lot of work.[152] And yet German authorities in fact took resolute action in a fair number of cases, forcing an improvement in the situation. For example, when complaints were filed in Wernigerode about "very vile-smelling gases" from a raw oil motor, the police administration swiftly demanded the procurement of a new motor, even though this entailed costs of around 10,000 Reichsmark.[153] Nor did the authorities shy away from the threat of punishment in this case, since the company in question, according to the police, "shows great indifference toward all official orders and can be prompted to follow them only through pressure."[154] A conflict over the soot emissions of a furniture factory in Waldshut (Baden) reached a similar outcome: after a variety of unsuccessful countermeasures, the company procured a steam boiler machine.[155] Even where their legal basis was shaky, officials did not simply give up. When residents complained about a business in Bochum that was burning ash-rich slurry coal, the factory inspector subjected the boiler plant to a close inspection—and what do you know: in order to burn slurry coal, the operators had added an artificial air supply, which could be interpreted as a substantial alteration of the plant, requiring a new permit, according to the industrial code. It goes without saying that the inspector would agree to approve this permit only "when the entrepreneur makes adequate provisions to remove the nuisance caused by fly ash."[156]

Undoubtedly, official lethargy was also obvious in some cases. For example, in 1928 a factory inspector in Düsseldorf considered it "unavoidable that at times a weak odor of acid will be noticeable in proximity to sulfuric acid plants; the neighbors will have to put up with it."[157] Also in 1928, a factory inspector in Triberg (Baden) demanded that a foundry "raise the height of the venting chimney" so that "the particularly bothersome smoke vapors generated by the casting process will be at least discharged over the neighboring buildings"; the following year, the same office noted that the raising of the chimney, "given the unfavorable location of the foundry in the middle of a valley basin," offered "no possibility of a remedy."[158] On the whole, the impression emerges that the core problem rested less with the officials than with the fact that virtually nothing got done without an outside impulse. "It remains to be seen whether further complaints will be lodged": this note, from the police administration in Bielefeld, pertaining to a remedial measure of dubious value, is a prototypical reflection of the mentality of German civil servants.[159] Even the Office of Factory Inspection (Gewerbeaufsichtsamt), which would become the most important air pollution control agency in postwar Germany, for the time being took on emission problems only when someone complained. But behind this stance lay not a deliberate policy of going easy on industry, but a bureaucratic path of least resistance. At any rate, German officials swung into vigorous action only when their honor was at stake. Even in industrial conurbations, officials were sensitive to the charge of negligence or bias; for example, the county commissioner (Landrat) of Dortmund explained, with respect to a conflict over emissions, that "immediate intervention from higher levels" was necessary so as to "cut the ground as quickly as possible" from the occasionally voiced opinion that the authorities "do not wish to take strong action against industry."[160] The fight against excessive air pollution was simply part of the normal bureaucratic routine.

Going through the surviving files, one gets the impression that the lack of a clear line did not pose a problem for officials. In fact, the lack of clarity suited them, entailing considerable flexibility: time and again, officials' daily bureaucratic routine reveals their clear desire to commit themselves as little as possible. For example, a 1924 licensing document from Dortmund states that "the cupola furnace should be set up and operated in such a way that harm and nuisance to the neighborhood and the workers of the factory should be avoided as much as possible"—a formulation that officials could interpret essentially as they saw fit.[161] The police could "take action against every commercial installation—whether licensed or not—as long as it was causing substantial dangers to the public, specifically to its health," explained the mayor of Würselen, near Aachen, referencing a decision by Prussia's Supreme Administrative Court (Oberverwaltungsgericht). In fact, the concept of danger in the Allgemeines Landrecht was almost a kind of wild card that the officials could play as needed; even the Reich Health Office declared in 1927 that where foul odors "impaired the well-being or damaged the appetite" or "the windows have to be kept shut . . . , a direct harm to health is undoubtedly present."[162] Officials could do something, provided they wanted to—and whether they wanted to was decided on a case-by-case basis.

This attitude meshed well with the fact that officials continued to count on the understanding of entrepreneurs and their willingness to cooperate. After all, it was to the advantage of both sides to deal pragmatically with problems rather than to communicate merely through official channels. "Practically speaking, the way it works today is that everyone who wants to set up a furnace, especially one that requires a permit, first gets in touch with the Office of Factory Inspection," the Reich labor minister explained in 1924.[163] The Prussian minister of public welfare also believed that "amicable influence and instruction by the officials of the office of trade supervision and by other expert organs of the bureaucracy often reached results better and more quickly than coercive police measures."[164] On another occasion, this same minister explained that one must not "go at it crudely, but with sensible deliberation and wise moderation."[165] To be sure, this did not mean that the administration was soft on industry; after all, officials still enjoyed a considerable array of formal and informal means to pressure industrialists into compliance. In fact, it was not rare for offers of talks to go hand in hand with threats of sanctions. In 1933, for example, one official of the Office of Industrial Inspection in Leipzig, in a conflict over a foundry business, declared his willingness to arrive "amicably at a remedy of the deficiencies"—only to proclaim in the same breath his belief that "the preconditions for coercive trade-police action are fully met."[166] But why use coercion if there was another way? From the bureaucracy's point of view, pragmatic solutions provided the least stressful solution.

Was the bureaucracy's case-by-case approach successful? Politically speaking, an affirmative answer could certainly be justified. The fact that lodging a complaint with the authorities was by no means a hopeless undertaking took much of the air out of reform debates, which, as a result, never got off the ground between 1914 and 1950. Anyone genuinely suffering from an above-average smoke nuisance had realistic chances of obtaining a remedy through lodging a simple formal complaint, and this reality had consequences: the officials' approach probably channeled—and in the final analysis neutralized—a considerable portion of the public protests. However, with a view toward the overall problem, a much more sober assessment is warranted. A policy on air pollution that essentially considered individual polluters in isolation and always shaped decisions around the special circumstances of individual cases was, in the end, an anachronism in an industrial society where emissions are not isolated special cases, but regularly occurring events. German officials continued to draw a fateful distinction between "excessive" air pollution, against which one had to take action, and "normal" air pollution," which individuals had nothing to complain about—and it did not occur to anyone that "normal" air pollution, too, could pose a considerable problem. The abstruse consequences of this bureaucratic logic became utterly clear in a program of dust-precipitation measurements that was launched in 1932: the goal of these measurements, undertaken simultaneously in a large city, an industrial conurbation, a residential suburb, and the countryside, was to "devise a yardstick to assess an area's degree of pollution from dust and other foreign particles precipitated from the air."[167] In other words, instead of pushing for a general improvement in air-quality conditions, the status quo was systematically measured, becoming the starting point for evaluating individual cases of pollution. "It must be our aspiration to find an unambiguous and objective expression for concepts that are anchored and laid down in existing law—as, for example, 'customary in a given location'—lest we get lost in a maze of problems," declared Wilhelm Liesegang, one of the most prominent air pollution experts of the interwar period.[168] In this perspective, consideration of the practical consequences of such a criterion became essentially irrelevant.

A Strong State

For students of the German political system, the unglamorous demise of the German smoke debate may seem an improbable turn of events. Didn't Germany have a bureaucracy with a legendary reputation for efficiency and autonomy? American reformers of the Progressive Era surely would have wondered about the dismal state of German air pollution control: Why was a state as strong as the German one acting so feebly and inefficiently in this field? But the contradiction was less pronounced than it might appear—and probably did not exist at all. After all, cooperation with industry was ultimately an expression of bureaucratic strength: in informal dealings with businesspeople, officials were aware that they had the power of the law behind them, and they made sure that the delinquents knew this too. If an industrialist was unwilling to cooperate, there was usually a chance to compel him by law and even to close down his enterprise altogether if all else failed. For example, the police shut down a limestone factory in Eschweiler in 1929 after it had produced excessive particulate emissions.[169] To be sure, cases of this kind remained rare, but this ultimately attests to the strength of the German bureaucracy: from a businessperson's point of view, it was important not to fall from grace with German officialdom.

Perhaps the best grounds on which to argue for the strength of the German bureaucracy was the array of expertise it could muster. In the interwar period, this array was further boosted by the fact that the Prussian State Agency for Water Sanitation (Landesanstalt für Wasserhygiene), the highest authority on all questions of water-related urban sanitation since the turn of the century, also began working on air pollution issues in 1923, henceforth known as the State Agency for the Sanitation of Water, Ground, and Air (Landesanstalt für Wasser-, Boden- und Lufthygiene, or "WaBoLu").[170] No comparable institution existed in the United States until well into the postwar period, and the possibility of officials having their own decisions declared authoritative by WaBoLu undoubtedly strengthened bureaucratic expertise yet further. However, the creation of this supreme scientific authority changed nothing about the structural problems of the German approach to cleaning the air: WaBoLu focused entirely on scientific topics and never considered its job to be coming up with political initiatives. At times, WaBoLu flexed its muscles about what it could and should do: "Where the nuisance is especially serious, and where it is possible to do so without an excessive economic burden on the businesses in question, the most rigorous action should be taken," one WaBoLu official proclaimed in 1932, at the nadir of the Great Depression.[171] On the whole, however, WaBoLu pursued a cooperative approach, which fit seamlessly into the existing system of air pollution control, with all its opportunities and problems. "Experience has shown that industry will not refuse its participation if it is persuaded that the measures do not entail unaffordable sacrifices for them," explained the president of WaBoLu in 1930.[172] And Wilhelm Liesegang, perhaps the most prominent WaBoLu expert, went so far as to offer a "vindication of industry" in one of his articles. While he admitted that technical problems continued to exist, as well as occasional negligence, emission problems were "surely never" attributable to a "ruthless pursuit of personal interests."[173]

The strength of the German bureaucracy is also revealed, however, in the simple fact that it could sustain its case-by-case approach. An American smoke inspector would hardly have been able to afford to pursue a strategy riddled with inconsistencies and flimsy compromises for any length of time. In Germany, on the other hand, officials could in fact work in this way for decades, without coming under significant pressure. German bureaucratic agencies were strong enough to rebuff any public oversight of their work, their inconsistency revealed only if one immerses oneself in the files—which were not accessible to the public. For decades, then, German agencies could assert, with the ring of true conviction, that they were putting a stop to all excessive pollution—without ever having to explain their definition of *excessive*. German agencies were strong enough to dispense with legitimizing their work through general rules: "There are only certain sanitary demands that should be assessed according to the circumstances of a given case," the Reich Health Office responded in 1927 to a question about general emission threshold values from a Bavarian County Administration, without having to worry about accusations of arbitrariness.[174] And German officials could afford to distribute jurisdiction over clean air measures in an almost Byzantine manner: one 1963 inventory of bureaucratic jurisdiction in the state of Lower Saxony revealed that the authority to keep the air clean was spread among fifteen divisions in eight departments across five ministries.[175] There is hardly a better demonstration of this entanglement of authority than the fact that the Ministries of Labor usually took the lead in the reform debates in the early years of the Federal Republic.[176] From a bureaucratic perspective, however, this made sense: this ministry was also in charge of factory inspection, involved in pollution control in a secondary capacity.

From this point of view, the German state was in fact a strong state when it came to air pollution control. The only problem was that this strength did not lead them to be tough on the problem itself, or at least not as a general rule. The German state could be strong if it judged pollution to be excessive, even more so if it believed that its directives were not being taken seriously, but these cases were rare exceptions to the rule of bureaucratic lethargy. The German administration, in other words, did not use its strength to solve the problem; instead, it focused on its own interest in processing complaints smoothly and without much ado. And this spawned an odd situation: the interwar developments that proved important over the long term took place for the most part without any contributions from the German bureaucracy; these developments include the rising interest in fuel economy, the upswing in dust control, and the consolidation of corporatist structures. During the interwar period, the foundations were being laid for the postwar upsurge in the clean air movement without contemporaries recognizing this shift— let alone taking advantage of it. Only in retrospect is it clear that the German clean air movement's ability to move so quickly into the fast lane in the 1950s vis-à-vis its American counterpart was largely the result of interwar developments.

Fuel Efficiency and Flying Ashes

In May 1899, one participant at a smoke abatement meeting held by the Association for the Promotion of Commerce and Industry ventured a look into the future: "I believe that the current rise in coal prices, which may continue for some time, will prove a more effective means for improving furnaces than all police ordinances."[177] In retrospect, this comment was almost prophetic: while American industrialists began to take an interest in optimized furnace operations because smoke inspectors and civic reformers were tirelessly reciting the creed that "smoke means waste," their German counterparts did so for the most part because coal was becoming scarce and expensive. "Rarely ever were the broadest circles of the population and the entire industry affected as disagreeably as they were by the coal shortage in the year 1900," an engineering handbook declared in 1904.[178] Similarly, a factory inspector in Unna explained in November 1901, "Given how large the procurement of coal looms in the budgets of most commercial installations, it comes as no surprise that in view of the steady increase of the price of coal by the Rhenish-Westphalian Coal Syndicate, more attention than before is paid to efficient furnaces and proper operation."[179] However, these efforts picked up broad momentum only through the conditions of the war economy after 1914 and the postwar coal shortage. Only these special circumstances created widespread awareness of the importance of the efficient use of coal; heating technology thus quickly advanced from a marginal topic to a key discipline.

In purely technical terms, this development was certainly favorable to antipollution efforts. A movement dedicated to the efficient use of fossil fuels invariably had to also be interested in reducing smoke and soot, the classic byproducts of incomplete combustion. But in Germany, unlike in the United States, the fight against smoke and the fight for efficient burning remained largely separate issues even after 1914. It was rare for interwar furnace-technology literature to even mention smoke and soot—and if it was mentioned, it was often only in connection with the attendant loss of heat.[180] De facto, many of the measures implemented during the war and in the postwar period to save coal had previously been called for in the name of smoke abatement, but nobody considered this noteworthy.[181] The smoke debate thus died out in the period after World War I, while many of its demands were met under the aegis of heating efficiency.

At the same time, however, the growing emphasis on furnace efficiency provided emphatic proof of what Joel Tarr calls the search for the ultimate sink: supposed solutions to problems often turned out to merely shift the problems somewhere else.[182] In this case, the quest for efficiency led to the construction of high-efficiency boilers with a previously unknown power density. The high stress on the burning process and especially the high gas velocity caused substantial amounts of ash to be pulled along and expelled outside through the chimney.[183] This problem became especially acute and explosive through the spread of powdered coal. From the point of view of emissions, powdered coal represented something like a worst-case scenario: the grinding of the coal created a particularly fine ash, while the blowing in of the fuel caused strong turbulence, as a result of which only a small part of the ash settled in the burning chamber. To make matters even worse, one of the advantages of powdered coal was that it allowed for the use of inferior fuels with a high percentage of incombustible substances.[184] Vociferous complaints from neighbors were the logical result: "Nobody can deny this ashfall," stated one complaint about the municipal Mark Electric Power Plant, while a teacher from Herbede attacked the emission of airborne dust from the Holland mine by noting that the situation was "so bad that the students are often forced to put on their hats during recess to get some kind of protection from the ash."[185] In the 1950s, the community of Kleinblittersdorf in the Saarland asked the authorities to intervene, "before our once flourishing town is turned into a Pompeii by this disastrous ashfall and all human and plant life suffocates in the dust."[186]

By chance, this development occurred at the same time as an upsurge in purification technology. Beginning in about 1910, there was a virtual boom in dust technology, triggered above all by the development of the first functional electrostatic precipitators by Frederick Cottrell in California and Erwin Möller in Brackwede near Bielefeld.[187] Compared to mechanical, cloth, and wet filters, the electrostatic filter had a number of substantial advantages: it had a very low filtration resistance and low operating costs; it was insensitive to high temperatures; and above all, it had an extremely high cleaning efficiency, since the electrostatic filter's capacity to capture dust did not depend on the size of the particles. Moreover, the resulting competition between the electrostatic filter and other dust-removal techniques quickly spurred a dynamic technical development.[188]

Two power plants became emblematic of the fly-ash plague and the difficulties of abatement: the Mont Cenis mine in Sodingen near Herne and the Klingenberg power plant in Berlin-Rummelsburg. In both cases, furnaces for powdered coal with a high fuel throughput had been built without any meaningful dust-extraction equipment, presumably out of sheer thoughtlessness—and in both cases, there was almost immediate outrage after they began operating.[189] In Sodingen, authorities openly threatened to close down the operation, while the Berlin victims formed a special organization and pursued a number of civil suits, finally settled by mediation in 1931.[190] Within the engineering community, the shock was profound: Klingenberg was the first German power plant to burn only powdered coal, a showpiece of contemporary power-plant technology—and now it was making headlines because of its fly-ash emission and because it was dragging the authorities into the business.[191] For once, antibureaucratic sentiment among German engineers, usually a burden on efforts to control air pollution, had a positive effect. As one publication put it in 1929, "It is in the interest of the public at large and desirable for the businesses themselves to achieve satisfactory results before the authorities find themselves compelled, because of individual complaints, to intervene in a comprehensive way, which could easily trigger a progress-retarding effect."[192] Never again would German engineers want to see supermodern furnaces being perfected through bureaucratic intervention.

However, engineers were driven not only by fear of further embarrassments but also by the inherent fascination of dust technology. After all, the developments of interwar period amounted to a classic case of science-based

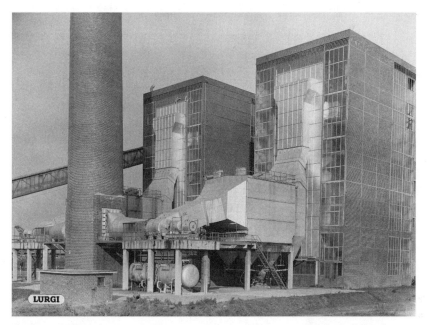

Figure 5. When electrostatic precipitators came into use in large coal-fired power plants during the 1920s, fly-ash control became a booming engineering field. Image courtesy of the Deutsches Museum.

technological progress. As late as 1917, a Swiss professor had declared that dust technology had "hitherto been limited to an empirical-practical adaptation to a given case," while a 1925 lecture about electrostatic gas purification before a local group of the German Society for Technical Physics had closed with the flowery assertion that "in the young forest of these theories many patches are still unplanted"—two comments that would soon be outpaced by rapid scientific developments.[193] The key attraction of dust technology was the interplay between theoretical calculations and practical adaptations: "It is a job that can only be solved through the proper organization of all available forces and resources, through the organized cooperation of theoreticians and practicians," one engineer noted.[194] In 1939, a technician reported on a centrifugal dust extractor for which around twelve thousand test series had been conducted over some ten years of development.[195] With this, there arose considerable need for technical expertise in the field of dust technology and thus a large number of job openings for engineers—especially since it became clear early on "that experiences with one installation generally cannot be simply transferred to another," which meant that every filtration system had to be adapted to the special conditions of each particular case.[196]

Many engineers were all but bursting with pride when they had a chance to talk about their special field. "The elimination or retention of industrial dust is an urgent demand of public health and not least of great economic importance, as well," proclaimed a popular scientific magazine in 1929, and others declared that the emission of dust was "among the most difficult things the engineer has to deal with."[197]

Of course, the counterpoint to the euphoria of the engineers was cost. In the case of the Mont Cenis mine, for instance, equipment costs for wet dust extraction amounted to around 70,000 Reichsmark, and operating costs were also significant: the installation used eighty-five cubic meters of water per hour, and even though two tons of chalk were added to the water every day to protect the construction against corrosion, annual repair costs still came to about 7,000 Marks.[198] Flue-gas cleaning at the Herdecke power plant was even more expensive, as existing mechanical filters did not achieve more than a paltry 40 percent efficiency. According to the plant operator, a bid from the Lurgi Company, then the leading manufacturer of electrostatic filters, came to around 500,000 Marks for the plant's six powdered-coal-fed boilers, plus some 200,000 Marks for the necessary construction work and a significant loss of income from the temporary shutdown of the boilers; the company, understandably, installed the filters only after long debate and under pressure from the authorities.[199] One engineer declared in 1931 that fly-ash control increased the cost of generating steam by 0.5–3 percent for grate combustion and by 2–10 percent for powdered coal combustion—and this did not even include the cost of waste removal.[200] In most cases, profitable utilization of fly ash was simply not in the cards: "A profitable use of the fine fly ash will hardly yield enough profit to service the debt for the dust extraction equipment," explained a Dresden engineer in 1931, and other experts expressed similar views.[201] To be sure, a few viable methods of utilizing fly ash were gradually developed, especially in the construction industry, but for the time being these methods could not shake off the stigma of being makeshift solutions; the "Basic Foundations of Planning for Flue Gas Cleaners," published in 1940, noted tersely that the utilization options for fly ash played "only a minor role at this time."[202] And even landfilling ash was rarely as simple as at the Essen-Karnap power plant, which mixed the ash with water and used it "to fill in the area around the plant."[203] This was especially true as power plants reached a daily fly-ash output of five hundred tons and more by around 1940.[204]

As a rule, a compromise generally was struck between what was possible and what it cost: when it came to flue-gas cleaning, the target was a degree of efficiency a few percentage points below what was technically feasible. "In the case of worthless dusts, as for example the flue-gas cleaning for power plants, we will have to be content with lower levels of retention for reasons of economic efficiency," one engineer explained in 1937. Other experts were also of the opinion that "a crude extraction" was good enough for fly ash.[205] In principle, this was also acceptable from the bureaucrats' perspective: "The minimum demand should be . . . 70 percent," WaBoLu noted in 1931.[206] Hardly anyone was asking for less anyway. For the most part, officials merely asked in general terms for "what is technically and economically feasible," a formula whose concrete meaning was "naturally dependent on the situation of the business using the furnace."[207] And so technicians, officials, and operators arrived with remarkable speed at a fundamental agreement: the cleaning of flue gases was indispensable above a certain boiler capacity—but the precise requirements had to be determined based on the circumstances of each individual case. An article in an engineering journal declared in 1930 that "the requisite degree of dust removal must be assessed according to population density, the type of plant growth, and so on."[208] The engineer Fritz Wellmann explained in 1936 that the crucial thing was "only the question of whether the fly-ash nuisance is eliminated completely through the installation of a dust extractor. Everything else is and remains theory."[209] Thus, officials maintained their classic German case-by-case approach, abstaining from general rules and regulations. For example, when a smelting plant asked in 1934 whether "any official regulations exist regarding the purification levels of dust extraction installations," WaBoLu responded by saying, "The supervisory agency decides case-by-case whether the installation of dust filters is necessary and what degree of dust extraction is required to avoid causing a nuisance to abutters."[210]

Needless to say, even a consensus like this left considerable room for differences of opinion. The issue, however, was no longer whether flue-gas cleaning should occur, but how it was to be done. Above all, the operators of boiler plants were intent on using mechanical dust precipitators, since they were significantly less expensive than electrostatic filters.[211] For example, at the beginning of the 1940s, WaBoLu published an expert opinion about a power plant in Upper Silesia, whose operators believed "that in the present case one should under all circumstances dispense with the installation of electrostatic filters."[212] However, serious conflicts arose only where the issue was expensive retrofitting.[213] The giant Goldenberg Power Plant in lower Rhenish Knapsack was probably the most spectacular case of this kind.[214] The operator of this plant had been able to forestall the installation of efficient dust extractors during the interwar period, but this turned out to be a rearguard action; in the early 1950s, officials, after intensive negotiations, pushed through the installation of efficient dust-extraction equipment. As a result, dust emissions from the Goldenberg Power Plant had dropped 76 percent from their 1936 levels by 1958, and the company was intent on cutting its fly-ash emissions by half again by 1960.[215] Flue-gas cleaning had simply become a matter of common sense for high-capacity boilers: "A dust extractor is no longer a matter of secondary importance," an engineer explained in 1941, and others noted that "hardly any large high-capacity boiler units are being erected without airborne dust precipitation."[216]

Dust technology even made it into the movies during the interwar years: a film produced in 1933 at the behest of the German Air Filter Construction Company was entitled "The Battle against Dust" (*Kampf dem Staube*). In addition to extensively promoting the company's products, the film also contained a feel-good depiction of the contemporary expert community. "Much has already been achieved . . . through the sympathetic cooperation of commercial sanitation, industry, and technology," the voiceover script declared.[217] Of course, the intention of presenting contemporary debates in the most harmonious light possible is readily apparent—but in its core the depiction is accurate: the emergence of flue-gas cleaning had in fact taken place without intense professional controversies. Power-plant operators were, needless to say, less than thrilled about the costs entailed by flue-gas cleaning, but they were at least willing to converse sensibly with officials about the matter. Thus, the overall picture had changed within a few short years. As late as 1925, Hanover's factory inspector had noted that the installation of an electrostatic precipitator for flue-gas cleaning "is too much to ask of any company today."[218] A little over a decade later, the installation of electrostatic filters was part of standard operating procedures.[219] All told, this provided emphatic proof of what was possible in Germany when officials and engineers for once joined forces. At the same time, however, this story raises the question of why this did not happen more often.

Corporatism Arising

The rise of dust technology also gave birth to a new scientific organization, the Expert Committee on Dust Technology (Fachausschuß für Staubtechnik). Its establishment, suggested by the Berlin District Association of German Engineers, was approved by the board of the association in November 1927 and carried out in February 1928.[220] "Its task is to bring together the efforts in its area and to provide support for urgent scientific work," explained Barkow, the first chair of the committee, at the founding meeting.[221] From the outset, the Expert Committee on Dust Technology had a decidedly interdisciplinary orientation. "The Committee should not be made up only of engineers, but should also draw on experts from other branches of science," Barkow declared, and Robert Meldau, elected as the official in charge of the professional literature, likewise emphasized "the need for cooperation between the experts in all areas having to do with dust."[222] And the committee in fact quickly turned into a lively meeting place for experts in the field; for example, its full meeting on March 3, 1937, was attended by around 260 members and guests.[223] As the most important result of the committee's work, apart from the definition of terms and concepts related to dust technology, the 1936 "Guidelines for Efficiency Tests on Dust Extractors" is notable—key for dust research, which was still in its infancy.[224] "A similar work exists neither in this country nor abroad," the managing director of the committee proudly noted in 1937, happily adding that there were "already several requests by American agencies . . . to translate the 'Guidelines' into English."[225] The committee's code of standards also earned approval and praise from others, and after a few years of work, its key role in the emerging scientific community was undisputed.[226]

Publicly, the Expert Committee on Dust Technology cultivated a stance of scientific neutrality. For example, in its 1927 founding appeal, the Berlin District Association explained that the plan was intended to promote "cooperative work" that avoided "duplication and a dissipation of energies."[227] When Barkow stepped down as chair in November 1931, he similarly declared that he had always been intent on "keeping together a society of scientists on the basis of scientific truth," and his successor was quick to express vigorous agreement.[228] Consequently, when one participant in the discussion following a lecture pointed to his own company's research results, Barkow immediately admonished him "not to name specific companies and to limit yourself here merely to personal experiences that are necessary for a thorough discussion of the lecture."[229] The members of the committee functioned "not as representatives of agencies, organizations, or companies to which they belonged . . . but as expert consultants," one VDI representative emphasized in 1931, doing so "even if they were recommended for participation by an agency or a company."[230] But these remarks notwithstanding, the Expert Committee on Dust Technology was indeed wedded to a political purpose: cutting the ground from any political initiative by promoting scientific and technical development. At a full meeting of the committee on March 19, 1930, Barkow, before moving on to the agenda, explained apodictically "that one cannot fight dust through ordinances or laws" and that the committee "does not exist to set the gears of legislation in motion"—

an explanation that was approved by those present without any further debate.[231] The aversion of engineers to a politicization of air pollution–control efforts was unmistakable.

Few noticed at the time that the establishment of the Expert Committee amounted to a gradual departure from the traditional German regulatory style, with its heavy emphasis on the state. While officials in Imperial Germany had displayed a marked preference for their own experts under the umbrella of the bureaucracy, a corporatist structure was now gradually developing, state organs and private organizations simultaneously taking on state tasks as well. This trend became even more apparent in the area of steam-boiler inspection associations, commonly called DÜVs (for Dampfkesselüberwachungsvereine). Since the foundation of the first DÜV in 1866, they had become pioneers of corporatist regulation in Germany, gradually taking over the inspection of steam boilers from the German state administrations.[232] But the state bureaucracies initially watched this rise of the DÜVs with some skepticism and feelings of competition. Prussian officials, in particular, had a hard time seeing in the DÜVs more than simple consultants on technical matters: for example, while the DÜVs in Saxony had been asked as early as 1882 to pay "special attention" during the inspection of steam boilers to ensure "that the burning takes place with the least amount of smoke possible and that neighboring property owners do not experience any damages or substantial nuisance from smoke, soot, etc.," a comparable regulation in Prussia was not incorporated into the boiler directive until 1909.[233] To be sure, a few Prussian DÜVs were consulted as experts on complaints about smoke emissions before 1914, but for the time being this was still the exception.[234] It was only after World War I that involving DÜV experts in the fight against air pollution problems became bureaucratic routine.

In part, this development was a byproduct of the previously mentioned rise of heating technology. Ever since the coal shortage during World War I, the DÜVs had been actively committed to promoting the efficient use of fossil fuels. For example, in 1921, the Aachen DÜV offered "courses on the most suitable design and operation of power and heating installations and on the proper implementation of an orderly inspection of operations."[235] The Württemberg DÜV, in its annual report, admonished the owners and operators of thermal power plants and heating installations "to have their businesses inspected as to whether they meet the requirements of the time with respect to heating economy."[236] But the DÜVs also played an important role in combating the fly-ash problem. In the 1930s, for example, WaBoLu and the Saxon-Thuringian DÜV agreed to cooperate in studying airborne dust in Bitterfeld: while WaBoLu would determine the extent of the damage through air measurements, the DÜV would inspect the technical conditions in the boiler house.[237] The DÜVs wanted to be a "truly independent, reliable, and objective advisor" on technical questions relating to firing and dust extraction, explained one DÜV engineer in an essay published in *Rauch und Staub*.[238] And the central association of DÜVs in Prussia quite matter-of-factly counted the fight against smoke among its "official responsibilities" in 1930.[239]

Thus the state and the organizations slowly came together, and both sides let go—at least gradually—of emotions that had previously prevented constructive cooperation. Officials got used to the corporatist structures and began to use the expertise assembled in the associations in their work; the VDI abandoned its purely negative attitude that viewed administrative officials as the natural enemy of the engineer. In the big picture, the interwar period thus presents itself as a period of transition: this was no longer Imperial Germany, where communication between engineers and administrators was limited, but it was not yet the era of the Federal Republic, when corporatist structures became a core element of the German regulatory style and the Expert Committee on Dust Technology the nucleus of the cooperative development of technical standards. Each side drew closer to the other, but the pace was glacial, even though they were in fact a perfect match: a strong and self-confident administration joined to the expertise of the German engineering community was, if anything, a promising combination. Once again the key dilemma of the German style of air pollution control is clear here: the lack of an aggressive fight against air pollution was a matter not of the absence of resources, but of the absence of thoughtful combinations of what the various sides could contribute. It was not until after World War II that policy brokers emerged who were able to bring all of these factors together, and the considerable advances since the mid-1950s provide ample proof of what German air pollution control could accomplish once it got its act together. Before 1945, however, neither administrators nor engineers were able to forge a similar alliance—and neither were the Nazis.

Cleaner Air under the Nazis?

The environmental policies of Nazi Germany have come into sharper focus in recent years. Initially, that a regime so obviously hell-bent on war and genocide would sponsor an environmental agenda created consternation and irritation, but sentiments of this kind reflect little more than a naive essentialization of good and evil. After all, the power of the Nazi regime relied to a significant extent on its remarkable ability to appease and befriend numerous groups that were not completely in line with its goals. Some of these groups pursued an environmental agenda, and here the Nazi regime allowed for some significant successes: a progressive animal-protection law was passed in 1933; foresters temporarily came to favor alternatives to traditional clear-cutting and monocultures; advocates of nature protection were jubilant after the Nazis passed a far-reaching national conservation law in 1935 that inspired almost feverish efforts at implementation.[240] Thus, it should come as no surprise that attempts were also made to push a clean air agenda under the Nazis. None of these attempts got very far, but they do deserve attention, shedding light on both the inner workings of the Nazi regime and the bureaucratic routine that held German air pollution control firmly in its grip.[241]

Before going into details, it is crucial to recognize the inconsistency of the Nazis' approach to environmental issues. Long before 1933, German environmental policies had been notorious for their fragmentation, making it difficult to trace widespread trends. Germany had nothing like the concepts of "conservation" and "preservation," which allowed some degree of coherence within the nascent American environmental community; air pollution control was different from urban sanitation, which in turn was different from energy conservation and nature protection, and so forth. As a result, German environmental policy was really a patchwork of mostly isolated approaches to a wide range of issues, and the Nazis' rise to power, if anything, increased this diversity of policies. Arguments that presume a consistent green policy in Nazi Germany, put forward by Anna Bramwell and others, have been so thoroughly disproved that it would be futile to discuss them again here.[242] In fact, policy fragmentation was a general trend in Nazi politics, with numerous groups competing over jurisdiction on a wide array of tasks and issues. Countless individuals and organizations sought to "work towards the Führer," as one contemporary remark put it, ultimately selling their own goals and interests as quintessential Nazi policies. The political scramble that took place in debates over air pollution was perfectly characteristic of the Nazi regime, bringing to mind a wide range of similar controversies in other political fields.

The debates over atmospheric pollution within the Academy for German Law, set up in 1933 in order to "Germanize" the judicial system, provide a case in point. These debates began with the Nazi notion of *Gemeinnutz vor Eigennutz*, or "the common good above the individual good," enshrined in point 24 of the platform of 1920.[243] The profit for the nation at large, rather than for the individual, was to be the central yardstick of legal affairs, a stance that was crucial for the boom in nature protection since 1935: with conservation interests deemed superior to individual property rights, numerous tracts of land were confiscated or purchased for token sums.[244] However, the debates over air pollution within the academy took a more theoretical approach, focusing on matters of legislation, specifically paragraph 906 of the Civil Code, or Bürgerliches Gesetzbuch (BGB), of 1900. In a nutshell, paragraph 906 ruled that citizens were protected from industrial emissions unless these emissions were either "insignificant" (*unwesentlich*) or "consistent with the local norm" (*ortsüblich*); for citizens seeking to sue a company for damages or an injunction, this paragraph was usually the key legal tool. It is characteristic that the Academy for German Law focused on paragraph 906, as this choice was clearly not driven by any concern over practical politics: the civil law had decreased in relevance in air pollution control since the mid-nineteenth century, replaced by state policies that the Academy for German Law never bothered to look into.[245] After all, the academy's motives were clearly ideological, as Dennis LeRoy Anderson points out in his history of the institution: "In many ways, the BGB was the symbol of liberalism, individualism, materialism, and legal positivism, all anathema to the National Socialist concepts of the Führerstaat and the Volksgemeinschaft."[246] Thus, the key issue was not so much air pollution as the general character of the German legal system, the Civil Code in particular.

Nazi jurists were emphatic in condemning paragraph 906 as a sellout to liberalism, incompatible with the principles of National Socialism. "The formulation of paragraph 906 of the Civil Code does not leave enough room (at least in its current interpretation) to take this point of view into account. . . . Only the National Socialist conception of property rights might provide some wiggle room to the provisions of paragraph 906," Ernst Eiser, a Berlin lawyer, wrote in 1938.[247] Similarly, Heinz Schiffer asked in 1936 whether paragraph 906 allowed a resolution of pollution conflicts "pursuant to principles that are in accordance with the legal philosophy of National Socialism," and his answer was "a resounding 'no.'"[248] But criticizing these provisions was far easier than specifying what should take their place. It was simple to state that the national interest should reign supreme, but who represented the national interest when an industrial facility's emissions damaged cropland? Heinz Schiffer urged all parties "to tackle the problem from the point of view of the community," asking both sides to "show consideration for each other" and to "make sacrifices"—vague recommendations that ultimately meant next to nothing, especially given the traditionally averse atmosphere in court battles over pollution.[249] Friedrich Klausing, a prominent member of the Academy for German Law, argued that it was important to find "a sensible balance of interests in accordance with the demands of the national economy." But in the end, formulations of this kind raised more questions than they answered.[250]

With the general direction of reform blurred, concrete proposals soon became documents of disorientation. For example, Klausing argued for "a more appropriate interpretation and use of the existing legal provisions on the basis of the new juridical philosophy."[251] Eiser called for "elastic jurisdiction," and Schiffer sketched a plan for special commissions of experts that could deal with complaints in a flexible way—opaque ideas that never evolved into anything even remotely resembling a legislative draft.[252] The confusion over ways and means reached its peak in a session of the Committee on Land Law (Ausschuß für Bodenrecht) of the Academy for German Law in February 1938, which at times took on the air of a brainstorming session. Lacking a clear guiding concept, everyone felt it was permissible to simply throw out whatever popped into his head. There were calls to consider "the idea of the folk character [*Volkstümlichkeit*] of the law" or "the local norms as a whole" (*Ortsüblichkeit im Ganzen*) and statements emphasizing the duty of landowners "to maintain their property in such a way as to keep the well-being of the nation [*Wohl der Volksgemeinschaft*] in accordance with the law and popular sentiment [*gesundes Volksempfinden*]."[253] Some speakers noted the vagueness of their statements themselves: one argued that people should not be able to prohibit emissions "if one could expect someone to tolerate it, taking into account all the details of the case, especially the local situation of the property in question," only to acknowledge moments later "that these provisions, taken by themselves, are not very meaningful"; he concluded with a proposal "to expand the present law and to make it more flexible."[254] Faced with this plethora of ruminations, the committee did not even come close to agreeing on a legislative proposal.

As a result, the ideological challenge to air pollution law spawned meager results. The most important output was a law of December 13, 1933, limiting the rights of citizens who lived or worked in the vicinity of facilities that were considered important for the physical training of the people. Neighbors of such facilities could not demand that they be closed, nor could they force the installations to mitigate their deleterious impact. In other words, neighbors were powerless to eliminate the nuisance itself; the only option they had was to apply for monetary compensation. However, the law did not offer blanket protection to all sports facilities. Rather, it gave special status to specific, highly important installations, a designation that could come only from the Ministry of the Interior. In granting this status, the ministry made the facility subject to certain conditions and regulations; it could repeal its decision at any time, in which case the facility would again fall under the terms of the Civil Code. Finally, the ministry also settled issues of monetary compensation to neighbors by incontestable decree.[255] This law was followed by a similar one on October 18, 1935, which extended this type of special status to hospitals, nursing homes, and similar institutions.[256] A renowned legal journal expressed jubilance about what it saw as a "crucial breach" in a heretofore dominant concept of property, but this remark was vastly overblown.[257] In reality, the impact of these laws was very limited. Even a governmental decree pursuant to the 1935 law stated that complaints about nuisances from hospitals and similar installations were "generally rare." It was more common for a hospital to *suffer* from industrial or nonindustrial emissions, but for such cases the new laws left everything unchanged.[258] Revealingly, an overview article published in 1936 only referred to the two laws "in passing."[259] And in 1937, Friedrich Klausing concluded that there had been "no fundamental changes" in traditional legal provisions as a consequence of these laws.[260]

A second challenge to existing air pollution–control policies grew out of a more practical concern, with the Reich Food Estate (Reichsnährstand), the farmers' organization in Nazi Germany, as the key player. Agricultural institutions had been supporting farmers suffering from pollution damage since the nineteenth century, offering expertise and moral support in order to extract compensation payments from the industrial corporations at fault.[261] Yet a number of documents suggest that the Reich Food Estate encouraged farmers to be more aggressive with damage claims. For example, when a German organization of heavy industry conducted a poll among its members about their experiences in dealing with air pollution problems, several companies mentioned activities of the Reich Food Estate during the 1930s.[262] At times, even the head of the Reich Food Estate, the so-called Reich peasant leader Richard Walther Darré, became active. In 1937, Darré sent a letter to the Gutehoffnungshütte in Oberhausen in which he urged an "amicable settlement" of pending damage claims.[263] And as late as December 1943, he tried to negotiate a deal in a conflict over the emissions of the Rütgers company near Dresden.[264] In 1941, a number of industrial associations agreed to set up their own "Research Institute for Air Pollution Damage" in order to counter the "strong and well-organized" Reich Food Estate, which had put industry "at a severe disadvantage" with its network of partisan experts.[265]

Some documents even indicate that the Reich Food Estate was planning to develop a full-scale policy initiative on the issue. In 1935, the Reich "peasant leader" asked the group's local branches to report pollution conflicts in their area. Mentioning a number of unfavorable court decisions during previous years, Darré took the initiative "to raise the question of compensation for damage caused by the emissions from industrial enterprises, with the goal of reaching a fundamental and permanent arrangement." He even indicated what he had in mind when he noted that it was imperative "to check whether a change of the existing legal provisions is necessary, specifically with a view to enlarging the obligation of liability for compensation pursuant to paragraph 26 of the commercial regulations."[266] The decree did not elaborate on the Reich Food Estate's motives, but they are quite plain in retrospect. After all, the Reich Food Estate had an uneasy relationship with farmers. The Nazis' agricultural policy, most prominently the Erbhof Law of 1933, was hugely unpopular among farmers, all the more so since the Reich Food Estate had shown in its implementation of the Erbhof policy that it was driven far more by ideological concerns about "eternal peasants" and "blood and soil" than by an actual interest in the situation of individual farmers.[267] Moreover, with the Four Year Plan of 1936, "the Reich Food Estate lost any sort of independence it may have enjoyed and simply became an outright instrument for the mobilization of the agrarian sector and its adaptation to a wartime economy."[268] In supporting farmers in conflicts over pollution damage, the Reich Food Estate was probably using one of its last remaining options to operate independently.

However, no legislative draft emerged in the end, probably because the matter was more complicated than it appeared at first glance. As the Reich Food Estate examined the issue more closely, it surely realized that higher compensation for crop damage had unwelcome side effects. If farmers received too much compensation, they might neglect their agricultural work, which in turn would conflict with the goal of higher agricultural production in preparation for war.[269] This was in fact more than a theoretical possibility, as the Reich "peasant leader" occasionally found himself in the interesting situation of arguing in favor of limited compensation: "From the point of view of the Reich Food Estate, it is not a desirable goal to breed so-called 'pensioners' on a large scale," he declared in his letter to the Gutehoffnungshütte.[270] In this regard, the position of the Reich Food Estate echoed that of the members of the Academy for German Law, who pointed to the negative side effects of monetary compensation. "In light of the interests of all parties involved, the 'pension farmer' in the environs of industrial plants is not a pleasant phenomenon," Eiser noted in a 1938 article. According to his account, when a new factory was built, neighboring farmers occasionally neglected their ordinary work, apparently keen to profit from the new enterprise as much as they could.[271] But how could the Reich Food Estate make sure that a farmer would use compensation payments productively? It is not difficult to imagine that the jurists of the Reich Food Estate pondered such questions and refrained from drafting a bill because they could not come up with persuasive answers. After all, the issue of compensation claims did not amount to a simple clash between evil industrialists and those suffering from their pollution, as the uninitiated sometimes saw it.

A third challenge to the dominant practices grew out of the exigencies of rearmament. Building up a powerful army was, of course, a key goal of Hitler's policies, and this entailed the construction of numerous plants for military production. This being Germany, the plants would need to request the customary license under the Industrial Code, and that spelled trouble, for the routine procedure required making the plans public and giving every citizen two weeks to file objections.[272] This requirement was obviously at odds with the Nazi goal of rearmament; after all, it meant that everybody could inspect the plans for a war-production plant. Consequently, the regime added a new paragraph (§ 22a) to the commercial regulations in the summer of 1934 that provided for a secret licensing procedure. The necessary requirement was that the construction of the installation in question had to be "in the public interest."[273] As the German Ministry of Trade and Commerce declared in a decree of October 30, 1934, this would regularly be the case "if we are dealing with plants producing military supplies."[274]

Interestingly, the decree did not include an expressed intention to exempt war-production plants from all regulations. On the contrary, the decree of October 1934 specifically ruled that the procedure pursuant to paragraph 22a did not imply a limitation on the usual examination of plans, since this would mean "a disadvantage to the workers, the neighbors, and the public at large."[275] Further, experience had shown that public reaction was generally unimportant for bureaucratic decision making.[276] Finally, and importantly, this decree clearly envisioned licensing procedures of this kind as the exception: the license had to be granted by the state government, with local authorities assuming a preparatory role only. Thus, there was no intentional negligence regarding air pollution from war-production plants. For example, the Ministry of Trade and Commerce issued a 1936 decree to inquire about problems with the awful smell emanating from viscose factories—a type of production that was highly important in the context of the Four Year Plan.[277] And when the ministry discovered that nine of the twenty-eight viscose factories were faced with complaints from neighbors, it issued a second decree that provided guidelines for mitigating this problem.[278] Still, it is important to remember that, in spite of such initiatives, the Nazi's rearmament program and the subsequent rise in war production inevitably meant more pollution. Even more, in spite of the emphatic call for screening plans thoroughly for pollution hazards, it is safe to assume that given the haste and the priority of rearmament efforts, bureaucrats were fighting an uphill job in mandating effective antipollution measures.

But still, even the special procedure in fact seems quite normal. After all, the procedure called for the kind of flexibility that had always been a hallmark of the German approach to air pollution control. Accustomed to looking at each case individually, with little concern for a general trend, German officials were in an ideal position to accommodate the specific demands of a war economy with tacit concessions that few, if anyone, would notice. In a way, paragraph 22a even marked a victory for the bureaucratic rank and file, in that the special procedure for war-production plants followed the standard procedure as closely as possible, forestalling alternative approaches that simply suspended all rules for ammunitions factories and the like. In fact, the paragraph was not seen as a quintessential Nazi law even after 1945, which allowed it to remain in force in the Federal Republic, as a conference of trading-law experts decided in 1958. However, jurists also found that the secret licensing procedure ought to remain limited to "special cases," and paragraph 22a was in fact rarely invoked in the postwar era.[279] In the highly industrialized state of Baden-Württemberg, only two companies had applied for a license pursuant to paragraph 22a by the early 1970s, one of them withdrawing its application after a modification of its production plants. The other application was refused by the state authorities.[280]

As a result, the dominant characteristic of air pollution control in Nazi Germany was its astounding normality. All three initiatives to change the law seem somewhat petty in retrospect, clearly inferior to the momentum of a bureaucracy that had long-standing routines for dealing with air pollution hazards. In fact, even the onset of World War II did not disturb these routines to any great extent, and air pollution control continued to show a surprising degree of normalcy for a number of years. In the spring of 1940, the Leipzig district organization of the Nazi Party filed a complaint against the licensing of an iron and steel foundry, and officials dutifully looked into the matter even though the plant was crucial to the war economy.[281] In 1942, a producer of artificial silk in the city of Krefeld built special cleansing towers in reaction to numerous complaints, although the company had some trouble finding the necessary workers, especially bricklayers. "The demands of the war are stronger than the goodwill of the company," an official commented on the factory's efforts.[282] Even the intensification of war production following the appointment of Albert Speer as minister for armament and ammunition in February 1942 did not spell the end to all efforts, and neither did the emphatic declaration of "total war" in early 1943.[283] A few months later, a conflict arose in Essen over a coffee-roasting plant that produced surrogate coffee for the military, ending with the company installing a special filter and agreeing not to process certain raw materials during morning hours.[284] In November 1943, a mining company in the Harz Mountains agreed to pay compensation of 15,578 Reichsmarks for forest damage.[285] And on May 19, 1944, half a dozen officials met in the city of Lüdenscheid in southern Westphalia to discuss the damage that the emissions from a local aluminum smelter were doing to the trees near the local stadium.[286]

In the end, even Auschwitz appeared on the radar screen of the German air pollution community, albeit awkwardly. In 1943, WaBoLu received a letter from the building department of the Auschwitz concentration camp (Zentral-Bauleitung der Waffen-SS und Polizei Auschwitz) asking whether it would be ready to compose the expert report mandated by local authorities; the project in question was "the construction of a heating plant at the Auschwitz camp."[287] WaBoLu, quite willing to provide such a report, asked, routinely, for a map that showed the environs of the projected plant within a radius of five kilometers.[288] Of course, the administrators of Auschwitz were in no position to provide such a map, and the correspondence died down. In fact, the "heating plant" may not have been a heating plant at all, as this letter was written at the same time as the SS started to use the large crematoriums of the Birkenau extermination camp.[289] This exchange of letters is arguably the most obscene type of communication imaginable in the field under discussion here, testifying to one of the most disturbing aspects of the Nazi era: the unsettling coexistence of monstrous crimes and bureaucratic routine. If there was any further need to demonstrate what Hannah Arendt has called "the fearsome, word-and-thought-defying banality of evil," the Auschwitz air pollution correspondence provides such a demonstration.[290]

However, the bureaucracy's momentum produced not only obscenities but also some promising trends. Strange as it may seem, the debate over reforming air pollution control between 1914 and 1945 was never more vivid than during the war years. For example, the Ministry of Trade and Commerce issued a decree on February 18, 1942, that effectively strengthened air pollution control: in licensing a number of plants, the ministry urged local authorities to consult with the experts at WaBoLu.[291] In doing so, the ministry followed a suggestion from the institute itself, shedding a revealing light on the motives behind the decree.[292] After all, the decree promised more work for the institute, underscoring the importance of its mission and demonstrating that none of its staff members was dispensable for military duty. Of course, this notion was never put into writing, but it can be inferred from the fact that this decree was merely one of numerous decisions, all of which implied more paperwork. For example, on June 25, 1941, the Ministry of Trade and Commerce took the initiative to revise the list of plants that required a special license—three days after the German attack on the Soviet Union.[293] And in 1944, work started on a new version of the "Technische Anleitung," a handbook for licensing plants that offered technical information and standards.[294] A 1942 WaBoLu memorandum spoke of "a reform of Paragraph 16 of the trading regulations [which dealt with the licensing procedure] after the end of the war."[295] And in a letter of January 1944, a member of the institute acknowledged that "the air in industrial areas is barely better now than a generation ago"; in earlier years, the institute had described the status quo much more positively.[296] "After the war, there will be an urgent need to control air pollution in industrial areas in a better way than heretofore," an official of the Ministry of the Interior wrote in a 1943 internal document.[297]

To be sure, these initiatives were a far cry from the discussions that would actually take place in the 1950s and 1960s. But nevertheless, it is noteworthy that World War II was indirectly a stimulus for reform initiatives. In spite of all the bureaucratic lack of interest in the problems of atmospheric pollution, this issue demonstrated that even in wartime there is still paperwork to be done. And when bureaucrats actually did the paperwork, it quickly emerged that German air pollution control was probably more deficient than had been thought. The direct result of the reform debate during World War II was thus marginal, if it existed at all, but the indirect result was significant: the debate spread a feeling among experts that something had to be done to improve Germany's approach to pollution control. That the reform debate emerged so quickly in the 1950s and led to a surprisingly enduring result may have been due to the fact that several of the officials involved had begun thinking about such reforms a decade earlier.

4 Beyond the Pall of Smoke

Both Germany and the United States saw growing public interest in air pollution control in the 1950s. As a result, administrative oversight began to grow, but that meant very different things in each country, with the position of industry making for the greatest contrast. American entrepreneurs were in many cases represented on boards of directors and were thus able to directly influence the relevant agencies, whereas German industrialists, shaped by constant contact with a self-confident and assertive bureaucracy, probably did not even dare to dream of such a strong position. To be sure, German officials, too, were by no means anti-industrial in their attitude, but on the whole there was a considerable measure of bureaucratic autonomy in Germany, something that was usually missing in the United States. In other ways, as well, German officials enjoyed a much more favorable position: German agencies seem to have had far less of a recruitment problem than their American counterparts; German air pollution control comprised the whole country, while the authority of American officials often ended at the city line. Further, unlike in the United States, where air pollution control had arisen out of the tradition of smoke inspection and where the fight against smoke was therefore long pursued with much greater vigor than the fight against other emission problems, German officials from the outset directed their attention at the totality of all pollutants. This allowed German authorities to act from a fairly strong position, while American officials were more or less dependent on the assent of the business community. When American industrialists demanded cooperation from the authorities, they were for the most part asserting authority. When German entrepreneurs spoke of cooperation, they were more likely acknowledging the unavoidable.

All in all, it seems that there was clearly more of a balance of powers in

Germany, and this was crucial for future developments. Before outlining the policy debates in detail, however, it is worthwhile examining more closely the actors involved: industrialists, engineers and researchers, officials, and the general public. In all these groups, postwar transformations implied new challenges for air pollution controls. The changes in policy grew from the ground up, from the interactions of myriad people on local, regional, and state levels. The time of great environmental master plans, with their promise to clean the environment through vigorous federal action, was still more than a decade in the future.

The United States around 1950

More than any other period, the two and a half decades following World War II can be seen as a self-contained chapter in American efforts to combat air pollution. Two sharp turning points bracket these twenty-five years: first, the spectacular campaigns in St. Louis and Pittsburgh and the subsequent broadening of the movement's agenda; and second, the passage of the Clean Air Act and the creation of the Environmental Protection Agency in 1970. The 1970 events, however, have exerted an enduring influence on the historiographic perception of the 1950s and 1960s: 1970 is generally seen as "year zero" of modern environmental policy in the United States. This perception is not without reason: 1970 did in fact see the establishment of a regulatory style that was new in several respects and that continues to shape American environmental policy. However, the unfortunate side effect is that the period before 1970 is commonly measured against the standard of this modern, ecologically inspired policy, appearing in a thoroughly negative light as a result. Thus, air pollution policies of the 1950s and 1960s are often viewed as a great failure, reaffirming in retrospect the need in 1970 for a turning point.[1] But when the activities of the late 1940s and early 1950s are viewed against the backdrop of contemporary perceptions, a different picture emerges: the postwar period was in fact characterized by an unprecedented upsurge in efforts at air pollution control, leading to significant improvements in local air quality. Thus, what seemed like a grandiose failure in 1970 was indeed quite an impressive success story in 1950.

Both perspectives, of 1970 and of 1950, contain a measure of truth, of course. If anything, post–World War II air pollution control presents an ambivalent picture. Between 1945 and 1970, American industry was willing to advance on the clean air front more energetically than before or after, but at the same time, this willingness brought air pollution control into a fateful dependence on industry's goodwill. As a result, the regulatory approach was a strange mixture of cooperation and cooptation, described here, for lack of a better term, as *pseudocorporatism*. It arguably takes such an oxymoronic expression to do justice to an approach so full of ambiguities: It was cooperative, but it gave tremendous influence to one party—industry. It was born out of the realization that smoke was only one issue in a wide range of air pollution problems—yet work continued to focus on smoke and dirt for quite a long time. It brought unprecedented investment in air pollution control—yet societal discontent by far outpaced any accomplishments. Perhaps the best way to highlight the contradictory nature of air pollution control between 1945 and 1970 is to point out that even after a quarter-century of progress, pseudocorporatism collapsed because it was not working.

The only way to make sense of these apparent contradictions is to consider the time scale. Around 1950, pseudocorporatist air pollution control was clearly effective—for the time being. But gradually, this approach led to a double crisis: an internal crisis, because of its own structural deficits, and an external crisis, because of growing environmental sentiments among the general public. Neither crisis was really addressed, let alone addressed convincingly—not for lack of good intentions, but merely because decisions made in 1950 had put regulators into a straitjacket. In a way, then, air pollution control in the 1960s was a curious mixture of past and future: the regulatory approach, dependent on industry's cooperation, drew on the Progressive Era's smoke inspection tradition, while civic protests provided a glimpse of the new age of environmental awareness. In the end, pseudocorporatist air pollution control simply collapsed in the heat of the environmental revolution, and activists and politicians alike searched for new approaches that did not smack of weak cooperation. Clearly, the 1950s style of air pollution control deserves more attention, not only for its own sake but because it allows a greater understanding of the events of 1970.

A New Urgency

In discussions of postwar changes in environmental policy, one name generally crops up sooner or later: Ronald Inglehart. His argument that postmaterial values gave rise to modern environmental awareness has found major resonance within environmental history, although it is controversial among sociologists. Notably, Inglehart casts the environmental movement in a flattering light, his thesis based on the assumption that the rise of the postwar generation, growing up in a context of peace and prosperity, led to a shift in dominant values. Inglehart proposes that those adults whose adolescence fell into the post-1945 period showed, because of their generational experiences, relatively little interest in classic material values—such as economic and political stability—and instead emphasized postmaterial norms and values.[2]

Obviously, this conception, independent of its empirical validity, possesses a normative attraction: if modern ecological awareness is based, in the final analysis, on a turning away from material values, this reflects fairly precisely the view that environmentalists hold of themselves as selfless idealists.

Examining the actual development of public opinion after 1945, however, a conceptual problem arises. According to Inglehart's thesis, the rise of air pollution as a political topic should have begun only several decades after the end of the war and should have been strongly linked to the political coming-of-age of a younger generation. However, observers were already noting a heightened public awareness of pollution shortly after World War II. "It is a fact that in recent years there are growing demands on industry in this direction," one article declared in 1948.[3] Robert Griebling, the managing director of the Air Pollution and Smoke Prevention Association of America, took a similar view. "Interest in air pollution control is increasing with every passing week," he proclaimed at the beginning of the 1950s. Journals like *Iron Age*, surely not pandering to the Zeitgeist, contained similar assessments.[4] Increased environmental activity was reported even in the rising cities of the Sunbelt.[5] And as early as 1951, one writer in the journal *Chemical Engineering* felt compelled to issue a warning against the public's excessive demands: "The public must learn to substitute patience for hysteria."[6]

This chronological inconsistency points to a fundamental problem in Inglehart's thesis: it directs attention toward a certain kind of criticism of air pollution—and in so doing ignores the fact that there were other perspectives on air pollution that had nothing to do with Ingleheartean postmaterial values. Thus, by looking intently for the origins of what would later turn into the environmental movement, scholars have overlooked the persistence of modes of perception that had occupied a fixed place in public awareness since the Progressive Era. Generally speaking, public criticism after 1945 still moved along fairly traditional tracks: the dominant theme was that air pollution was dirty and caused a nuisance—with the small difference that these ill effects were now attributed not only to smoke but in principle to all air pollutants apparent to the senses. It was therefore no break with traditional patterns of perception when foul-smelling gases emerged as an issue alongside dirt and dust. For example, one complaint about a chemical factory in Cleveland noted, "Last Sunday the fumes from the Grasselli plant almost stopped the Ten O'Clock Mass." The air pollution committee of the League of Women Voters of Eugene, Oregon, was fittingly named "Committee for Smoke and Stench Control."[7] Furthermore, it was still widely accepted that the harmful effects of air pollution did not extend substantially beyond the hazy atmosphere surrounding large cities. In many cases, the horizon was

Figure 6. This Walt Disney Cartoon of 1955 still associates air pollution with dirt, rather than with invisible hazards that would soon dominate the environmental discourse. Image © Disney Enterprises, Inc.

narrower still: it was not rare for the perspective to be initially limited to a specific part of the city. The establishment of the Cleaner Air Committee of Hyde Park–Kenwood in 1959 grew out of the dismay that Laura Fermi—the widow of the physicist Enrico Fermi—felt when confronted with the dust and grime of her Chicago neighborhood after her return from Europe.[8] At times, protests against dust and noxious gases merged with protests against other neighborhood problems. The Southeast Community Council in Cleveland, for example, noted tersely in its minutes: "Coke Ovens are biggest nuisance. They are destroying our neighborhood. Younger people moving out of neighborhood. Neighborhood being invaded by Puerto Ricans, W[est] V[irginians] and other undesirable people."[9]

This is not to say that the Inglehart thesis has no explanatory power at all, although the fact that Inglehart himself emphasized the limited validity of his thesis when it came to environmental issues is usually overlooked. At any rate, the dichotomy between material and postmaterial values Inglehart favored is not documented here with the kind of clarity he claimed on other topics.[10] The rise of postmaterial values, in other words, did not lead to the discovery of the problem of air pollution, but, at most, to the transformation of an already existing awareness, which then invigorated the modern environmental movement. What was truly new around 1950, then, was less perception than a new urgency. In the period before World War I, the smoke nuisance—the vehemence of civic protests notwithstanding—had played a comparatively secondary role in public perception. Even the Syracuse Chamber of Commerce, the driving force behind the reform efforts in the city, had declared in 1907, "We hesitate to designate the problem as a 'vital' one."[11] But this status had changed: air pollution could now safely be proclaimed the key problem confronting society, which in turn indicates that the decisive cause behind the change was probably not perception itself, but social context. Many of the problems that had preoccupied reformers at the turn of the century were now seen as having been solved: the global economic crisis, the dominant problem of the 1930s, had been overcome; and the danger of a new war could be judged as low, at least until the Korean conflict. While air pollution had long taken a backseat to more pressing problems, there was now plenty of room in the public arena for a more intensive debate of the issue. The rapidly progressing suburbanization after 1945 also likely contributed to the new prominence of air pollution: in the new suburbs, the contrast with the polluted metropolises became particularly glaring.

This new social context met with two developments in the specific field of air pollution control. First, the campaigns in St. Louis and Pittsburgh provided striking proof of the possibilities for mitigating the problem—and of the resultant broadening of the air pollution agenda. Once the smoke nuisance, for decades the dominant air pollution problem, had been brought under control, other emission problems came into view that had long been overshadowed by smoke. Second, the much-cited air pollution disaster in Donora, Pennsylvania, intensified public focus on pollution. In this town on the Monongahela River, some thirty miles southeast of Pittsburgh, at least twenty people died in the fall of 1948 when emissions from a zinc works became trapped under an inversion layer and accumulated to toxic levels; an even greater number of fatalities was prevented only by the courageous actions of local emergency personnel.[12] This was without question a shocking catastrophe, but its impact was probably increased by already heightened public attention to air pollution problems.[13] How easily an air pollution disaster could evaporate politically had been demonstrated nearly twenty years earlier by the "fog catastrophe" in the Belgian Meuse valley, which, in Wilhelm Liesegang's verdict, had "no general significance."[14] The times when such callousness seemed acceptable were obviously over in postwar America.

As part of this new context, the social composition of protest groups was significantly broader than during the Progressive Era. That was already evident during the Pittsburgh campaign at the beginning of the 1940s: although the city's upper class continued to be the driving force behind the efforts, public interest clearly extended beyond these circles. It took eight pages of a 1941 committee report to list all the organizations that had come out in support of the antismoke drive: the spectrum ranged from the Disabled Veterans of the World War and the International Brotherhood of Firemen and Oilers, to the Pittsburgh Council of Catholic Women and the Retail Grocers Association.[15] A poll conducted toward the end of the 1940s revealed that 75 percent of Pittsburgh's residents supported the authorities' efforts; even among low-income individuals, who had initially been skeptical of the endeavor, approval had risen to 62 percent.[16] Consistent with this picture, unions now also became involved in the drive against air pollution for the first time.[17] To be sure, a social bias remained, but it was far less pronounced than in earlier decades. Against the background of the Progressive Era's exclusive leagues, civic protest against pollution problems now looked increasingly like a mainstream endeavor.

However, it is crucial to recognize that the consequences of this development were double-edged. As the social basis of air pollution efforts broadened, the selective attraction enabled by the socially exclusive definition of the problem faded away; it was this definition that had supported the antismoke movement and allowed the general interest in clean air to be organized as the specific problem of the civic upper class. Public protest against air pollution in the postwar period was therefore highly subject to the paradoxes of collective action described by Mancur Olson. This is why the public's new interest found only limited expression in organized action and remained politically underrepresented—a state of affairs that would essentially prevail until the ecological revolution. With few exceptions, public discontent during the 1950s and 1960s was always stronger than its institutional representation.

It should also be mentioned that the protests continued to remain free of anti-industrial overtones. Indeed, the behavior of the organized public repeatedly demonstrated a submissiveness toward industry that occasionally verged on an embarrassment. For example, on the first anniversary of the local ordinance, the Pittsburgh United Smoke Council gave an "Award of Merit" to ninety-five companies that had done nothing more than follow the law.[18] "Industrial self-interest, stimulated by the pressure of public opinion, is still the most enlightened approach to the mitigation of air contamination," declared the Southeast Air Pollution Committee in Cleveland, even going so far as to add, "Directly and without the aid of public officials, it is often possible to accomplish much more than with them."¹⁹ When the steel company Jones & Laughlin eliminated a notorious emissions problem in 1956, the *Cleveland Press* commented, "The community is grateful for the company's public-spirited action."²⁰ Given this broad confidence in industry, the public discontent of the postwar period left considerable latitude for the political response; when it came to strategies, the public's ideas were and remained flexible, if not vague. Of course, this was because officials in many places often reacted very quickly to the growing unhappiness of citizens. Rarely, then, did public organizations find themselves in a situation in which effective laws and their implementation had to be considered; as a rule, groups could limit themselves to giving friendly nods of approval to the proposals and measures put forth by the regulatory establishment.²¹ And the lead on such measures was often taken by the one party from which environmental historians would least expect it: industry.

The Role of Industry

Perusing reports on air pollution problems in industrial journals of the late 1940s and 1950s leads to a striking discovery: in the postwar period, these journals were among the most ardent champions of clean air efforts. "Keep ahead of public opinion," recommended one article in the journal *Coal Age*, while *Chemical Engineering Progress* declared that industry had "an obligation to the community to do everything reasonable to minimize atmospheric pollution."²² Other organs addressed their readers directly: "How many of these suggestions is your company following now?" asked *Dun's Review and Modern Industry* at the end of a list of suggested improvements. *Factory Management and Maintenance*'s advice was short and to the point: "Better move fast."²³ "Objections from the citizenry of the community around a chemical plant should never be lightly dismissed," counseled *Chemical and Engineering News*, while an article in *Chemical Engineering* announced its position in its title: "Pollution Control—an Industry Must."²⁴ Even the catastrophe in Donora was by no means downplayed by industrial journals: "The Donora 'death smog' could happen in your town—given the right combination of smoke, hilly terrain, and foggy weather," *Business Week* noted when scientific inquiry into the incident had only just begun.²⁵ And a whole host of contemporaries from all walks of life attested that industry did not ignore such published admonitions.²⁶ Indeed, the postwar period was marked by a historically unprecedented increase in industrial activities to control emissions.

Several motives came together in this remarkable industrial stance. First, corporate public relations were increasingly important. World War

II had led to a massive expansion of public relations (PR) departments in American companies, and the dominant leitmotif of the 1940s was to present industry as a "good neighbor."[27] Needless to say, such an image was credible only if industry also strove to cut its emissions, and thus the obligation toward one's own neighborhood became the most popular justification for measures to reduce air pollution. In a survey conducted by the journal *Mill and Factory*, fully 91 percent of industrialists reported that they felt that the advantage of clean air measures was above all in "improved community relations."[28] Procter and Gamble explained its motivation by saying, "We want to remain in good standing in the communities in which we operate"; other companies said much the same.[29] One PR firm even suggested seeing the matter in a positive light: "Air pollution problems can be turned into community-relations opportunities."[30]

A second reason for the interest shown by business was the desire to avoid courtroom battles. For American companies, legal proceedings represented an essentially incalculable risk. The problem was not only the long duration of such proceedings and the resultant expenses; in many cases, it was also entirely unclear on what legal basis the verdict would be rendered. Unlike the German BGB, which at least articulated the criteria of significance and local norms and offered grandfathering to companies licensed according to §16 of the Industrial Code, American emissions law was a vast and nearly unmanageable conglomeration of regulations, at times with peculiar touches.[31] For example, anyone who had caused an emissions problem in New Jersey for twenty years thereby acquired the formal right to continue producing these emissions.[32] "There is perhaps no more impenetrable jungle in the entire law than that which surrounds the word 'nuisance,'" a 1941 legal handbook declared—a state of affairs that frequently turned legal proceedings over emissions into a judicial nightmare.[33] Industry therefore had good reason to prepare for the eventuality of lawsuits by taking preventive measures—especially since it was reported as early as 1950 that awards for compensatory damages could reach $250,000 in some cases.[34]

In addition to civil suits, industry feared that public displeasure might lead to what businesspeople called "restrictive legislation" or "impractical laws."[35] "If industry doesn't police itself—right now—politicians and aroused citizens will take over," the journal *Factory Management and Maintenance* declared. The San Francisco Chamber of Commerce warned, "Great care should be exercised in the consideration of proposed legislation to insure that the emotional approach to air pollution control is eliminated."[36] This fear of public wrath is remarkable not least because it did not quite match the behavior of the organized public, which, as mentioned earlier, was for the most part still rather tame during this period. But in spite of emphatic declarations about a sense of personal responsibility, industry continued to see public opinion as an element of uncertainty, monitoring its stirrings with a skepticism bordering on distrust. As early as 1952, a representative of the Manufacturing Chemists' Association cautioned against the demagogical approach to dealing with emission problems: those who took this approach, he warned, "are sometimes stirred up . . . by crusaders who inject fear into their minds."[37] Other industrialists noted in a similarly pessimistic vein, "As long as the passions of the men refuse to conform to the dictates of reason and justice, we may expect more stringent laws and more severe penalties."[38]

When looking at all these measures, it must be borne in mind, moreover, that American industry had been experiencing a sustained upswing since the end of the war. Notoriously dirty sectors such as steel, chemicals, and energy therefore did not have to worry all that much about the costs of air pollution measures, especially since substantial improvements could often be achieved with little effort and expense at this early stage. To be sure, keeping the air clean was nearly always a money-losing proposition: in a survey conducted in the mid-1950s, only 2 percent of companies indicated that the measures they had taken had paid off through the extraction of valuable materials.[39] Still, investments were substantial: the annual outlay at the beginning of the 1950s was estimated at between 100 and 400 million dollars.[40] "Obviously, air-pollution is no small item in the budget of even a large plant," one journal noted already in 1950, and as the postwar years unfolded, expenditures rose steadily.[41] In the mid-1960s, the New York utility Consolidated Edison declared that it had invested 104 million dollars in pollution control over the previous twenty-five years.[42] In 1966, the Public Health Service cited an estimate according to which expenditures for the construction and operation of air control installations amounted to 850 million dollars annually.[43]

After a few years, when pollution control had become established in many places, these three motivations were joined by a fourth: the desire for certainty in planning. Since investments of some kind were unavoidable, industry wanted to ensure that it would not be forced to engage in costly retrofitting a few years down the road. This desire was reflected, for example, in the fact that as early as 1960 Florida's industrialists regularly communicated with authorities about what measures had to be taken, even though they were not legally required to do so.[44] In Cleveland and Pittsburgh, industry was given ten years to get a grip on certain emissions problems, while in Milwaukee the parties agreed on a five-year plan.[45] To be sure, these moves did not necessarily put off the problems. In 1965, for example, Cleveland authorities moved the target date up by six years because industry had solved the issue much earlier than expected.[46] In these and other arrangements, industry was concerned first and foremost with criteria that would allow it to assess what air pollution meant to a company over the long term and to plan accordingly: "Industrialists want to know of the various requirements which will be placed upon them by State and local governments," the Public Health Service explained in a 1966 report, going so far as to surmise "that most industrialists would rather locate in an area where air pollution control programs and requirements are well developed and based on long-range plans, rather than in an area where the matter is poorly handled or in a state of evolution. The uncertainty of the latter situation could result in deteriorated air quality and frequent, costly changes in emission control requirements."[47]

Clearly, the constructive attitude of many industrialists stemmed from the fact that a certain degree of air control activities was simply the path of least resistance. In this way, the risk of expensive civil suits could be minimized, the company's image burnished, and the political reins tightly held. Of course, this left considerable leeway for the individual company to determine the extent of its measures. Companies like Du Pont generally placed great importance on clean air efforts and responded to complaints with statements like, "We have a common interest."[48] But other companies reacted to stringent requirements by threatening to relocate.[49] Each individual company's position depended strongly on the local circumstances and the attitudes of management, which made it exceedingly difficult for industry to come up with a uniform stance on state or local regulations. In this respect, industry found itself in a classic dilemma: on the one hand, it could hardly deny that official efforts were legitimate; on the other hand, overly strict control could not be in its interest. Even companies that were receptive to clean air measures had to be interested in keeping official jurisdiction as narrowly defined as possible; after all, the "good neighbor" policy was persuasive only if the measures companies took were indeed voluntary. A statement made by the chair of the Manufacturing Chemists' Association's committee on air pollution control provides a prime example of this ambivalence: "This committee believes in self-regulation in air pollution as in every other problem, though it recognizes that in some areas conditions may require concerted action under government leadership."[50] More than once, representatives of industry found themselves in the paradoxical situation of having to approve a law that would in fact have been unnecessary if only all industrialists had followed the wishes of their representatives.

If a certain measure of official action was thus unavoidable, industry believed that it should most definitely meet a number of conditions. First, it should focus primarily on those pollutants that were directly perceptible to the senses. Since companies were interested chiefly in protecting their immediate neighbors, industry was little concerned with solving problems these neighbors were unable to detect. Second, industry demanded that companies be subject to the fewest possible restrictions when it came to choosing technology. The work of officials should focus on whether or not a remedy was called for, not on how it was achieved, an attitude that found expression above all in strong opposition to a formal licensing requirement for new installations. When the Public Health Service came out in favor of such a requirement in 1959, the Manufacturing Chemists' Association reacted with criticism that was unusually harsh for the time: "We feel that a system of permits unreasonably interferes with the free choice of a manufacturer to abate air pollution in the manner which seems best to him."[51] Third, the behavior of officials should be as predictable as possible; in many cases this meant that they should be subject to direct control by industry. How unabashed industry was in this regard is revealed by a model law published by the ASME in 1949, which proposed the establishment of an "Advisory Board" to advise top-level officials: three members were to be experienced technicians, while the other two spots were to be reserved for representatives of industry.[52] Since such conditions might sound a bit harsh if put too bluntly, industrialists preferred to formulate their demands as an offer: we are ready and willing to cooperate with the authorities. This was an offer authorities just couldn't refuse.

The Administrative Response

Around 1950, authorities were generally subject *to* events, rather than subjects *of* events. This was because smoke inspectors, entrenched for decades and still the only established authorities in the area of air pollution control, suddenly found themselves confronted with radically different circumstances after 1945. First, the unambiguous public demand for more stringent protection against emissions now went far beyond the framework of traditional antismoke measures. Second, industry now had good reason not only to respond receptively to recommendations from officials, as it had done since the Progressive Era, but also to take the initiative on emission problems and come up with policy ideas. Third, in the wake of the campaigns in St. Louis and Pittsburgh, the municipal antismoke movement was at the height of its fame, which meant that officials could hardly ignore the new tasks. As a result, broadening the range of activities was the order of the day, and even more—in essence, it seems that officials really had no choice. As early as 1948, one speaker at the annual meeting of the Smoke Prevention Association declared, "If the smoke abatement officials had not voluntarily reached out to embrace these other pollutants of the atmosphere, they would have been compelled to do so by the weight of public opinion."[53]

On paper, the resulting activities look quite impressive. According to a survey published at the beginning of 1951 in the journal *Iron Age*, thirty-seven of fifty-six large American cities had revised their local ordinance since the end of World War II. Further, as a rule, regulations now no longer pertained only to the smoke nuisance but also to other types of airborne dust, the majority of laws even containing regulations on gaseous pollutants; a similar surge of innovation in municipal air control had not occurred since the Progressive Era.[54] Not least, and remarkably, smaller cities were now also enacting local ordinances: according to data from the American Municipal Association, by the mid-1950s there were only four cities of more than 25,000 inhabitants in the entire country that had no regulations of any kind.[55] However, the administrative reality behind these numbers was often much more prosaic. Especially in smaller cities, frequently nothing was done beyond passage of an ordinance. In Florida, for example, there were sixty-two cities with ordinances around 1960, but only three of those had any meaningful enforcement program.[56] And even where such programs existed, they were often weak: in 1962, only thirteen supervisory agencies had more than ten employees; 58 percent consisted of only one or two persons; and 69 percent of municipal governments employed fewer than two persons per 100,000 inhabitants.[57] "Municipalities usually lack the financial resources to support the technical facilities and staff required for effective air pollution control," noted a study by the Legislative Research Council of Massachusetts with an air of regret, and this was not a singular opinion.[58] Even Pittsburgh's agency, which enjoyed an excellent reputation far beyond the city, was forced to concede in its annual report for 1958 that it had been unable, because of a shortage of personnel, to respond appropriately to all complaints and had thus resorted in these cases to an appeal to civic consciousness by sending out a "'Be A Good Neighbor' post card."[59] In 1965, Chicago's Department of Air Pollution Control had fewer inspectors to ferret out violations of ordinances than the Department of Smoke Inspection had had in 1910.[60]

Of course, the weakness of local agencies in terms of personnel was not unintended. After all, it was very much in line with what industry wanted: an agency should be just strong enough to admonish individual businesses that were negligent, but not so strong that it could, for example, take the initiative on more stringent and systematic oversight.[61] The smaller the agency, the easier it was to keep under control: at times the dominance of industry was in fact institutionally secured in a way that exceeded even the generous ASME recommendations. In the 1950s and 1960s, the rule was that industrial polluters furnished a substantial number of the members of committees and supervisory bodies. At times, the representatives of industry and business virtually ran the show. For example, the Air Pollution Advisory Board of Milwaukee was made up of representatives from the Wisconsin Electric Power Company, the Wisconsin Gas Company, the Consolidation Coal Company, the Pelton Steel Casting Company, and Industrial Combustion Inc.[62] The situation was hardly better on Chicago's Air Pollution Control Committee, where Laura Fermi was the only representative of the concerned public, alongside seven men from various sectors of the economy and one union representative.[63] In other instances, the background of industry representatives spoke volumes: the Allegheny County Smoke Control Advisory Committee, responsible for air pollution control in the area surrounding Pittsburgh, comprised the president of Duquesne Light, the local utility; the general manager of the Baltimore & Ohio Railroad; the vice president of the Pittsburgh Consolidation Coal Company; and the president of U.S. Steel.[64] Clearly, this gave industry de facto veto power—and nobody knew this better than the Allegheny County Bureau of Smoke Control, which noted in one report, "Were it not for the cooperation of industry, this Bureau might as well turn out its lights and close its doors."[65]

However, this certainly did not mean that cooperation in clean air efforts took a negative tone during the postwar period. On the contrary, cooperation was widely seen as the key to success. "Industry as good corporate citizens already has demonstrated a willingness locally, regionally and on a statewide basis to cooperate wholeheartedly in air pollution control programs," the New York State Air Pollution Control Board declared at the beginning of the 1960s, and many other agencies echoed this sentiment.[66] At least until the end of the 1950s, civic organizations were generally convinced of the value of cooperation between the authorities and industry. As a resolution by the Izaak Walton League of America put it in 1959, for example, "There is a great need for continued and improved cooperation between industry, government and other agencies concerned in the study and solution of air pollution problems." In the Hudson River Conservation Society, cooperation was even enshrined in the society's Certificate of Incorporation as of 1961.[67] The prevailing view was that the penal provisions of the laws should be applied only to a small minority: those who stubbornly resisted cooperation.[68] Sometimes cooperative programs even ran successfully without any legal basis.[69] Even within the Public Health Service, which later took an increasingly skeptical stance toward cooperative clean air efforts, there was still widespread optimism throughout the 1950s: "With . . . a spirit of mutual confidence and unity, we shall be able to cope with the problem before a threatened saturation impels desperate action."[70] Time and again, cooperation by all parties was invoked as though this were cooperation between equals—though, admittedly, the most emphatic voices usually came from representatives of the regulatory establishment.[71]

Widespread consensus also still existed on a second point: air pollution was a local affair. "Most people agree that air pollution should be regulated by the lowest level of government capable of dealing with a particular problem area in its entirety," declared Jean Schueneman of the Public Health Service as late as 1962.[72] However, the reasons for this view were quite varied. The organized public most likely opted for local control because it still understood air pollution as a geographically limited problem that did not extend much beyond the borders of a large city. Industry, meanwhile, seems to have been interested chiefly in exerting as much control as possible, which was generally easiest to achieve at the local level. Among the authorities, support for local control was a given, since any form of supraregional control would have invariably entailed a partial loss of power for municipal agencies. In spite of differing motivations, there was therefore a shared tendency to focus the reforming zeal entirely on local organs. For the time being, the general belief was that control programs on a state or federal level were unnecessary. Thus, the first state program to combat air pollution was established not in one of the highly industrialized states of the Manufacturing Belt but, of all places, in rural Oregon.[73]

Oregon presents an interesting contrast model, elucidating the characteristics of the postwar regulatory mode. Oregon's case reveals in particular how important the tradition of the fight against smoke was to reform efforts after 1945. Since the primary source of energy on the West Coast was not coal but oil, Oregon lacked the tradition of smoke inspection so firmly entrenched in most American cities.[74] As a result, discussions in Oregon were initially characterized by an exceptionally shrill tone. For example, representatives of industry vehemently attacked a local ordinance for the city of Portland because they believed it gave authorities the option of resorting to drastic measures; between the lines one could sense near panic at the thought of an irascible official who would be able to harass local industry with arbitrary directives.[75] In other cities, such concerns were all but unknown: after all, decades of antismoke efforts had made it abundantly clear to industrialists that it was possible to cooperate splendidly with municipal officials. The journal *Modern Industry*, for example, raved: "Industry can count itself extremely lucky in the men who head up smoke-abatement bureaus in many cities. Men like J. H. Carter of St. Louis, Sumner B. Ely and David M. Kuhn of Pittsburgh, and Frank Chambers of Chicago, have a good understanding of industry's problems, and a wide knowledge of pollution-control equipment."[76] A similar sentiment was voiced by the representative of a chemical factory: "I have found most of the officials to be reasonable men who recognize that the arbitrary exercise of their powers would, in many cases, be even more harmful to the community than the discomfort or inconvenience resulting from atmospheric pollution."[77] When the Chicago Association of Commerce and Industry sought to set up a special committee, it even brought the now seventy-eight-year-old Thomas Donnelly out of retirement to recount the memorable campaign of 1907—an event that, in the assessment of those in charge, played no small part in gaining the goodwill of the members.[78]

In fact, one dominant characteristic of the transition phase after 1945 was that it was marked almost throughout by a decidedly placid atmosphere. Whereas the problem of domestic fires had still been the topic of lively and at times bitter debate during the interwar period, there was now an almost eerie agreement on all sides: neither the strategy of cooperation nor the local basis of clean air efforts became a topic of any significant controversy. Local authorities, by default, ended up charged with the new tasks, since no one had expressed any serious reservations about such a solution. Municipal officials thus confidently set to work. There was, then, a simple reason why midcentury Americans were content with a simple reform of classic smoke inspection: it was the most obvious solution. But only time would tell whether it was also the best.

The Rise of Scientific Research

If there was one demand raised even more frequently during the postwar period than the calls for cooperation and local control, it was the demand for more research. Today, when billions are invested annually in scientific research on air pollution, this does not seem remarkable; around 1950, however, such a demand had almost revolutionary ramifications. After all, the smoke abatement tradition did not include major research efforts. Crucial to earlier effective antismoke measures was the systematic application of technical knowledge, most of which already existed, meaning that the need for scientific explanations was fairly low. For example, the motivation behind the smoke investigation headed by Pittsburgh's Mellon Institute was at least as political as it was scientific: "The formation of an enlightened public opinion upon the smoke problem is one of the principal objects of the whole investigation," Raymond Benner, the head of the study, openly declared.[79]

Moreover, the scientific value of the extensive measurements of dust precipitation carried out in many cities was controversial: "Such data do not bear on the solution of the air pollution problem," a 1929 article declared.[80] Further, the comprehensive studies conducted during the New Deal, which had gathered an enormous amount of data at considerable cost without any clear sense of its usefulness, were all too readily identifiable as mere job-creation programs.[81] Those working in the trenches of clean air efforts rarely saw the lack of scientific research as a genuine problem.

This changed, however, after World War II: there was suddenly an urgent call for scientific studies of air pollution problems. Many of the subsequent early research projects were associated with control agencies rather than academic institutions, thus underscoring the extent to which the demands posed by regulation were calling the shots.[82] One 1954 survey of research began by noting, "There has been a significant increase in technical interest in all phases of atmospheric pollution . . . , and there is little sign of it diminishing." This was not an isolated view.[83] July 1951 saw the launch of the first scientific journal about air pollution problems under the title *Air Repair*; renamed the *Journal of the Air Pollution Control Association* in 1955, it was long considered the most important publication in this field.[84] The year 1955 also saw the first issue of the *APCA Abstracts*, published by the Air Pollution Control Association—the former Smoke Prevention Association—which offered an overview of the steadily growing number of publications.[85] Soon, then, the cardinal problem for researchers was no longer a lack of information but too much of it: a survey commissioned by the Public Health Service in 1965 revealed that researchers, officials, and other experts bemoaned first of all the large number of information sources and the lack of authoritative syntheses.[86] Beginning in 1955, scientific research was also supported by federal funds, which additionally boosted already booming research efforts, especially since the federal program grew rapidly: from $1,516,000 in 1956 to $9,637,000 in 1963.[87]

It quickly became impossible to discern a common thematic thread in all these efforts. Already the U.S. Technical Conference on Air Pollution, held in May 1950 in Washington, which for the first time brought together the various currents from all parts of the country, indicated the breadth of research interests: the spectrum ranged from effects on plants and animals, to questions of measuring technology and health effects, all the way to meteorological aspects.[88] In 1957, the United States was home to sixteen scientific associations of supraregional importance that maintained special committees on air pollution problems; the more important ones included the Air Pollution Control Association, the American Industrial Hygiene

Association, the American Public Health Association, the American Society of Mechanical Engineers, and the Manufacturing Chemists' Association.[89] Since 1945, the conventional wisdom was that "the subject is so complicated and has so many facets that the co-operative skills and experience of many professions are needed."[90] Research on air pollution, then, was interdisciplinary long before that attribute emerged as an academic buzzword.

It would be beyond the scope of the present book to detail the various branches of research. However, with a view to regulatory policy, it must be emphasized that the project of keeping the air clean was now, for the first time, conceived as a progressive one: it was generally expected that new research findings would be directly incorporated into ongoing abatement efforts. The Chicago ordinance of 1958 even contained the explicit provision that its content was to be revised regularly in light of the latest research. In 1951, the head of the Division of Air Pollution Control in Cleveland, speaking about some unresolved problems, noted matter-of-factly, "As research goes on, these will be solved eventually."[91] All this made the job of local authorities significantly more difficult: not only were they supposed to solve a host of new problems for which the inherited knowledge of the smoke inspectors was generally inadequate, but they were also expected to integrate yet-to-be-generated knowledge about pollutants into their work. Obviously, this complicated officials' work enormously, all the more so since the great majority of emission problems were soon regarded as manageable from a purely technical point of view. In many cases, then, it was not the invention of technological countermeasures that was the key problem, but their practical application.[92] It was anything but clear whether local officials, induced to pursue a rather casual pace by the heavy influence of industry interests, would be able—in spite of scanty manpower—to keep pace with booming scientific research.

A Coalition for the Time Being

A cooperative approach to air pollution in the United States probably never had better prospects of success than in the years following World War II: if one examines the intentions and interests of the parties involved, one finds a degree of mutual agreement that is rare in politics. On the one side, the general public was calling for better protection against emissions, yet their demands were not linked to a fundamental critique of industry; on the other side, many companies were now willing to make considerable investments in clean air efforts in order to present themselves as "good neighbors." The existing regulatory agencies also favored a cooperative solution: after all, municipal smoke inspectors had been firmly committed to cooperation with polluters for decades, and there was no sign that anyone planned to compete with local authorities. The obvious approach, then, was to set up institutions that brought together the various parties to concur on mutually agreeable solutions. There is a standard term in modern political science for such an arrangement: *corporatism*.[93]

Corporatism is generally considered a very un-American phenomenon, and this is also evident in this case. What in fact was established in the late 1940s and early 1950 was not classic corporatism, but a lopsided, "biased" corporatism—a pseudocorporatism of sorts. In theory, corporatist regulation presupposes the equal representation of all involved parties: only when all parties communicate openly with one another and search for a middle ground that offers all sides a balanced mix of advantages and disadvantages can one speak of corporatism in the true sense of the word. However, American industry very rarely went along with such negotiations, as it usually sought a dominant position that often amounted to a de facto veto right. Cooperation, according to industry, was good and helpful—but it had limits. And industry alone should decide what those limits were.

Given the conditions at the time, was pseudocorporatism the best strategy? Revealingly, this question was hardly ever raised in the immediate postwar period. Controversy only rarely surrounded questions of strategy; rather, the climate was dominated by mutual agreement that emerged without much discussion—almost magically, it seemed. "The administration of these programs is taking on a definite pattern of maturity and the Government, industry, and the public are adopting attitudes which are conducive to further accomplishment": such harmony marked the general climate of debate—all the more remarkable in view of the fact that decisions with far-reaching consequences were obviously being made.[94] However, around 1950, hardly anyone was thinking about the long-term future; few, for example, were wondering whether cooperation from industry would be guaranteed even when ongoing research suggested a much more rapid approach. For now, there was little room in the regulatory establishment for longer-term perspectives—a position that would present the champions of industry with new and difficult problems in the 1960s.

But this is the wisdom of hindsight. Contemporary statements displayed a much more positive, at times even enthusiastic, viewpoint. "Allegheny County is on the way to being the cleanest industrial community in the country," declared Pittsburgh's United Smoke Council as early as 1953, while the *Cleveland Plain Dealer* noted in 1964 that the decline in dust precipitation over the previous sixteen years provided "dramatic evidence of what can be accomplished through a cooperative program."[95] Continuous measurements in Dayton showed a 60 percent decline in dustlike emissions during the 1950s, and Cincinnati was able to achieve a 67 percent reduction between 1948 and 1966.[96] Even Morton Corn, who was spearheading the environmental critique in Pittsburgh in 1969, paid tribute to the successes of the previous years. "We have worked according to a gentlemen's agreement," he declared, and he certainly did not mean this disparagingly. The cooperative approach, he believed, "did wonders in the beginning."[97] Finally, a comparison between the interwar years and the postwar period reveals that the accomplishments of clean air efforts around 1950 should not be underestimated: many industrial enterprises were investing in control measures on a far greater scale than ever before, and scientific research, after years of neglect, was surging ahead on a broad front. To be sure, the pseudocorporatist strategy that most cities had settled on did not lead to an optimal implementation of what was technologically possible at the time: Los Angeles's approach demonstrated, after all, that a far more effective campaign was entirely possible from a technological point of view. However, whether the organized public would have supported tougher measures against industrial polluters seems doubtful. On one occasion, for example, the New York City–based Citizens Union approved the easing of an ordinance with a telling explanation: "We oppose anything which may discredit smoke control in the eyes of the public."[98] In any case, the real problem with pseudocorporatism was not so much its immediate effectiveness as its potential for development. For as time went on, it became more and more obvious that a series of structural problems existed that increasingly restricted the reformist potential of the local approach. The pseudocorporatism that was institutionalized around 1950 allowed a coalition for the present, but was it also a coalition for the future? Doubts were growing as to whether this question could be answered in the affirmative.

Germany in the 1950s

It should come as no surprise that the German debate lagged behind its American counterpart for a number of years. While American cities were debating the merits of more comprehensive air pollution control almost as soon as the war ended, the defeated Germans were struggling with the bare essentials of life. With most German cities largely destroyed, and millions displaced, homeless, or unemployed, air pollution was obviously a lesser worry. But when the postwar reconstruction gained momentum, paving the way for what would later be called Germany's economic miracle, the situation began to change, and a new kind of urgency crept into debates, much as it had in America. The 1950s were a time not only of burgeoning mass consumption but also of a new awareness of the dangers posed by polluted air.

Needless to say, this change in public awareness did not happen overnight. As late as 1951, a medical official from Duisburg asserted in the Health Committee of the German Association of Towns and Cities (Deutscher Städtetag) that the population "has generally come to terms with 'its' industrial air, willingly or not," and in December of the same year, the city council of Frankfurt passed a decree that stated that the existing nuisances "must be accepted in the interest of the necessary industrialization and in consideration of the reduction in unemployment that comes with it."[99] But public debates soon began to follow a different theme. As early as 1954, the director of the Ruhr Area Federation for Regional Planning (Siedlungsverband Ruhrkohlenbezirk) cautioned "that the wishes of the population are pushing more strongly than before towards the elimination of nuisances."[100] An article about "daily gas warfare" reached a similar conclusion three years later: "There are mounting complaints by the defenseless about the pollution and fouling of the air, especially in the Ruhr region, in Hamburg, Hanover, Salzgitter, Frankfurt, and Mannheim; in short, wherever there is a concentration of industry and people."[101] "The problem of air pollution has become a question of the day," the journal *Gesundheits-Ingenieur* opined in 1956. Noting the growing public discontent soon became a fixed point of reference in publications.[102] "The question of industrial exhaust fumes is, without a doubt, one of the most serious of our time and especially of our future," asserted Robert Meldau. One member of the Sanitation Institute of the Ruhr Region went so far in 1957 as to declare air pollution a "global problem of the highest order," and nobody felt compelled to proclaim such statements crude exaggerations.[103] Even the Association of German Industry (Bundesverband der Deutschen Industrie, or BDI) conceded matter-of-factly in a letter to the minister of the interior: "The public demand for measures to keep the air cleaner is increasing."[104]

On the whole, complaints were for now still dominated by fairly traditional patterns of argument. For example, one complaint from Lünen was directed "against the intolerable soot and coal dust pollution . . . of our apartments, gardens, and parks."[105] A woman from Augsburg was unhappy that "the balconies, windows, and sills are constantly soiled heavily with soot flakes and soot dust."[106] In Itzehoe, in Schleswig-Holstein, complaints were lodged about "nuisances of the vilest kind" and about "gasings and filthiness" from cement dust.[107] Residents of Beckum claimed that emissions from a nearby cement plant had rendered "cleaning . . . completely pointless" and that "our wives are actually afraid of washing days."[108] And in Ottobrunn, near Munich, a "Working Group of Ottobrunn Associations and Organizations" attacked "the stench" from a waste oil refinery, which was "spreading and becoming more and more unbearable."[109] "Complaints about mere nuisances prevail," an article on the health dangers of air pollution noted as late as 1957.[110] Just as in the United States, health did not become the prime concern in air pollution disputes until the 1960s.

The sources do not reveal a uniform social profile of the population members lodging complaints. For example, the Conference of German Physicians (Deutscher Ärztetag) criticized the pollution of the atmosphere, as did the workers' representation of the Ewald-Fortsetzung mine in Oer-Erkenschwick, which blamed the exhaust fumes from the mine's own power plant for "visible damage to fruits and vegetables," as well as for "severe headaches, frequent nausea, and laryngitis" among workers at the coking plant.[111] "When the preservation of the health of our nation is at stake, the pursuit of profit and the economic interests of companies must take a backseat," proclaimed a press release from the German Trade Union Federation (Deutscher Gewerkschaftsbund) in 1958.[112] The Federation for Heimat Protection (Deutscher Heimatbund), too, regarded the fight against air pollution as the "moral duty of the originators."[113] The Central Association of German Home and Property Owners (Zentralverband der Deutschen Haus- und Grundbesitzer) and the League of German Civic Associations (Verband Deutscher Bürgervereine) received much acclaim for a joint "Resolution on Air Pollution" dating from September 1957. Among other things, the resolution called for "enforcement of the guaranteed right to life and freedom from bodily harm, without consideration for economic interests."[114] Even Alfred Müller-Armack, the mastermind behind West Germany's postwar economic policy, emphasized "the task of keeping the air clean, which has already been undertaken but which must surely be continued with much greater intensity in the future."[115]

Of course, the 1950s were still far from the unprecedented level of ecological mobilization that would characterize the West German public in later decades. But the broad spectrum of voices shows that the demand for better pollution control was already commonplace a decade after the end of World War II. Above all, it is striking that protests also drew noticeable support on the local level. For example, Kirchlengern, in eastern Westphalia, saw a protest rally in November 1954 after repeated complaints about fly-ash emissions from a local power plant had fallen on deaf ears.[116] It was not rare for such discontent to be formally organized now, the organizations describing themselves, quite revealingly, as "communities of interests" (*Interessen*-

gemeinschaften). One example is Krefeld's "Interest Community of Those Harmed in Their Land, Property, and Health" (Interessen-Gemeinschaft der Flur-, Sach- und Gesundheitsbeschädigten).[117] Looking back, one should not underestimate the significance of such organizations: a community of interests against fly ash from a power plant near Wetter an der Ruhr was able to collect more than two thousand signatures right off the bat in support of its demands, and an Emergency Association in the village of Kleinblittersdorf in the Saarland, located downwind from a large French power plant, comprised—by its own account—98 percent of residents.[118] However, no umbrella organization brought these local initiatives together, which naturally limited their effectiveness. Supraregional protest was centered almost exclusively in existing associations whose primary tasks were only indirectly concerned with air pollution. In spite of all the civic activism, air-control efforts lacked a well-organized lobby during all of the 1950s and 1960s, a deficiency that some more progressive officials regretted more than once. During a meeting with representatives of the city of Duisburg, an official from North Rhine–Westphalia's Ministry of Labor asked despairingly in 1961 "whether a central association of the civic organizations could not be found as a serious representative for the defense against emissions in opposition to the polluters' lobby."[119]

This organizational weakness was, however, more than compensated for by aggressive reporting in the press. Even the most skillful lobbyist could hardly have put as much pressure on the German administration as a media community that had committed itself, across all political camps, to the fight against air pollution. "Cancer lurks in the air," proclaimed one headline in *Welt der Arbeit* in October 1958.[120] The "misappropriated sky" was also a topic in *Christ und Welt*, while the *Süddeutsche Zeitung* noted sadly that "the white flowers are gray" in the Ruhr region, and the *Frankfurter Allgemeine* reported with a shudder that in areas with a concentration of heavy industry, "the rule still holds today that the bright white shirt that the businessman wears to the office in the morning must be replaced with a fresh one the next day."[121] "Our air needs to improve," declared the *Westfälische Rundschau* as early as February 1954. The *Kölner Stadtanzeiger* published—under the sensational title "Breathing Forbidden—Danger!"—a series or articles with no fewer than twenty-three parts. The *Neue Ruhr Zeitung* reported, following air measurements in Mülheim by the TÜV Essen, that the air there was "virtually contaminated with poisonous carbon monoxide!"[122] Since the spring of 1956, at the latest, when the influential weeklies *Stern* and *Spiegel* for the first time reported extensively on air pollution, a staunch antipollution stance had been considered good form in the German press—and this usually left little room for subtlety. "All children in large cities are living unhealthy lives," the *Frankfurter Neue Presse* declared.[123] In fact, the media commented even on vigorous initiatives with pronounced impatience: when North Rhine–Westphalia's minister of labor, Konrad Grundmann, declared that "within ten years, at the latest, the pall of smog [in the Ruhr region] will have been reduced to a tolerable level," the *Rheinische Merkur* noted dryly that this was "a long, barely acceptable period."[124] This comment in turn drew a disappointed note from one ministry official: "That is not what we meant!"[125]

But while the German public was calling in no uncertain terms for tougher measures, it was entirely unclear how such demands were to be implemented. For example, the *Ruhr-Nachrichten* demanded "that Father State must finally intervene with drastic measures lest we 'die in the filth,'" while the Federation for Heimat Protection called for the "speedy passage of effective laws," and the Conference of German Physicians demanded a "revision of the legal regulations" in which "technical and economic considerations must definitely be ranked behind medical requirements."[126] Obviously, such formulations left the agencies concerned with very broad latitude. The debate did not even include a clear mandate for action. While a poll conducted by the Allensbach Demoscopic Institute in 1959 showed a solid two-thirds majority in support of tougher measures against emission problems, the response to the question of who, specifically, should take on this task showed a broad range of opinions: 45 percent of those polled went with the government in Bonn, 38 percent with local authorities, 26 percent with state governments, and 24 percent with industry; even unions and churches were mentioned, respectively, by 7 and 4 percent of those surveyed.[127] Similar to what had happened during the smoke debate around the turn of the twentieth century, authorities now found themselves confronted with diffuse public discontent that did not provide any clear directives for palliative measures. As a result, discussions in the 1950s may well have led to a similarly unedifying result as had the corresponding debates in Imperial Germany, had there not been, in certain places, a very open response to the new urgency of air pollution control.

The Role of the Authorities

Without a doubt, the most remarkable thing about the administrative reform debate of the 1950s is that it took place at all. After all, agencies could easily claim good reasons for responding to public unhappiness placidly. Was there not a firmly established bureaucratic routine for dealing with air pollution problems, a routine that the broad public had tolerated for decades without much grumbling? Did this not justify the expectation that it would probably be sufficient to respond to public criticism by giving the problems just a little more attention? In fact, such thoughts were indeed still bouncing around in the heads of many officials. For example, a report to Bremen's Senate in 1958 asserted that "detrimental effects in individual cases . . . can be adequately countered with existing legal provisions," and the factory inspection for Ludwigsburg noted lethargically that it was not necessary "to accord all that much weight" to the exhaust fumes of local industrial enterprises with respect "to their objective effect on the population." As late as 1963, a Bavarian county official referred a petitioner reflexively to the path of a civil suit.[128] But at least one state faced up to the new public demands and, in the process, took the lead in German reform efforts for decades to come: North Rhine–Westphalia.

In order to understand the sequence of events, it is crucial to note that the most eager reformers were working on the state level, especially within the Ministry of Labor and Social Affairs, which was traditionally in charge of air pollution.[129] For example, in 1957, one ministry official proclaimed that the existing regulations were "undoubtedly outdated."[130] The officials at the Ministry of Labor also liked to point to the continuously growing discontent among the population to shore up their own demands. "The concern of the public . . . is directed at the widest possible elimination of nuisances from air pollution," declared a ministry official named Franz Oels during an interministerial meeting in 1956. On a different occasion, a ministry official spoke of "growing alarm among the public and the members of parliament."[131] When Oels was invited as a guest to a meeting of the working group Smoke Damages and Dust Collection Methods, of the Association of High-Performance Steam Boiler Owners (Vereinigung der Großkesselbesitzer, or VGK), he emphatically took the side of the suffering population: "The public and the administration maintain that too little has been done up to now."[132] What made it easier for ministerial officials to respond to public protest was the fact that they were simultaneously defending their own bureaucratic turf: the draft bill that the director of the Ruhr Area Federation for Regional Planning, Sturm Kegel, had presented in 1952 was a direct challenge, from the Ministry of Labor's perspective.[133] After all, the concept of a cooperative solution propagated by Kegel, discussed in greater detail below, amounted to a reduction in the ministry's own range of tasks. Thus, when officials from the Ministry of Labor appeared to be strikingly motivated, this had at least as much to do with the consolidation of their own authority as with the discontent of the broad public.

At the level of agencies lower down in the hierarchy, a somewhat more mixed picture is evident, in line with the nature of the work being done at this level. At a time when all decisions were made on a case-by-case basis, the day-to-day practice of air pollution control consisted primarily of constant negotiation over ever-new compromises; it was thus perfectly natural that these officials spoke in much more measured tones. But here, too, some motivated officials pushed vigorously for a reduction in emissions. For example, during August 1954 negotiations over the maximum fly-ash emissions from brown-coal power plants along the lower Rhine, government representatives lent added weight to their demands by warning of imminent "'emergency' measures [taken] by the district administration under pressure from public opinion."[134] The factory inspection in Soest also demonstrated remarkable courage in quarrels with the cement plants in its district, even going so far as to solicit the help of the state prosecutor when needed to force the companies to adopt the desired measures.[135] And when the Mark Electric Utility, after critical newspaper reports about its Cuno electric power plant, sought the help of the district administration in Arnsberg in 1956, emphasizing soothingly that they had been "in contact for many years with the factory inspection in Hagen," the officials in charge responded with a determination that had been rare in earlier times: "As a matter of principle, we must hold to the fact that the air, as a common good, must be kept as clean as possible." They proceeded to curtly deliver a scathing indictment: "The countless complaints and the repeated meetings are evidence, are they not, that this fundamental requirement was not adhered to at the Cuno plant."[136]

In the 1950s, however, such committed officials saw themselves confronted with a double handicap. First, a large number of licensing documents had been destroyed during the war, which made it impossible for officials to point to the requirements they had laid down; the minister of labor in North Rhine–Westphalia, for example, estimated that about a third of the licensing paperwork had been lost.[137] Second, the regulations of the Industrial Code were still in force, and they generally prohibited retroactive additions to licensing requirements: the authorities could impose additional demands only when the installation in question was undergoing substantial alterations.[138] Of course, this did not completely rule out improvements after the fact: officials continued to have at their disposal an arsenal of formal and informal sanctioning tools they could use to strong-arm an entrepreneur into cooperating. But if a company simply refused to be intimidated and stubbornly insisted on its rights, even the most agile official was invariably stymied.

However, the Industrial Code was federal law and thus beyond the purview of a state government. Fortunately, the federal government proved receptive to the state's problems, not least because of a private backchannel between North Rhine–Westphalia and the federal government in Bonn: the aforementioned Franz Oels was the brother of Heinrich Oels, who worked on the same issues in the Federal Ministry of Labor and later in the Federal Ministry of Health.[139] But in Bonn, too, an external impetus was needed to push the Ministry of Labor into dealing with the issue aggressively. One such impetus came from the German parliament, which in January 1957 unanimously adopted a proposal by the Social Democratic Party (SPD) that obligated the federal government to prepare a "comprehensive report about the pollution of the air by industrial enterprises and other causes."[140] A second impetus stemmed from the fact that the federal government was being challenged by the work of an Interparliamentary Working Group (Interparlamentarische Arbeitsgemeinschaft für naturgemäße Wirtschaft, or IPA), an organization of members of parliament at the state and federal levels.[141] Soon after its creation in 1952, this group took on, as one of numerous issues, the problem of air pollution. The formation of the VDI Clean Air Commission, in which the IPA played a crucial role, took place without direct participation by federal officials.[142] So when the officials of the Federal Ministry of Labor responded constructively to ongoing reform efforts, they were driven not only by public discontent but also by a desire to keep things in their own hands.

Here it must be emphasized that such an official response did not require a great deal of courage. In the 1950s, air pollution control was by no means the political minefield it would become in the age of environmental politics; even the changes that industry lobbyists were asking for were still noticeably moderate at the federal level. Nor did the officials in question have reason to fear opposition within the federal government in the 1950s—least of all from the chancellor's office: when Chancellor Adenauer was presented with a memo in October 1958 about "legal measures to keep the air clean," he noted in his own hand that he considered "the inclusion of roadway traffic exhaust" an urgent necessity and went on to ask whether it might be possible to "incorporate measures against noise."[143] A reassuring letter from Minister of Transportation Hans-Christoph Seebohm about automobile exhaust did not satisfy Adenauer: Seebohm's contention that "the existing legal regulations are sufficient to combat avoidable smoke and noise generation in road and waterborne traffic" drew Adenauer's exhortation that Seebohm ensure, "through more stringent control of vehicles and traffic participants, that the regulations are in fact observed."[144] It seems that these were spontaneous initiatives of a chancellor known for his solitary decisions, but they demonstrate that the growing activities of officials met with favor even in the top echelons of Germany's young democracy.

It would surely be an exaggeration to presume that Bonn and Düsseldorf always saw things the same way. But as early as December 1956, at a meeting between the federal and the state ministry, there was "agreement on the view that... an initiative should be developed in order to be able to impose subsequent requirements on old installations"; moreover, there was agreement to revise the index of installations, which required prior approval.[145] Officials elsewhere were still more cautious at this time: a contemporary interministerial meeting in the Saarland, for example, concluded that "for the time being one should be content with the already existing legal options."[146] But outside of North Rhine–Westphalia, as well, the notion slowly began to spread that a reform of legal regulations was unavoidable. When measures to combat air pollution were discussed in the municipal working group Rhein-Neckar in 1957, participants came to the unpleasant realization that in the end everything would depend on the goodwill of industry: "Given the existing legal regulations, it is likely to be extremely difficult to enforce remedies against the will of recalcitrant enterprises."[147] In 1958, Munich's factory inspection told, in connection with complaints from neighbors, of the "hopeless situation" of many agencies, which could not "do anything because they lacked a legal basis."[148] Even veteran experts were slowly starting to have their doubts: while Arnold Heller, a scientist at WaBoLu since 1929, had declared as late as 1950, in the journal *Gesundheits-Ingenieur*, that "it would be adequate for now if the previous regulations in this areas were applied throughout Germany as they were intended," a mere four years later he wrote in the same journal that there was an "urgent need for new regulations."[149] For the time being, these were still isolated opinions, which would solidify into a genuine debate over strategy only as the years went on. But the 1950s saw only the beginnings of such debate—primarily because the crucial decisions had already been made. By the time the need for reform was recognized outside of North Rhine–Westphalia, the solution was already at hand in Düsseldorf and Bonn.

The Role of Industry

It is not difficult to understand that the growing public discontent was extremely inconvenient to German industrialists. Tougher steps against air pollution problems would bring additional costs; public protests endangered a bureaucratic practice with which industry in general had been able to get on rather well; and for an entrepreneur it was, needless to say, unpleasant

Figure 7. Smokestack of the Völklingen steel mill in the Saar region in 1962. Reluctantly, German industry came to accept the imperatives of air pollution control, although it usually obtained generous timetables for compliance. Image courtesy of Ullstein Bild.

when the press wrote, for example, of an "ash war at Lake Harkort."[150] Consequently, many documents mirror a sense of discomfort among German industrialists during the 1950s. For example, a representative of mining in the Ruhr complained in the *Industriekurier* that the public was taking a "strong, sensationalistic" interest in the health dangers of air pollution, "even though public health experts who specialize in this area maintain that everything is still unclear."[151] The VGK also felt compelled to note that the public debate "has in part taken on forms that are not suited to promoting the concrete work in this field, which still needs more thorough investigation."[152] When Sturm Kegel's cooperative idea dominated the debates, the Association of German Industry published not one but two pamphlets to shed light on the "problems and difficulties that are, upon closer inspection, attendant upon the implementation of the proposal, both in economic-technical as well as in legal terms." Behind closed doors, there was even talk of "demagoguery" and remarks that "calm . . . is the citizen's first duty."[153] If it had been up to German industry, the whole debate over reforms probably would not have happened in the first place.

Against this background, it is all the more remarkable that industry at no time dug in its heels on a fundamental position. From the beginning, industrial interest groups, in particular, were eager to prove that, for all their criticism, their attitude was a constructive one. Consequently, in its response to Kegel's proposal, the Association of German Industry did not simply reject it but simultaneously declared its willingness "to support, in both technological and legal regards, all measures that are suitable [to] bringing about an improvement in air quality.[154] And when a short time later changes to the Industrial Code and the Civil Code were under consideration—changes that meant a considerable weakening in the legal position of industry—the Association of German Industry deliberately refrained from principled criticism: in a letter to the parliamentary committees in charge, the federation explained that it had come to accept "that there is no denying the essential justification of the intended measures, in spite of the burden they will impose on individual companies in the interest of general welfare."[155] Likewise, the Association of German Chambers of Industry and Commerce (Deutscher Industrie- und Handelstag) presented itself as emphatically willing to compromise: the draft law, it noted to the federal minister of economics, represented "a measured attempt to delineate the contradictory concerns of industry, on the one hand, and of the general public, especially agriculture, on the other."[156] Representatives of notoriously polluting sectors were also eager to avoid the reputation of being obstructionists. The VGK presented its own efforts "as an expression of a self-evident obligation undertaken on our own responsibility"; the managing director of the Lower Rhenish Chamber of Commerce in Duisburg emphasized that Duisburg emitters had "behaved from the outset with an open mind"; one representative of the cement industry declared that it had to "remain our primary task to do everything to make life in our industrial regions bearable and pleasant for working people by trying everything to push emissions down to the lowest practical level."[157]

To be sure, this course was by no means uncontroversial in industrial circles. "Unfortunately, anyone who has to deal with dirt in the name of industry will make few friends," a BDI expert complained to a reporter from the *Industriekurier*.[158] But if industrial associations endeavored—internal resistance notwithstanding—to find a compromise between their own interests and those of the public (incidentally, unlike the German Association of Farmers), there are probably three main reasons for their behavior.[159] First, at a time when the most important pollutants were still readily apparent to the senses, it was often impossible to deny that many places were experiencing considerable nuisances. The BDI therefore declared matter-of-factly that "many of the measures demanded by state agencies, if looked upon objectively, must be recognized also by industry as justified in the interest of the common good."[160] Second, when analyzing the behavior of industrialists, it must be recalled that there were also industrial sectors with a special interest in clean air: for example, a dairy objected to the construction of a waste incinerator in Rosenheim.[161] Third, industry clearly did not wish to find itself too much on the defensive. Sturm Kegel's legal initiative, which had caught industrialists off-guard, provided a constant reminder to the business community that it ran the risk of unpleasant surprises if it remained as passive as it had been in earlier decades. If reforms were inevitable, the business community at least strove to gain as much input as possible.

Industry did not like the debate, but what it abhorred even more were experiments in regulation. It therefore pushed for cautious changes to improve the existing system of air pollution control, without being confronted by the imponderables of fundamental reform. As early as 1954, Karl Oberste-Brink, the chair of the BDI emissions committee, declared that it might be helpful to examine "to what extent the Industrial Code is still adequate for today's needs, given the development that has occurred in the intervening period."[162] The following year, the BDI was already more concrete: in a printed statement, it proposed "a review of the catalog of installations that require licensing and are in need of oversight in accordance with the regulations of the Industrial Code," as well as a "revision of the 'Technical Manual for the Industrial Code' and the 'Steam Boiler Operating Instructions.'"[163] In addition, industrialists wanted to make sure that they would not be the only ones taking measures to combat emissions: few representatives of industry forgot to point out that domestic furnaces, small businesses, road transportation, trains, and ships contributed in no small measure to atmospheric pollution.[164] Finally, industry was eager to strengthen the "close cooperation between companies and industrial associations, the Association of German Engineers, the Association of High-Performance Steam Boiler Owners, the factory inspection, the agencies for technological inspections, and the mining bureau." Put simply, as long as competent technicians were in charge, industrialists felt fairly safe. After all, the BDI was absolutely convinced "that the issue of dust and exhaust is primarily a scientific-technological problem."[165]

What made it easier for the BDI to put forth such demands was the fact that efforts in this direction were already under way. On the problem of air pollution through flue gases from steam boilers, the VGK, in April 1955, presented a comprehensive "Report about the State of Technology," whose spectrum of topics ranged from the effect of airborne dust and questions of measuring technology, all the way to the design and construction of dust collectors.[166] The Association of German Cement Plants (Verein Deutscher Zementwerke) also financed extensive research projects, the findings of which were passed on to the commercial supervisory authorities, so that—as one representative of the cement industry put it—"the licensing requirements for new plants will not incorporate unacceptably high numbers, as occasionally happened in the past."[167] Of course, no one could entertain any illusions about the bias of such publications: at a meeting in July 1957, a factory inspector from Recklinghausen commented that the cement industry's target numbers reached "levels that we, as the supervisory authorities, can in no way regard as the goal that should be aimed for today"; at best, the values could be seen as the "maximum framework."[168] Still, the positive value of this research work should not be underestimated: at a time when technical norms were generally scarce in the area of air pollution control, even proindustry publications represented valuable points of orientation. A ministry official from North Rhine–Westphalia's ministry of labor declared quite frankly that the research work by the cement industry had the basic "advantage for us that the industry on its own has named figures, and figures that one can talk about."[169]

In describing the attitude of industry, however, it is imperative to emphasize very clearly that there was a considerable difference between the behavior of the associations and that of individual entrepreneurs. For while the former were generally intent on constructive cooperation with the authorities, individual companies showed themselves in many cases much more reserved.[170] Especially in areas of high industrial concentration, where the authorities acted most aggressively, for obvious reason, there were often intense conflicts between officials and entrepreneurs about antipollution measures. For example, the factory inspection in Nuremberg-Fürth reported in October 1958 that entrepreneurs were generally "intent only on achieving the primary technical purpose of their manufacturing processes" and would therefore reject installations to reduce emissions as "too expensive and uneconomical."[171] Mannheim's mayor also had some frustrating experiences at the beginning of the 1950s when he sent out a circular calling upon all the businesses in the city "to keep smoke emissions as low as possible": only a third of the businesses even deigned to respond, and the majority of those that did simply declared "that everything is already being done."[172] "At large meetings and congresses," explained a factory inspector, "large polluters repeatedly profess their willingness to undertake thorough measures to combat air pollution. In daily practice, however, the situation is usually different." Some enterprises were "taking every opportunity ... to gain time through a constant stream of new objections, some justified, some not."[173] The factory inspector in Soest could only concur: the assertion by the Association of German Cement Plants "that the cement industry is willingly pursuing neighborhood protection on its own accord" contradicted "the experiences in nearly all enterprises in this district."[174] And in Munich, too, authorities found that industrial emitters were "rarely willing to remedy the situation without coercive means."[175]

To be sure, one must be careful not to jump to conclusions by generalizing such comments. The authorities and industrialists for the most part still got on fairly well, especially in regions where air pollution control was still not high on the agenda.[176] Moreover, even in heavily polluted conurbations, there were companies that looked upon the fight against air pollution as a question of honor: the Mannheim Public Utility, for example, "considered it its self-evident duty to keep the emission of smoke as low as possible."[177] Beyond that, a smart entrepreneur would be intent, simply out of self-interest, not to antagonize factory inspectors willfully in spite of countless conflicts over details. On the whole, however, there was a considerable chasm between associations and individual companies: while associations emphasized their willingness to compromise, industrialists were often much more reserved on the ground. And this situation had significant consequences for the course of reform, as the ministerial bureaucracy, the essential player in shaping the reforms of the 1950s, had more favorable experiences with industrialists than did the lower agencies. Anyone working on day-to-day air pollution control knew that only a self-confident posture and a good dose of suspicion brought success; the notice from one official of industrial inspection in Augsburg, who believed that "a visit to the operator of the boiler is imperative, especially since more can be accomplished through personal negotiations," indicates the extent to which respect played a role in everyday control practices.[178] On the level of state or federal ministries, on the other hand, the matter looked far more positive: officials continually confronted by the conciliatory rhetoric of industrial associations were, naturally, much more inclined to give industrialists the benefit of the doubt. Of course, confidence was not unlimited: as we shall see, ministerial officials initially took a critical view of the activities of the VDI Clean Air Commission. But it spoke volumes about the image that North Rhine–Westphalia's ministerial bureaucracy had of the entrepreneur when the Ministry of Labor, for example, informed Sturm Kegel "that it is indispensable in the view of the administration to win over industry's positive stance on the draft law," or when Heinz Seiler, speaking on behalf of the same ministry to representatives of the VDI, declared that the tasks at hand "can be solved in a much more promising way on the basis of mutual trust by those involved . . . than with legal coercion."[179] It remains doubtful, to say the least, whether German ministerial officials would have ventured upon the path of cooperative air pollution control had they been aware of the full extent of the conflicts that would ensue over implementation.

The German Rise of Research

In August 1954, WaBoLu officials recommended purchase of the *Proceedings of the United States Technical Conference on Air Pollution*.[180] They noted that this book offered "a very good compilation of the progress that has been made in the United States in all areas of air sanitation." In fact, from a scientific point of view, it was a good idea to take a look at what was being done on the other side of the Atlantic: "While relatively few findings could be collected in Germany during the last ten years in the area of protecting abutters against air pollution, because of the war and the postwar situation, there is a lot of modern literature in the United States of America."[181] This assessment, however, seriously shortchanged German research. Of course, after 1945 German researchers had a more difficult time than their American colleagues, for obvious reasons. But Germany did have a number of research traditions from the interwar period that were more or less absent in the United States. In any case, the United States did not have general technological superiority—with the exception of the special problem of photochemical smog, where the United States' pioneering role was beyond doubt. This became apparent when Americans in Plittersdorf, near Bonn, complained to German authorities about the emissions from a nearby cement plant: after a tour of the plant, in which a U.S. expert participated, the Americans were forced to concede, shamefacedly, "that dust control is far better than in industrial plants in the United States."[182]

The most striking characteristic of the research landscape in Germany was its highly fragmented nature, evident even upon superficial examination. The research activities of the owners of large boilers and of the cement industry, mentioned earlier, began in the early 1950s. In addition, however, in 1956 a number of companies from different sectors of the industry established the Advisory Institute for Water Management and Emissions (Beratungs-Institut für gewerbliche Wasserwirtschaft und Emissionen), headquartered in Cologne. One reason behind its founding, at least according to a BDI circular, was the notion "that bureaucratic measures, experience has taught us, are in many cases best prevented by voluntary measures."[183] Those who considered such an approach too conciliatory could turn, for example, to Dr. Helmut Berge and his Agricultural-Chemical Institute in Heiligenhaus, near Düsseldorf. According to a note from the North Rhine–Westphalian Ministry of Labor, this group was "exceedingly pro-industry in its activities of rendering expert opinions."[184] But the other side, those damaged by emissions had their own experts as well, and some even worked for the government. Other institutions, for example, the Associations for Technological Inspections (Technische Überwachungsvereine, or TÜV) were eager to maintain neutrality for the simple reason that they were close to the state.[185] After 1953, WaBoLu, as well, continued to exist as a department within the newly created Federal Health Agency.[186] Finally, it is likely that a considerable number of university institutes dealt with various aspects of air pollution, although the lack of contemporary studies makes a comprehensive overview difficult. Still, we do know, for example, that in the field of agronomy and forestry alone, there were, according to a report by the Federal Ministry of Agriculture in February 1957, about ten institutes at six different universities that had "become known through scientific work in this field."[187] Even though the institutionalization of scientific research in the 1950s cannot compare to what followed in subsequent decades, there was still considerable activity that met with remarkably broad interest. The first large conference on air pollution, organized in December 1955 by the Association of German Engineers, in partnership with the VGK and the TÜV Essen, attracted no fewer than 841 participants.[188]

However, no scientific institution was as important to the development of air pollution control after 1945 as the Expert Committee on Dust

Technology within the Association of German Engineers. This committee, which, as we have seen, had already attained considerable prominence during the interwar years, was revived at the end of 1947. Its purpose remained the same: "It is the task of the Expert Committee to systematically research the scientific basis of the dust abatement processes, to the extent that they have not yet been clarified, with the goal of further perfecting the dust precipitation apparatuses and boosting their efficiency," explained a letter from the VDI in November 1947.[189] This mission evidently still drew wide support: soon the Expert Committee again had five working committees that dealt with themes from "dust sanitation" to "measurements in dust technology"; as early as 1950, it was reported that the Expert Committee had "resumed its practical-scientific work on a large scale."[190] The committee's meetings evidently provided a welcome forum where experts could meet and exchange ideas. For example, one WaBoLu official commented in his report about the 1951 meeting in Goslar: "Far more important than the lectures themselves were personal meetings with professional colleagues from industry and the bureaucracy." Meldau reported that since 1947, each meeting had been drawing "320 to 450 members and guests."[191]

For all that, the work of the Expert Committee was still not nearly as unpolitical as was purported by constant invocations of the slogan of "cooperative work."[192] To be sure, the blatant partisanship within the VDI still did not stand a chance: when the discussion of Kegel's draft law got under way, Robert Meldau noted with satisfaction that the Expert Committee, thanks to a BDI memorandum, "has been taken out of legal questions and can now devote itself again to its true area of work, namely technical questions."[193] At the same time, however, the Expert Committee was also interested "in preventing for its part too vigorous an initiative in the area of legal regulation"—every politicization of air pollution control invariably threatened the position of technical experts.[194] At first glance, this seems like a staunchly obstructionist stance, but through a series of unexpected events, it would turn into a major asset of German environmental policy.

5 Going Local, Going National
THE POSTWAR DIVERGENCE OF ENVIRONMENTAL POLICY

The Silent Crisis of Municipal Air Pollution Control in the United States

It is rewarding to compare the postwar period of institutional change with the Progressive Era. In principle, the situations were similar: in both cases, new public opinion demanded reforms in order to gain control of a pending problem—and in both cases these reforms actually came about. But the way in which these reforms were discussed reveals a far-reaching difference. The Progressive Era saw a broad public debate in which civic leagues, entrepreneurs, and scientific experts participated and that eventually led to a consensus about a strategy that promised benefits to everyone involved. The situation after World War II was very different: at most there was the beginning of a strategy debate. Instead of discussing and weighing various alternatives, the parties involved decided without much ado on the model most readily at hand, which was essentially little more than a modification of the tried-and-true smoke inspection approach. This happened even though, from an institutional point of view, there was plenty of room for a fruitful debate: postwar municipal air pollution control displayed a number of problems that gradually curtailed its effectiveness—and these problems could have already been foreseen more or less clearly around 1950. The enduring crisis that plagued municipal clean air efforts from about 1960 on was, then, the direct result of the amicable agreement of 1950.

The Limits of Municipal Air Pollution Control

It was essentially by default that municipal governments took charge of air pollution control during the Progressive Era. With the federal government never envisaging more than an advisory role in this field, and state legislatures showing little concern due to the predominance of rural interests, it was left to cities to deal with smoke. Solving an urban problem through city policies seemed like a natural and obvious solution, but this approach, in spite of its seemingly self-evident nature, began to diminish toward the end of the Progressive Era, when American industry began to spread beyond city limits.[1] Obviously, this spelled trouble for an approach that centered on municipal authorities: slowly but surely, the field of action for city smoke inspectors began to erode. In spite of this, the interwar period saw only two initiatives that responded to this development with new institutional forms. First, in Boston, the fight against smoke was already in the hands of a state agency before World War I, and authorities adjusted the sphere of action in 1928 to reflect the growth of the city. Second, after 1931, at least one county tried to take steps against the smoke nuisance—Hudson County, New Jersey.[2] But these approaches hardly resonated anywhere else at the time, experts offering little more than pipedreams. In the case of boundary-crossing pollution, Henry Obermeyer recommended "friendly negotiations" with neighboring municipalities, while Harry B. Meller of Pittsburgh's Mellon Institute dreamed about the establishment of so-called air hygiene districts that were to be set up without regard to state or municipal boundaries.[3] While this did demonstrate an awareness of the problem, it did little beyond that.

To be sure, the suburban problem in the 1920s and 1930s should not be overestimated. After all, Raymond Tucker's St. Louis campaign gave emphatic proof of what could be achieved even within the limited framework of a municipal antismoke drive.[4] However, the situation grew critical after World War II, when the process of suburbanization accelerated noticeably. This essentially meant that an ever-larger number of polluting businesses were located outside of the boundaries of large cities and could therefore no longer be controlled by municipal officials; municipal air pollution control thus increasingly became a living anachronism. According to a 1961 survey, more than 60 percent of American cities had an emission problem whose location put it beyond the reach of municipal authorities; in cities of more than two hundred thousand inhabitants, fully three-quarters of the officials polled reported such problems.[5] In the final analysis, though, such a study merely demonstrated the obvious. "The fact that much of the new industry is being developed near, but outside of, metropolitan areas limits the ability of the municipal authorities to deal with the resultant air pollution," the Minnesota Health Department noted in 1959, while Cleveland's Southeast Air Pollution Committee observed matter-of-factly that "pollution knows no boundaries."[6] The only question was what should be done.

At first glance, the best response was to shift the administrative competencies to the next-higher level of the bureaucracy. There was even a shining model that impressively demonstrated the merits of such a solution: Pittsburgh. After that city's legendary campaign against coal smoke had more or less settled into a routine, it became increasingly clear that a control program was also necessary for the surrounding communities. To this end, the Allegheny County Bureau of Smoke Control was set up in 1949, responsible for keeping the air clean throughout the county, with the exception of the city of Pittsburgh itself. But Pittsburgh also showed that creating an agency on the county level was not a simple matter. First of all, the state legislature had to pass a special law in order to create the legal basis for such an institution; this took two attempts, because of the resistance of the railroad lobby. Afterward, activists had to persuade the Board of County Commissioners of the importance of air pollution control, and this required, among other things, the collection of more than fifty thousand signatures.[7] If all this effort alone did not exactly foster enthusiasm for such an approach, subsequent events gave municipal officials especial pause: after officials from the city and the county had worked side by side for several years, the municipal Bureau of Smoke Prevention was dissolved in 1957, and the authority of the Allegheny County Bureau of Smoke Control was expanded to include the city.[8] Any municipal official who supported air pollution control by county agencies therefore risked his own job over the medium term. This was probably a key to the fact that air pollution control by counties generally remained the exception. Of the eighty-five agencies with an annual budget of at least five thousand dollars in 1961, a mere fifteen were at the county level, and of those, seven were in California, where the smog of Los Angeles had led to a different regulatory approach.[9]

Of course, there was no lack of attempts by municipal officials to solve the problem on their own initiative. For example, in the greater Philadelphia area, air pollution problems were discussed during a Regional Conference of Elected Officials, and in 1962 the Northeastern Illinois Metropolitan Area Air Pollution Control Board was set up in Chicago. Meetings of this kind, however, led to little more than a noncommittal exchange of information and vague agreements.[10] In 1964, Denver created a Regional Air Pollution Control Agency, but it was ultimately an advisory organization dedicated to promoting and coordinating local activities.[11] Of course, the value of such institutions should not be underestimated at a time when cooperation was an almost magical concept, but in the end it was obvious to everyone involved that over the long term such groups could not take the place of formal, supraregional controlling agencies. To be sure, in a 1961 survey, no fewer than 46 percent of cities with sources of emissions beyond the city boundaries claimed that they were cultivating a "working agreement" (whatever that meant) designed to solve these problems.[12] But in most instances, these agreements were most likely voluntary in nature, since there were only nine states at the time where air pollution control was even legally possible across city and county lines.[13] And even such arrangements could end in failures of almost tragic proportions. For example, when a Metropolitan Air Pollution Control Program was set up in Cincinnati in 1957, with seven suburbs signing an agreement that gave Cincinnati's Bureau of Air Pollution Control oversight also in these municipalities, Charles Gruber, the head of the city's air pollution efforts, hailed this as the first step toward a comprehensive solution: "While the Program is still in its infancy, it is hoped that the addition of other villages, now considering entry into the organization, will substantially aid in its development to maturity."[14] However, this hope proved illusory: twelve years later, the number of cities and towns that were cooperating with Cincinnati still stood unchanged, at seven.[15]

This entire problem could have been avoided from the outset had air pollution control been declared the responsibility of state governments—except that the creation of state-run programs had been blocked precisely because incumbents nearly unanimously favored a local approach. As a result, state legislation differed enormously: "State approaches to air pollution have probably shown the greatest variation of all levels of government," a 1961 study declared.[16] At that time, thirty-three states did have laws to combat air pollution, but only seventeen invested a minimum of five thousand dollars annually in relevant programs, and even these programs focused mostly on research and technical support. "Depending on the exact meaning given the word 'enforcement,' four to six States 'enforce' air pollution regulations," noted a 1963 study for the U.S. Senate Committee on Public Works.[17] The New York State Air Pollution Control Board, for example, which had enforcement authority from a legal point of view, rejected the use of its powers on principle and focused entirely on promoting local control programs.[18] Even when states took a more vigorous role, their efforts were not necessarily intended to compete with local initiatives. "It is probable that the State will have to carry much of the load until city and county air pollution control agencies grow," the Florida State Board of Health noted in

1960.[19] When the 1963 Clean Air Act created the possibility of supporting air pollution control programs with grants, the Public Health Service assumed, without much ado, that the lion's share of the available money would go to local agencies.[20]

For all these reasons, the activities of various states were in most cases hopelessly outdone by the efforts of municipalities. In 1968, the budget of the Illinois Air Pollution Control Board, for example, was only about one-sixth the budget of Chicago's Air Pollution Control Department. In Michigan, a comparison of the state's expenditures to those of Detroit and Wayne County elicits a similar result. While a total of 876 individuals were working in local control programs in 1961, state agencies had merely 148 full-time and 29 part-time employees, and of these more than a third were working in California.[21] What these numbers meant in concrete terms is evident, for example, in a letter from the New York State Air Pollution Control Board to the Hudson River Conservation Society: the board bluntly informed the conservationists that one could likely begin the process of developing formal regulations in ten or fifteen years; however, the board was perfectly willing to advise local communities on how to come up with their own ordinances.[22] Even New Jersey, which ran one of the best programs in the country, could not achieve systematic oversight. "Principal emphasis of the air pollution program . . . is directed toward correction of specific nuisances in the vicinity of particular sources of pollution," declared a 1966 report by the Public Health Service, which recommended a tripling of state funds.[23] "State legislation is the only fair method for uniform air pollution control," the Michigan Department of Public Health concluded in 1967, and this was certainly more than merely disguised self-praise.[24] Still, the Council of State Governments was forced to acknowledge in March 1967: "Local governments . . . are the leaders in the air pollution control efforts that have been made to date."[25]

A more vigorous and activist policy on the part of the states would have been desirable for a second reason, as well: it was the only way to guarantee an effective fight against air pollution problems in smaller municipalities and the countryside. For if a city was not big enough to afford a local control program—contemporaries put the threshold at a population of about 150,000—the successful model of the 1950s was invariably stymied, and the same was true for polluters in unincorporated areas.[26] Moreover, officials in small industrial cities, which often had the biggest emission problems, were particularly dependent on the goodwill of local industries. A case in point, presumably not untypical, was the steel town of Struthers in northeastern Ohio. In a private conversation with an official from the Public Health Service, the mayor of Struthers revealed that representatives of the local steel plant had threatened to relocate some parts of its operations if the town took measures against the company—an argument to end all arguments in a city of 14,000.[27] In situations like this, however, communities generally could not count on any help from state agencies. "By far the great majority of States are not even serving those communities which are too small to operate their own local programs but are nonetheless affected by serious air pollution problems," declared Arthur Stern, assistant chief of the Public Health Service's Division of Air Pollution, before a committee of the U.S. House of Representatives in 1966.[28]

However, there was one problem even state agencies could not have solved: emission problems that reached across state lines. A nearly insurmountable dilemma stood in the way of institutions that were a match for such problems: if the authorities were weak, they had other problems to worry about, but if they were strong, they were disinclined to relinquish their competency to a superior agency. And such problems were by no means rare: in 1963, nineteen states reported emissions that crossed state lines, a total of about 38 million people living in conurbations with this kind of problem.[29] Yet only a single institution in the entire country represented something like a regulatory agency above the states: the Interstate Sanitation Commission, in greater New York.[30] But even this institution hardly provided an encouraging model. For one thing, the Interstate Sanitation Commission had not been created specifically to fight air pollution (it had been responsible for fighting water pollution since 1936).[31] For another, the expansion of its agenda had succeeded only after a long and nerve-wracking legislative process, mostly due to resistance from preexisting control agencies in New York and New Jersey.[32] But if it took considerable effort to institutionalize cooperation across state lines even in an area with an established tradition of joint regulation, one can only imagine the problems in other urban areas.

It was perhaps conceivable to declare that it was the task of the federal government to deal with boundary-crossing pollution problems. But this solution was more theoretical than real; in general, most people believed that a limited federal role was desirable.[33] Even the Air Quality Act of 1967, whose passage clearly mirrored a sense of disillusionment concerning the local approach, still regarded air conservation explicitly as a task of state and local agencies.[34] And this created a constellation that was rife with the potential for crisis: on the one hand, it was obvious that municipal air pollution control was increasingly reaching its limits, while on the other hand, the very persistence of municipal air control ensured that institutions capable of regional oversight could not establish themselves—or could do so only to a limited extent. City officials, who had once spearheaded the American effort to combat air pollution, were now no longer part of the solution, but part of the problem.

New Frontiers

Seemingly, the broadening of the definition of air pollution after 1945 marked a dramatic turning point. After all, the transformation in the tasks at hand could hardly have been more radical: after decades of pretending that there were no air pollution problems beyond smoke, municipal officials were now responsible for the totality of all emissions, at least as a matter of principle. However, if the bureaucratic reality behind the grandiose laws is examined, it is evident that no similarly profound change came about in the consciousness of the authorities. Instead, administrative practice was generally characterized by a peculiar "business as usual" mindset: although coal smoke became less and less of a problem, it continued to draw the lion's share of attention. Take, for example, Pittsburgh: even though the much-vaunted ordinance of 1941 allowed a fight against all pollutants, little of this was evident in the work of the authorities for a long time.[35] It was only when public discontent was expressed in a growing number of complaints about gaseous emissions that officials began to wonder. "Complaints of smoke are lessening," noted Pittsburgh's Bureau of Smoke Prevention in its annual report for 1953. "Now that it is greatly cleared, people are becoming more conscious and interested in complaining of disagreeable odors, noxious gases, fumes, etc." And this had consequences for the day-to-day work of the bureaucracy: "The character of the Smoke Bureau's work is somewhat changing," the report observed.[36] Apparently, Pittsburgh's officials did not become aware of their own shifting agenda until the transition was well under way.

Other agencies, too, were rather sluggish in responding to the new challenges. Some even refused to accept them at all. In Columbus, for example, the work of the authorities was focused exclusively on the smoke nuisance as late as 1958.[37] In Chicago, legal authority was expanded as early as 1948, enabling the Department of Smoke Inspection and Abatement to rename itself the Department of Air Pollution Control, but it was not until 1958 that rules and regulations were passed that made the law effective in everyday practice.[38] Even in Cleveland, where Herbert Dyktor, as the head of the Division of Air Pollution Control, had early emphasized the importance of other industrial emissions, abatement activities long continued to be focused on smoke; at the end of 1967, there were six employees in the Bureau of Industrial Air Pollution, as compared to fourteen in the department in charge of fighting smoke. And even this number was the result of a reform that year that increased Bureau of Industrial Air Pollution personnel by 50 percent.[39]

And in Allegheny County, the director of the Bureau of Smoke Control stated candidly, "Our problem is solids first and gases afterward."[40]

The lack of open debate makes it somewhat difficult to pinpoint the reasons behind this bureaucratic inertia. But if one considers the dominant role that engineers trained in furnace technology had played for decades in municipal antismoke efforts, it is hard to avoid a certain suspicion: in all likelihood, smoke inspectors harbored an antipathy toward the expansion of the pollution agenda that was as deep as it was unspoken. Since these engineers usually lacked the technical knowledge to take on other pollutant problems, they could not have a personal interest in the reorientation of bureaucratic priorities. The new problems required specialized knowledge that chemists, industrial hygienists, and other scientists possessed to a much larger degree than mechanical engineers. Smoke inspectors, whose scientific monopoly had been practically uncontested for decades, were now in danger—to use terminology proposed by Andrew Abbott—of losing professional jurisdiction in questions of air conservation.[41]

The debate over renaming the Smoke Prevention Association became a nearly exemplary manifestation of the smoke inspectors' hesitation. The national professional association of smoke inspectors had considered a change in name as early as 1939, on the grounds that the title should indicate their expanded spectrum of tasks; in the end, however, the proposal failed to garner majority support.[42] When the topic was raised again in 1949, the discussion again revealed a clear tendency to preserve the traditional name.[43] At the next conference, in 1950, members voted for a compromise solution: the organization would now assume the rather clunky name of the "Air Pollution and Smoke Prevention Association of America."[44] However, no one was particularly happy with this solution, and members were debating yet another change in name only two years later. Nevertheless, the suggestion forthcoming from a special committee—Air Purification Society—was disqualified by the simple fact that its acronym, APS, was not a pleasant one.[45] Eventually, at the suggestion of Charles Gruber, the name "Air Pollution Control Association" carried the day, and it was subsequently adopted as the association's permanent title.[46] In 1946, the Smoke Prevention Association had even directly opposed the Zeitgeist by emphatically rejecting ongoing efforts to revise local ordinances: "It is the mature judgment of this Association that the need of the present period is for more enforcement and public support of existing agencies than for new ordinances," a resolution of smoke inspectors declared.[47] It was only when an Air Pollution Control Conference was formed in October 1946 as a competing organization, under the chairmanship of Charles Gruber, that a reform process got under way within the

Smoke Prevention Association.[48] After a change to the bylaws, its board of directors was made up of nine officials (including the president) and eight representatives of industry, which meant that the omnipresent cooperation was enshrined within the group's internal structure.[49] "The main purpose of the revitalized Association ... is to co-operatively seek the solution to the air pollution problems," explained Gruber, who had been president during the transition phase.[50] Smoke inspectors had boarded the air pollution–control train in the nick of time.

Of course, the "old guard" of air pollution control was merely fighting a rearguard action. The long-term trend was clearly toward experts who understood and could combat the totality of all emission problems. Over the medium term, however, engineers could certainly act as an impediment that slowed things down: when municipal agencies systematically battled the smoke nuisance with elaborate bureaucratic programs while going after other emissions only haphazardly, this clearly amounted to a grotesque misallocation of scarce bureaucratic resources. And so the inertial response of municipal officials to their changed tasks fits into the picture earlier encountered in small cities and the suburbs: the pioneers of the Progressive Era now stood in the way of urgent reforms.

Recruitment Problems

As air pollution control was booming throughout the postwar period, insiders were quickly bemoaning the shortage of qualified personnel. For example, in the mid-1960s, the Public Health Service reported that "competition for personnel with any experience at all is very active."[51] As a professional field, air pollution control suffered from a dual handicap in the 1950s and 1960s. First, the work, by its very nature, was not prestigious: "Hanging over this profession is the relation with dirt and filth, with smelly rivers, sewers, and dead fish," explained Arie Jan Haagen-Smit, one of the field's most renowned researchers, in the journal *Research Management*.[52] Second, municipal authorities were generally not able to lure capable people with attractive salaries. According to the National Science Foundation, the average income for the head of a municipal control program was $8,300 around 1960, while the average income for all engineers was around $10,500.[53] The complaint was thus constantly raised that the usual salaries were simply too low to attract good people.[54] The Toledo Division of Air and Water Pollution Control once lost no less than 40 percent of its employees to industry in a single year—"for the usual reason, higher pay."[55]

Recruitment problems were not entirely new to American cities, as the antismoke campaigns of the interwar years amply demonstrate. In Cleveland, the average tenure of the commissioner of smoke inspection was less than two years between 1910 and 1927.[56] This was an exceptionally high turnover rate, but it reflected a weak identification with a bureaucratic job that was by no means unusual: "To many, the job is just a job," lamented Victor Azbe, a veritable crusader against St. Louis smoke.[57] Municipal authorities were essentially offering poor pay for a difficult job, since a competent smoke inspector had to fill several roles: he had to have engineering experience, he had to give speeches and secure support from politicians and organizations, he needed tact and sensitivity to coax the necessary cooperation out of entrepreneurs, he needed administrative skills—and for all that he should not be reluctant to get his hands dirty.[58] An engineer with all of these qualifications naturally came at a price. At least $10,000 had to be budgeted for a capable department head, Harry B. Meller explained in 1924 in a conversation with the city manager of Cleveland—much to the surprise of the women of the Women's City Clubs, who had arranged the meeting and thought they had no chance of getting that kind of money authorized.[59] Of course, there were also established, dominant figures who sustained their cities' smoke-abatement drives for decades and ultimately became seasoned veterans: Frank Chambers in Chicago, John Lukens in Philadelphia, Charles Poethke in Milwaukee, Harry B. Meller in Pittsburgh, Gordon Rowe in Cincinnati, and William Christy in Hudson County. On the whole, however, they were an exception: of thirty-one officials who were responsible for the antismoke drive in one city in 1924 and were therefore listed in the handbook of the Smoke Prevention Association, only twelve were still in office eight years later.[60]

Twenty-five years on, this situation had hardly changed. Even renowned agencies like the Allegheny County Bureau of Smoke Control complained that they were simply unable to attract individuals of the requisite caliber for its tasks: "There are only two places to look for men: (1) from the ranks of those who have been unsuccessful in industry and are looking for what they think is a soft berth; and (2) young men just out of college who consider the Bureau an opportunity to broaden their experience and their acquaintanceship with a view to finding something better later on," a review of the Pittsburgh program noted in 1951.[61] Throughout the first decade of its existence, the Los Angeles County Air Pollution Control District was unable to actually fill all of its authorized positions. For example, on October 31, 1955, it had 252 employees and 107 open positions—even though the salary problem was comparatively marginal in California.[62] One official who switched in the mid-1950s from Cleveland to Orange County doubled his salary at a single stroke.[63] To be sure, the federal government did try to ameliorate recruiting problems by generously funding training programs alongside research—in 1965, for example, around $2.3 million was available for this purpose.[64] But in the end, these programs barely kept pace with the growing demand for manpower. In its annual report for 1968, California's Air Resources Board was still complaining that even experts with no practical experience were hard to come by.[65]

Of course, it is difficult to assess the consequences of the recruitment problem. In the end, it remains a matter of speculation whether the authorities would have pursued a more effective policy with better-qualified personnel. In light of the other structural problems of local agencies, which even the most skillful department head would have had to accept, the human factor should not be overemphasized. It is striking, though, that there were hardly any charismatic and widely respected heads of municipal agencies during the 1950s and especially the 1960s. If official reports all too often reflect an uninspired "working by the book," this was not least due to the fact that local authorities tended to be run by second- and third-rate personnel.

The Quest for Rules and Regulations

The broadening of the pollutant agenda invariably brought with it a greater need for technical regulations that provided the officials in charge with guidance on how to deal with the new problems. In this respect, as well, the postwar period saw a clear break with the smoke abatement tradition, since smoke inspectors had been content with fairly simple methods of establishing norms. As a rule, smoke inspectors had used the so-called Ringelmann Scale, a table with various shades of gray that could be compared to the emissions from smokestacks; as soon as a certain shade of gray was exceeded, it was considered a violation of the local ordinance. Although this method was notoriously imprecise, it was generally adopted because it was easy to use.[66] In the postwar period, the Ringelmann Scale was even applied to other pollutant problems, but this was a makeshift solution: other standards were simply not available.[67] At a time when the scientific study of air pollution was just getting started, the search for standards opened up an abyss of ignorance.[68]

When drafting local ordinances, it was still possible to avoid the entire problem by describing tasks in the most general terms. For example, the 1941 Pittsburgh ordinance prohibited "such quantities of soot, cinders, noxious acids, fumes or gases . . . as to cause injury, detriment, or nuisance to any person or to the public"—a formulation that could hardly have been more elastic.[69] Of course, such vagueness merely shifted the problem elsewhere: establishing specific threshold values could no longer be avoided when concrete problems were being confronted. In fact, the problem would have become dramatic if it had not been for industry cooperation. If industry, in other words, had been willing to take measures only where clear standards existed, bureaucratic work would likely have ground to a halt very quickly. Once again, cooperation thus proved an opportunity and a limitation at the same time. Remarkably, representatives of industry were by no means the only people calling for a cautious approach: "Extreme caution should be used at this time in drawing up rules and regulations, particularly regulations placing maximum allowable limits on the amount of a contaminant which could be liberated by an industry or which may be tolerated in the atmosphere," a Public Health Service (PHS) official noted in 1953.[70]

Against this backdrop, there was little hope in taking pollutants' effects as the starting point. To be sure, occupational medicine certainly did have threshold values for the concentration of toxic substances in indoor air, but the general perception was that they were not usable for protection against emissions: the standards of occupational doctors were set for the temporary exposure of healthy male workers—and thus much too high for the purposes of air pollution control, where the young and the sick also had to be considered.[71] Some people proposed deducing threshold values from the nuisance effect of pollutants, but the results were not very convincing. "The question, is, then, whether the dust or odor or other pollutant is sufficient in amount to detract seriously from the amenity of living in that particular community," one such attempt concluded—not a very useful formulation in a specific case.[72] In practice, people simply used common sense—or what passed for common sense—to assess emission problems. "We usually rely on personal experience and opinions of a few of our associates," stated an internal memorandum of the Public Health Service regarding procedures on foul odors in the vicinity of refineries.[73] Since the attempt to derive threshold values from the effects of emissions was therefore bound to fail, given the deficiencies in scientific knowledge at the time, officials had no choice but to take technological possibilities as their starting point. In the end, this amounted to defining problems in such a way that they could be readily eliminated with the available means—an approach that sounded much better in the language of the day: "It is necessary to operate on the basis of accepted engineering practices in evaluating discharges," the Public Health Service declared.[74] An Allegheny County case will serve to highlight the ensuing problems: two large factories created a serious fly-ash problem in their environs—yet they were clearly operating within the given threshold values.[75] But what was the alternative? "To prepare standards, it will require engineering, legal, chemical, meteorological, hygienic and general scientific knowledge to do a complete job," Abraham Wolk explained in a 1955 letter to the attorney general of Pennsylvania.[76] And such a research program was beyond the abilities of chronically understaffed local authorities.

Nor could municipal authorities count on help from supraregional organizations—at least not in the form of technical norms that deserved the label "neutral." For example, the Manufacturing Chemists' Association published an "Air Pollution Abatement Manual" with a great deal of useful information as early as the beginning of the 1950s, but the association's bias was plain.[77] Other organizations were barely any better. One revealing example is a model law created and published by the ASME, which appeared in May 1949 and served as a guideline for many cities.[78] The first and most obvious drawback of this law was that it had been in the making for an exceedingly long period of time: the Model Smoke Law Committee of the ASME had formed on December 4, 1940, and met twice a year for deliberations, even during the war years.[79] Moreover, the model law avoided taking a clear position on the question of whether the coal market had to be regulated in order to solve the domestic furnace problem, clearly a serious omission at a time when the campaigns in St. Louis and Pittsburgh were being widely discussed.[80] Finally, the model law formulated excessively permissive standards when it came to the fly-ash problem. Purification efficiency in new plants was to be a maximum of only 85 percent, in existing installations a mere 75 percent.[81] These values could have been easily attained already in the early days of dust precipitation: even the Klingenberg power plant in Berlin had achieved an 85.2 percent rate of dust collection in the interwar years.[82] In the face of such values, independent experts like Raymond Tucker were left with a sense of powerless disappointment. "It would appear as though they have weakened the dust clause to the point where it will be of little improvement over any that are now in existence," Tucker wrote to the chair of the ASME committee.[83]

And this was not an isolated case. In 1966, when a vote was to be taken in a similar committee on a "Recommended Guide for the Control of Dust Emission," Arthur Stern, as the representative of the Public Health Service, refused to approve it. Among other things, Stern pointed out that the electrostatic precipitators that had been installed in the country's power plants during the previous two years already fulfilled all of the proposed standards. "In my opinion, it makes ASME look foolish to publish as the basis for a standard a figure on which the operative lines are outside the area of real installations," Stern wrote to the chair of the Air Pollution Standards Committee.[84] ASME's reaction was even more revealing: even though the vote produced a clear majority in favor of the threshold values (fifteen to two), the association put pressure on the two dissenters to change their vote or at least abstain.[85] When it came to Stern, however, the ASME was up against an immovable obstacle: "I do not believe that it is in the best interests of ASME to adopt this standard," Stern informed the committee chair.[86] But beyond that, there was little that Stern could do.

It is important to note that such a farce was not simply an accidental result of individual failure. The role industrial representatives played as members of the ASME committees was sure to forestall a tougher approach: the 1940 Model Smoke Law Committee included, among others, a representative of the Pittsburgh utility Duquesne Light. Du Pont even provided the chair of the 1966 Air Pollution Standards Committee.[87] This was, in fact, standard practice: for example, the chair of the Committee on Criteria for Community Air Quality of the American Industrial Hygiene Association came from Eastman Kodak.[88] For an industry intent on having the final say about the possibilities and limits of air pollution control, the process of establishing technical norms furnished an ideal starting point from which to exert effective control. The handling of technical norms thus fits into the general picture of a pseudocorporatist regulatory mode that allowed industrial interests to reign supreme.

The price of this situation was that no routine for dealing with technical norms was established. No agency was capable of providing authoritative clarifications, and no regular way of defining norms emerged—one that would have provided at least procedural backing for threshold values. The definition of technical standards thus invariably became a pawn in negotiations: "In the absence of established facts from which to derive standards, it is perfectly proper to adopt values which represent simply a reasonable compromise between the ideal cleanliness of country air and the high degree of pollution to which some of our cities are now exposed," declared two members of the Industrial Hygiene Foundation. Such an attitude obviously opened up considerable leeway in negotiations.[89] If discussions of technical norms in the 1960s gradually took on the air of a carpet sale, this too was a result of the amicable accord of 1950.

The "Insider Perspective"

It comes as no surprise that the environmentalists of the early 1970s had nothing good to say about the pseudocorporatism of the preceding decades. It is not without irony, however, that these environmentalists at the same time complimented officials, though unintentionally. More than once, they plunged into existing procedures with the goal of proving that authorities had been corrupted by industry—only to come up empty-handed. For example, the *New York Times* examined the state of bureaucratic activities in late 1970. Even though there were thirty-five states at the time where industrialists and other interests belonged to the bodies charged with overseeing air pollution–control efforts, the *Times* was unable to name a single case in which the decision not to adopt a particular measure could be traced back to the representative of some interest group—a finding that evidently did not satisfy the author of the article. The reporter thus decided to declare industry cooperation scandalous in and of itself, titling his piece "Polluters Sit on Antipollution Boards." Over several columns, he proceeded to report, with an undertone of outrage, which companies had their representatives in which agency.[90] It was only toward the end of the article that the author conceded that this massive industrial presence seemingly had no effect at all: "The few states that have panels composed of engineers, professors, pharmacists, housewives and other disinterested citizens . . . give every evidence of getting along just as well as boards with members from the pollution sector."[91] The great corrupting of air pollution control, which this article seemed at first glance to document, was simply nonexistent.

At the same time, however, ample evidence exists for a much more subtle kind of corruption. For while an intensive working relationship developed between the authorities and industry, growing alienation pervaded the relationship between officials and the public, little remaining of the reciprocity that had characterized the Progressive Era. To be sure, this was not only the fault of the officials: since public discontent in the postwar years was for the most part more vociferous than it was organized, it was not easy for officials to develop a viable relationship with representatives of the public. At a 1963 meeting in upstate New York, the state air pollution commissioner, Arthur Benline, was delighted to see some new faces: "Most meetings of this kind are attended by the same people—mostly air pollution control officials who find themselves talking to other air pollution control officials."[92] Still, the dominant impression is of officials' notable lack of interest in an active citizenry: it might have been nice to know that one enjoyed the public's support, but few sought to reach out to civic leagues; contacts with industry were usually seen as far more important. In any case, it hardly seems to have occurred to officials in the postwar years to build up a separate power base by communicating intensively with affected citizens.

This attitude probably found its clearest expression in the institution of "Cleaner Air Week," which was called into being by the Smoke Abatement League of Cincinnati and, thanks to support from the Air Pollution Control Association, quickly established itself as a nationwide event.[93] The very nature of this event was revealing: a carefully planned-out "action" week was not exactly the best way of starting a conversation. For the most part, the public functioned here merely as a mass of interested but ignorant laypersons, who had to be informed of their own duties and the wonderful work of the authorities. A typical example is the official program for the 1960 Cleaner Air Week. When it described the tasks of the common citizen, the program consistently highlighted his or her obligation to take certain measures. For example, on "Home Heating Day," furnaces were to be checked for proper operation, while on "Automobile Day," people were to watch out for vehicles trailing a cloud of soot from their tailpipes. When it came to industry, by contrast, a very different aspect was front and center, the pamphlet recommending that participants "arrange industry tours for air pollution control progress of your community plants." Officials evidently considered it unnecessary to admonish industry within the event's framework.[94] The 1966 program was even more blunt: "The entire drive is toward proving to the individual citizen that he has a large stake in the community's program to attain cleaner air and convincing him that he must give that program wholehearted support."[95] As a result, cooperative rhetoric came to dominate Cleaner Air Week declarations. "It is our goal to promote sound cooperation among public officials, representatives of industry, and interested citizens," emphasized the chair of the Cleaner Air Week Committee in Cincinnati in 1969.[96] Without any apparent qualms, the interests of the public were thus equated with those of the regulatory establishment—even though an ever-deeper chasm was opening up in the 1960s, at the latest. And where this chasm could eventually lead was made clear in Pittsburgh in 1969, when the president of U.S. Steel was awarded the Good Neighbor Award during Cleaner Air Week, only months before environmentalists launched a fierce attack on him.[97]

To be sure, the dominance of the regulatory establishment during Cleaner Air Week had limits. Especially toward the end of the 1960s, Cleaner Air Week was increasingly shaped in a number of cities by individuals who were part of the budding environmental movement.[98] But even at that time, events were often still characterized by a friendliness toward industry that was at times so blatant as to be almost touching: first prize for participants in the 1968 poster and essay contest of Pittsburgh's Cleaner Air Week was two shares of PPG Industries.[99] "The public needs to know that industry is making strides toward cleaner air," declared an industrialist at Chicago's Cleaner Air Week in 1964, meaning this as a plea for a better-informed public.[100] An event that seemingly solicited the cooperation of all parties thus became de facto a PR initiative for the regulatory model of 1950; the regulatory establishment's perspective usually came across as the

Figure 8. Cartoon from the 1956 annual report of the Allegheny County Bureau of Smoke Control. In earlier times, air pollution–control officials had emphasized the crucial role of active citizens, rather than poking fun at their presumed ignorance. Image courtesy of the Allegheny County Health Department.

only rational—indeed the only possible—perspective. It was thus perhaps no coincidence that the participation of women reached a historic low during these years. There was, of course, the occasional exception. In Pittsburgh, for example, Jean Nickeson became chair of the Cleaner Air Week Committee, with "splendid cooperation from the Public Relations department of United States Steel."[101] In general, though, the organization of Cleaner Air Week seems to have been firmly in male hands, for only males belonged to the regulatory establishment. And the fact that in some cities younger women could compete for the title of "Miss Cleaner Air" did not necessarily improve the general picture.[102]

If there was any communication at all between officials and the public in the postwar period, it was usually one-way: officials informed interested citizens about the details of their measures and then expected, as a matter of course, that the public would respond with appreciative applause. For example, Dayton's Bureau of Combustion Control (as the municipal agency for air pollution control in this Ohio city was still revealingly named in 1959) cultivated its contact with the public chiefly through lectures and a weekly radio show, and cities like Chicago and Denver even had separate departments for Public Information and Education in the 1960s.[103] But efforts of this kind often went along with strikingly insensitive, previously unexhibited behavior. Sometimes language itself gave officials away: Robert Griebling, for example, attributed what he regarded as an exaggerated complaint to the work of "certain emotionally unstable residents."[104] Similarly, the Allegheny County Bureau of Smoke Control set a revealing cartoon at the beginning of its annual report in 1956: it showed two smiling housewives talking about how well-suited the grill was for burning the household garbage—while a pall of thick smoke spread over the neighborhood.[105] And when the electrostatic precipitators in a Pittsburgh steel plant had to be shut off for seven days in October 1960 because of work on a high-voltage electric power line, as a result of which an enormous amount of dust was dispersed over the neighborhood, officials saw residents' protests as evidence of their own success. After all, didn't the angry protests demonstrate how well the neighbors were usually protected against emissions?[106] A word of regret or even apology for such unavoidable suffering is nowhere to be found in officials' statements.

To be sure, the alienation between officials and the public should not be exaggerated. On the whole, complaints from those affected by air pollution were taken seriously, and officials did not deny their public mandate. "It is only for the benefit of the people that the control of air pollution is inaugurated," declared Cleveland's Herbert Dyktor.[107] Rather, what developed in the minds of officials was a mentality that Joseph Sax would attack in 1971 as evidencing an "insider perspective": a way of thinking shaped by internal bureaucratic considerations, which often seemed quite irrational from the public's point of view.[108] While the authorities cooperated intensively with industry and often more or less adopted its way of seeing things, they had no direct line to the public and thus often no understanding of the sensitivities of technical laypersons. "All too often, we have tended to take a view of air pollution control that is bounded by narrow partisan interests," Vernon MacKenzie, chief of the PHS Division of Air Pollution, declared to a group of engineers in 1966.[109] This left officials very badly prepared for the process of transformation that public opinion underwent in the course of the 1960s: they lacked not only channels of information but also ways of integrating these new perspectives institutionally. It was therefore perfectly logical that the new, ecological definition of the problem was largely displaced into a realm that lay outside the regulatory establishment. The split of the air pollution discourse into "insiders" and "outsiders" was the logical result of what had initially been a productive alliance between the authorities and industry—a logical result that, in the long run, would prove disastrous.

The Unspeakable Crisis

All in all, there are a host of good reasons to speak of a crisis of municipal air pollution control as early as the 1950s. Local agencies were generally understaffed; they had considerable problems attracting qualified personnel; they were slow to react to the broadening of their range of tasks, smoke inspectors lacking the necessary expertise. What had long been an intensive relationship between officials and the public was now replaced with a state of growing alienation. The lack of useable threshold values forced officials to take a hands-on approach to problems, the crucial factor not the effects of pollutants but the available technological options. Finally, suburbanization was constantly eroding the official's sphere of action—not to mention the situation in smaller towns and unincorporated areas, where local industrialists could often decide the fate of air pollution control as they saw fit. Even though the accomplishments of pseudocorporatist air pollution control should not be underestimated, it was clear that under these conditions, official efforts necessarily fell behind the technological and economic possibilities. After all, these were not simply the kinds of temporary problems expected during any period of rapid change; rather, these were structural, built-in defects that would not simply resolve themselves over time. This was true even for the problem of threshold values: although there was reason to believe that cognitive uncertainty would shrink as scientific research advanced, the lack of established procedures for setting threshold values was clearly an enduring burden. The future of pseudocorporatist air pollution control was therefore anything but rosy.

What made these problems truly explosive, however, was the fact that contemporaries could discuss them only to a very limited extent. The striking absence of public debate about the structural problems of municipal air pollution control certainly had a logic of its own: municipal officials, who made up the majority of experts in the postwar period, had good reasons to fear debate over reform. After all, a municipal official who pointed all too insistently at the problem of unregulated enterprises in the suburbs was indirectly arguing for the abolition of his own agency. The situation was similar with regards to recruitment: officials who were vehement in complaining about the lack of qualified personnel ultimately discredited their own profession. In any case, the smoke inspectors' timid response to the broadening of the pollutant agenda was a rearguard action: they did not wish simply to stand by and observe the erosion of their professional jurisdiction. Other problems were by their very nature not amenable to constructive debate. The growing alienation between officials and the public was grounded in an insider perspective of which officials were for the most part unaware. And there was no way to talk about emission problems in small industrial cities—not in the cities themselves, where industrial enterprises could act like they owned the place, and not in the higher agencies, which usually had problems of their own. Municipal officials could have begun a debate about the definition of threshold values, but they were short on arguments for doing so: if local industrialists had demonstrated their willingness to cooperate constructively with officials, on what grounds could they now be blocked from having a hand in defining the technical norms? From the perspective of municipal officials, therefore, there were good reasons to dispense with an open discussion—especially since local industrialists, on whose cooperation officials depended, could hardly be interested in such a discussion. The result was a discursive situation that could hardly have been any more critical. It was increasingly apparent that the time of municipal air pollution control was past—but from the inside perspective of officials, there were no genuinely serious problems. The prospects for incremental progress, whereby a more productive regulatory approach would gradually replace pseudocorporatism, were therefore exceedingly bleak. It would appear that, on its own, municipal air pollution control would hardly be able to resolve its structural problems.

Reforming Air Pollution Control in Germany, 1952–70

It is generally not easy to pinpoint the precise beginnings of a strategy debate. For the most part, such debates begin modestly, with small corrections to established procedures or scattered complaints that condense into a real debate only over time and finally lead to reforms. In this regard, the reform debate in Germany in the 1950s was the exception to the rule: it began abruptly in 1952 when the director of the Ruhr Area Federation for Regional Planning, Sturm Kegel, presented his concept of air pollution control cooperatives (*Luftreinhaltegenossenschaften*). Kegel's proposal immediately triggered a lively debate, and it is possible to draw a straight line between that debate and the reforms that were eventually implemented. Therefore, it is hard to exaggerate the importance of the concept of cooperatives to the discussion in the early 1950s: In the beginning, there was Sturm Kegel.

In fact, Kegel's concept was from the very outset not simply an idea, but a detailed legal draft with thirty-eight paragraphs.[110] At the heart of the proposal was the formation of special cooperatives for air pollution control in the industrial regions of North Rhine–Westphalia. In a given region, these cooperatives were to bring together industrial enterprises, local municipalities, and the "local regional planning community"—in the Ruhr region it was Kegel's Ruhr Area Federation—in order to "jointly implement measures that are necessary and suitable according to the current state of technology to prevent as best as possible the polluting of the air in the area of the cooperative."[111] The organs of the cooperatives were to be a general meeting and a seven-man board, which had to include at least two representatives each from industry, the municipalities, and the planning community; in addition, a special appeals board was to resolve disagreements.[112] However, while the formal organization of a cooperative was addressed in exhaustive detail, down to its dissolution, the draft's guidelines on finances were quite open to interpretation: contributions from those members who were themselves adding to air pollution should be determined "in accordance with the degree (amount and nature) of their emissions in the area of the cooperative"; the other members should be "assessed according to standards determined by the board."[113] In addition, "expenditures made by individual members to attain the goals of the cooperative must be adequately considered in their favor." Finally, there was obligatory reimbursement for air pollution–control installations that already existed when the cooperative was set up, although this reimbursement must not "exceed the costs that would have accrued to the cooperative from setting up installations that would do the same."[114] How these general regulations were to be put into practice was something that Kegel believed the members should figure out among themselves.[115]

The most remarkable thing about Kegel's proposal was that it influenced debates for more than a decade, even though none of the groups involved could muster any real enthusiasm for it. It has already been noted that the Ministry of Labor was anything but thrilled about this proposal; after all, the cooperatives were intruding on its administrative turf.[116] The Interparliamentary Working Group, too, could not warm to the concept, even though its members were in principle sympathetic to a reform of the regulations.[117] Technical experts were also skeptical: Karl Schwarz at the TÜV Essen considered the proposal "technically not feasible" for the time being; Alfred Löbner declared on behalf of WaBoLu that the current path had proved "highly successful" and that "a new organization is therefore not necessary"; and Robert Meldau, speaking for the Expert Committee on Dust Technology, also expressed reservations, declaring that the Expert Committee could "only represent something that is solidly backed up technically, is reasonable, and holds promise for the future."[118] And opposition from industry was certainly not hard to miss. In February 1955, a statement from the

Association of German Industry identified "problems with the assessment of contributions"; criticized "overorganization and work duplication by air pollution control cooperatives"; and registered "concerns about setting compulsory contributions," as well as "concerns with respect to constitutional law."[119] Not even those harmed by emissions backed Kegel's proposal. "Participation by agriculture and forestry is undoubtedly too small in the current draft," the North Rhine–Westphalian Ministry of Agriculture commented, while the city manager of Mülheim an der Ruhr saw in Kegel's proposal merely "a new and substantial burden for the communities of the Rhenish-Westphalian industrial region," which would be compelled to contribute even though they were "not responsible for the pollution of the air from dust and ash."[120] And if papers like the *Westdeutsche Allgemeine* took Kegel's side, they probably did so primarily because his draft bill signaled that at least something was happening.[121]

In retrospect, it is in fact hard to deny that there was some truth to the remarks of these contemporary critics. The draft law did show considerable flaws in several respects, flaws that would have proved obstacles to implementation under the best of circumstances and fatal under the worst. The most obvious was Kegel's uncertainty over how to deal with technical expertise, documented in several proposals that were presented in various forms and in quick succession.[122] In addition, Kegel said little about how to assess financial contributions. If one considers the unhappiness of many industrialists about growing public pressure, and the concern of many cities in the Ruhr region about additional expenditures, it is easy to imagine that negotiations over the distribution of costs would have quickly degenerated into a hopeless melee. At any rate, setting contribution rates would have required extensive measurements and complicated calculations, and cooperatives would have presumably wasted a good deal of their time on unproductive administrative tasks.[123] Moreover, it was by no means certain that the communities would in fact have formed the intended counterweight to the polluting industries, because as the operators of municipal enterprises they also represented the emitters' side. Furthermore, there was the open question of whether existing institutions like licensing procedures and oversight through factory inspectors should continue. Finally, it was not even clear that Kegel was seeking a quick solution to pressing problems, for he readily conceded, as early as November 1952, that it would probably take the cooperatives "many decades to carry out their work."[124] In all likelihood, the cooperatives would have created more problems than they would have solved.

It is also worth noting that Kegel's penchant for this particular organizational form had little to do with the exigencies of effective air pollution control. Kegel's motivations became apparent, for example, when the conservative CDU party's Committee on Municipal Politics, at Kegel's urging, wrote to "important members of the CDU party group in the state assembly," pointing out "the political importance of the bill with regard to the so-called subsidiarity principle."[125] What Kegel merely hinted at here was openly expressed in an essay by Fritz Rabeneick, who was among the few who explicitly defended Kegel's idea of cooperatives.[126] According to the principle of subsidiarity, Rabeneick explained, "the highest state authority should not arrogate those tasks that could be carried out better, or at least equally as well, by subordinate communities." He went on to say that "this recognized principle of any liberal order probably received its most trenchant formulation in Christian social philosophy"—and then referred baldly to the papal social encyclical "Quadragesimo Anno."[127] Since Kegel himself never spoke quite so frankly about what inspired him, the question about his motives must remain open. It is conceivable that other cooperative models, like that overseeing German industrial safety, figured into Kegel's thinking, although the surviving documents offer no direct proof. In any case, there are clear signs that the question of how to most rapidly clean up the air in the Ruhr region was not paramount in Kegel's thinking. In an essay published in the CDU's journal on municipal politics, Kegel called for a "change in the mental attitude of all those involved" and a "new basic attitude" that had to proceed from "the common good."[128] Against this backdrop, the problem itself seemed, in the end, almost arbitrary.

It seems doubtful in retrospect whether Kegel's concept of cooperatives had a realistic chance under these conditions. The best argument for Kegel's idea was, in the end, not its specifics, but the condition of the air in the Ruhr region. Since the result of the bureaucratic efforts was widely regarded as unsatisfactory, the concept of air pollution–control cooperatives at least stirred a vague hope for change. But when it became quickly apparent that there was broad support for certain reforms to existing approaches, this argument also became less persuasive. As early as April 1955, Arnold Heller declared, with the ring of conviction, that "this bill will hardly have any chance of passage."[129] By that time, it must have been clear to Sturm Kegel himself that the prospects for his proposal were fading; at any rate, his supplementing his proposal in June 1955 with the draft of a federal law, even though he had far fewer connections on the federal level than he did on the state level, certainly seems like an act of desperation.[130] The definitive end to any chance of its realization can be dated to around the end of 1956, when the members of the interministerial Clean Air Working Group in the state of North Rhine–Westphalia declared unanimously that they regarded "the bill proposed by Kegel as not feasible."[131] When the CDU member of parliament Karl von Buchka, during a parliamentary debate on January 1, 1957, remarked, regarding Kegel's concept, "that doubts have arisen as to whether this is the right approach when it comes to the air," it was, in the final analysis, nothing more than a first-rate funeral.[132]

The ministerial officials in North Rhine–Westphalia must therefore have been surprised when Kegel suddenly began to tout his proposal again around 1960.[133] By this time, important steps had been taken at the federal level, with the establishment of the VDI Clean Air Commission and the 1959 passage of a law to change the Industrial Code. Moreover, the Ministry of Labor, after lengthy preliminary work, had drafted a state clean air act that promised to close the remaining legal loopholes.[134] Officials at the Ministry of Labor, however, could no longer simply brush the proposal aside. Formally, it no longer came from Kegel himself, but from the CDU's Committee on Municipal Politics, which invoked "a mandate from state prime minister Dr. Meyers to present a bill about air pollution control associations."[135] In the resulting dilemma—unable to reject the concept of cooperatives but not wishing to support it—the Ministry of Labor opted for delaying tactics. A May 1961 memo pointed the way: although "a few basic ideas of the bill and the elaboration of details concerning procedural laws are certainly worth considering," ministerial officials simultaneously pointed out that "extensive (possibly many years of) deliberations about the tasks of self-governing bodies in the area of air pollution control are likely to be required."[136] The Ministry of Labor was therefore tireless in pointing out that the cooperatives concept still contained many unresolved problems and was thus on the whole not ready for passage: the state clean air act should be enacted first—and this, of course, would "in no way prejudge" Kegel's proposal.[137] In the end, officials found this approach successful: eventually, even Kegel himself conceded that passage of the state clean air act should not be delayed unnecessarily.[138] Although Kegel continued to solicit support for his bill, he was now fighting a losing battle.[139] The critique of the proposal's details assiduously put out by the Ministry of Labor was so effective in the end that even members of the CDU party group in the state assembly raised concerns about Kegel's bill.[140] When the decision was made in favor of the state clean air act, a de facto decision was simultaneously made in favor of implementation by existing agencies, and the cooperatives concept was laid to rest once and for all.

But even if Kegel eventually failed with his cooperatives concept, he undoubtedly rendered a great service to the cause of air pollution control—though very differently than he had intended. The 1952 bill had the productive side effect of pressuring officials, industry, and scientific experts into taking action. The debate about Kegel's bill functioned like a litmus test of the regulatory tradition: since few dared to deny that Kegel was responding to a very real problem, it was obvious that hardly anyone still considered the established way of doing things sufficient. Even though Kegel's cooperatives concept must be judged in retrospect as a dead end—an idea that would presumably have hurt emission control more than it helped—his activism ensured that reform of air pollution control was being discussed in North Rhine–Westphalia as early as 1952.

The Reforms of the 1950s

Sturm Kegel's initiative thus acted as an unexpected wake-up call for North Rhine–Westphalia's officials. They had been dealing with emission problems for decades on a case-by-case basis, without any meaningful challenges, but this bureaucratic practice now had to justify itself overnight. But no matter how much officials tried to expose the weaknesses in Kegel's concept, there was no indication that they were coming up with a creative counterproposal in the process.[141] When officials attacked the cooperatives concept, they also implied a general critique of fundamental change. From the outset, their efforts were aimed merely at improving, through certain modifications, the existing system of air pollution control to the point where it would be up to the new challenges. When an official from the federal Ministry of Labor noted "general concerns" about a change to the licensing procedures—"there is no way to predict what the repercussions of such a step might be"—his words were a fitting expression of broader bureaucratic sentiment toward reform.[142] In spite of the challenge Sturm Kegel had laid down, officials were still thinking along traditional lines, and they never sought an open debate over regulatory strategies. It was thus soon possible to achieve a consensus within the administration about the necessary next steps: without much discussion, an "agreement" was reached at an administrative conference on occupational safety in June 1955: "a revision of §16 of the Industrial Code and the Technical Directives should be accelerated and undertaken under the aegis of the Federal Ministry of Labor." Participants hoped for access to "notes about the last deliberations of the Technical Deputation" during World War II, which had supposedly been in the possession of Wilhelm Liesegang, who had died in 1953. Subsequent work should then be "taken on by working committees that will be formed for the various specialized areas from experienced factory inspection officials and experts from industry and air pollution–control practice."[143]

It is tempting to speculate what would have happened if an actual attempt had been made to implement this resolution: the draft for new technical directives, found in the possession of Liesegang's widow, proved exceedingly meager. A memo from the federal Ministry of Labor noted laconically as late as the early 1960s that, "given the current manpower" in the relevant departments, the ministry was "in no position to come up with the general administrative regulations without help from other agencies." The ministry's attempt to develop new technical directives on its own could well have turned into yet another example of bureaucratic mismanagement in the history of German air pollution control.[144] It was thus probably fortunate that the federal government quickly lost the initiative. At the beginning of May 1955, Sturm Kegel was scheduled to lecture at the plenary meeting of the Interparliamentary Working Group; Robert Meldau, as the chair of the Expert Committee on Dust Technology, and Heinrich Grünewald, the director of the VDI, were also invited. Although there was no time after Kegel's lecture for thorough discussion, the participants at least knew each other now.[145] At the same time, Karl Mommer, an SPD member of parliament, was preparing a proposal for deliberation in the Bundestag, which the daily *Die Welt* reported about in a short article on May 28.[146] This article gained the VDI's attention: Mommer's proposal to appoint "an independent commission of experts" amounted to direct competition for the Expert Committee on Dust Technology; moreover, professional self-interest made the VDI opposed to any politicization of air pollution control.[147] Within a week, Grünewald and the secretary general of the Interparliamentary Working Group, Wolfgang Burhenne, arranged a meeting. While the VDI ran into a brick wall with Burhenne regarding its desire to rule out changes to the law in principle, the "idea of personal responsibility" seemed to hold some appeal for the IPA; at the end of the meeting, the VDI was asked to present "a draft of about three typed pages with positive proposals."[148] The ensuing report, discussed at a second meeting between Burhenne and Grünewald on June 14, was formally a proposal from the VDI, but it clearly bore the marks of a compromise offer, as when it noted, with emphatic caution, that "it must be reviewed . . . whether a new legal foundation must be created."[149] The question of a change to the law was thus postponed for the time being, the memorandum focusing entirely on the technical side of the issue: "As an important prerequisite for air pollution control there must exist technical rules and guidelines that are objective, adequate, and universally applicable, so as to assure adherence to them in individual cases," the VDI explained.[150] The Expert Committee on Dust Technology should thus take on the task— joined by experts "in voluntary, personal responsibility"—of developing technical norms, which would then be "obligatory for all parties involved in the interest of keeping the air clean."[151] And this is what happened: in a special meeting on July 5, 1955, the IPA described "the draft of technical rules and guidelines of the VDI Expert Committee on Dust Technology" as "an urgent matter and one worthy of support in every way," declaring the reform of legal regulations a "task of the secondary level of urgency."[152] That same month, a Guideline Committee was established within the Expert Committee on Dust Technology, with the mandate of "drafting suitable guidelines within a year."[153] In a supplemental move, the VDI, in November of the same year, set up a Clean Air Committee within the Expert Committee on Dust Technology. In March 1957, this committee became the separate VDI Clean Air Commission.[154]

By its very nature, the VDI Clean Air Commission was thus initially nothing more than a nonbinding suggestion. Neither the government nor parliament had officially asked the VDI to draw up technical norms, and the IPA, too, had not entirely given up on a legal initiative. For now the parties were simply going to wait and see what the agreement would yield; Meldau's warning at a board meeting of the Expert Committee on Dust Technology in June 1955, that it was imperative to prevent the committee from being "seen as the extended arm of industry," clearly mirrored a sense that it was being watched closely.[155] "The question about the independence of the VDI from industry ... was briefly touched on," the Ministry of Labor of North Rhine–Westphalia noted following a meeting with representatives of the federal Ministry of Labor in July 1955.[156] Ministry officials became even more suspicious when the VDI began planning to set up a working group on "publication issues" with the goal of "countering the public's mistrust on questions of keeping the air clean and, through objective information, creating a sphere of trust between emitters and those affected."[157] An internal ministry letter documented "the sense that the VDI is exploiting the entire issue of its mandate ... to influence the public in a one-sided way."[158] The president of the Federal Health Agency (Bundesgesundheitsamt) even responded to an invitation to attend the founding meeting of the VDI Clean Air Commission with the snippy comment that while it was "the perfect right of every citizen" to propose a bill, scientific bodies in general should merely "concentrate on research ... whereupon the state, with its agencies and auxiliary institutions, in conformity with the provisions of the constitution, condenses into laws the imperatives to be derived from what has been discovered."[159] The initial skepticism of contemporaries makes perfect sense in retrospect: for example, the process by which norms were established in the American Society of Mechanical Engineers, the U.S. equivalent to the VDI, makes it clear that the danger of drawing up lopsided threshold values was very real.

In truth, such concerns eventually proved unfounded. From the very beginning, hegemonic claims by industry representatives were consistently rebuffed within the VDI Clean Air Commission. When a representative of the BDI, at a meeting of the commission, boldly demanded that "future proposals [from the Expert Committee on Dust Technology] be discussed with him," since this question, after all, "had not only a technical but to a quite considerable extent also an economic side," the response from the members of the commission was cool. Grünewald ostensibly called for "a parity-based composition" of the working committees and declared the work to be "to achieve agreements and not impose regulations on anyone." Another member followed up by remarking that above all else, the VDI had to be recognized "as a neutral body," and to that end it had to submit "well-thought-out findings."[160] The suggestion "to have the findings of the committee worked over by lawyers," and to use for this purpose chiefly the "circle of jurists of the BDI," was likewise brushed aside with the simple comment that "the work of the committee must be objective."[161] And a representative of the nitrogen industry, who declared at one meeting that "it is necessary to set up barricades against agriculture, which was stepping forward in even greater numbers with claims for emission damage," was immediately reprimanded by the chair of the VDI commission, Heinrich Lent, who admonished him "for a breach against objectivity"—even though Lent, as VGK chair, was certainly no partisan of farmers.[162] Lent was therefore naive only at first glance when he boldly proclaimed that "the members of all committees are appointed not as lobbyists for the individual sectors of industry, but as independent experts."[163] The constant invocation of "teamwork" established a rather constrictive framework for the commission's internal debates: anyone who did not argue the facts merely isolated himself.[164] Obstructionism and blatant lobbying therefore had rather a limited chance in the VDI Clean Air Commission. Helmut Berge, whom the BDI steered into a working group of the commission, became an example of where the limits of acceptability lay. When he launched a polemical attack against a large experiment on the sulfur dioxide problem in Biersdorf in the Siegerland, calling instead for extensive studies throughout Germany, which would have delayed addressing the issue by years, the committee opposed him unanimously. Berge, humiliated and stripped of influence, subsequently withdrew from the working group.[165]

The tradition of "teamwork" thus ensured the VDI Clean Air Commission a much greater degree of independence compared to the corresponding committees of the ASME. And yet in retrospect there is reason to doubt that this alone would have prevented the commission's work from being infiltrated by lobbying efforts, had it not simultaneously come under outside pressure. Although the VDI's guidelines rapidly became the basis for the work of federal bureaucracies, in and of themselves they as yet had no legal relevance. The States' Committee for Pollution Control (Länderausschuß für Immissionsschutz), founded in 1964, which brought together the state ministries responsible for emissions, noted that the guidelines "do not, in legal terms, contain a binding and definitive definition of the state of technology in the sense of §§16/25 of the Industrial Code, but are merely statements by expert agencies about the state of technology."[166] In other words, every VDI guideline was thoroughly reviewed yet again by experts in the administration before it was given the force of law; if it did not meet with approval, it was simply not adopted. For example, the Technical Directives of 1964, substantially based on the VDI Commission's work, laid down an emission threshold value of 0.4 mg sulfur dioxide per cubic meter, despite the fact that a previous VDI guideline had set 0.5 mg as the upper limit.[167] Even vigorous opposition from industrial circles was unable to change this.[168] The contrast to the American situation is patent: while Arthur Stern, as the representative of the Public Health Service, could do no more than register an unavailing protest when the ASME passed its lopsided threshold values, German agencies had reserved for themselves a kind of veto right. The VDI Clean Air Commission therefore had good reason, out of pure self-interest, to fear creeping corruption of its work: as soon as one of its guidelines failed bureaucratic review, it ceased to have any practical relevance, and the VDI Commission was made to look bad.

Of course, all this does not mean that the work of the VDI Clean Air Commission was beyond criticism. Its first guidelines for dust emission, still substantially based on the earlier work of the Association of High-Performance Steam Boiler Owners and the Association of German Cement Plants, drew especially vocal criticism from various authorities.[169] Even the official annual report released by North Rhine–Westphalia's factory inspection noted on one occasion that "because of already existing high levels of dust and fumes in the industrial areas, requirements had to be imposed that went far beyond the upper level of the VDI guidelines."[170] Still, compared to the work that was being done in the United States at the same time, there was a qualitative difference. While the fight against sulfur dioxide emissions in the United States had been put on the back burner for the time being, this issue was quickly placed on the German agenda: "The most severe issue . . . is the sulfur dioxide problem," Lent declared as early as 1957.[171] Furthermore, the VDI Clean Air Commission did not evidence the kind of complacency that was rampant within proindustry circles in the United States: "What has so far been done and could be done is only a small beginning," Lent modestly admitted in 1962.[172] And since the work of the VDI commission was intended to continue indefinitely, there was now an institution that was able to keep up with progress in technology; the fate of the 1895 Technical Directives, which were not updated after the turn of the century and were therefore soon regarded as "hopelessly outdated," had thus been preempted.[173]

Incidentally, the VDI Clean Air Commission also presented a good solution from a financial point of view: in December 1967, Karl Schäff, the chair of the VDI, calculated in a meeting with Federal Minister of Health Käte Strobel that it took net expenditures of about 70,000 DM to come up with one guideline.[174] "If the government wanted to have this work on guidelines done by contracting it out, the expense would be several times the figure I have given," Schäff argued, and he was presumably right.[175] Last but not least, the VDI commission also proved to be a very flexible organization capable of learning: only a few weeks after a spectacular smog episode in the Ruhr region in December 1962, the VDI commission established a subcommittee on "Clean Air Warning Systems."[176] And this case was unexceptional. From the very beginning, the VDI Clean Air Commission, like the Expert Committee on Dust Technology before it, was open to representatives from various scientific disciplines; physicians and veterinarians worked there alongside engineers and representatives from polluting industries; agriculture and forestry could send experts as easily as the top municipal associations.[177] And after the 1964 introduction of the so-called green-print procedure, which called for the publication of a draft version followed by a generous period in which objections could be raised, it is difficult to see how the process by which the VDI arrived at threshold values could have been any more transparent.[178]

As a result, a lack of technical norms was probably the least of all worries for officials in the field. The only problem was that implementation of these norms still faced the obstacle presented by provisions of the Industrial Code, which prohibited any additions to an operating license once it had been granted. This issue was addressed by a bill from the Interparliamentary Working Group, introduced in the federal parliament in March 1958. Essentially, its goal was to make it obligatory for all enterprises to keep up with the current state of technology when it came to cleaning flue gases, regardless of when they had been licensed.[179] Such a demand was not particularly controversial: as early as December 1956, officials of the federal Ministry of Labor had arrived at an agreement with their North Rhine–Westphalian colleagues to pursue such a change in the law.[180] As a result, deliberations within the administration were rather unspectacular.[181] The bill was passed shortly before Christmas 1959 without substantial changes as the "Law on the Change of the Industrial Code and Supplementation of the Civil Code"; the remaining legal loophole for enterprises that were not subject to mandatory licensing was left to state legislation and closed with the North Rhine–Westphalian clean air act of 1962.[182] The Interparliamentary Working Group thus saw its mission fulfilled: when Chancellor Adenauer, in a speech to the Bonn Civic Association (Bonner Bürgerverein) in November 1960, declared with respect to air, water, and noise pollution that "the enactment of laws is necessary to protect people against these unpleasant side effects of technological progress" (incidentally a formulation that, according to Adenauer's personal advisor Franz Josef Bach, the chancellor had personally inserted), the IPA chair responded in a letter that "the legal work is all but completed." It was now "up to the bureaucracy to actually implement these directives."[183] The reforms that Adenauer had wanted to push for had, unexpectedly, already been achieved.

A Bureaucratic Revolution

To be sure, the reforms of the 1950s were rather unimpressive on paper. The "Law on the Change of the Industrial Code and Supplementation of the Civil Code" radiated little charm in its cumbersome name alone, nor did the drafting of new technical directives by the VDI Clean Air Commission seem like a revolutionary step: earlier technical directives, published in 1875 and 1895, had had no lasting impact on political debate.[184] The 1950s reforms would presumably have petered out without significant effect, had they not been something like the tip of the iceberg—the legislative expression of a general change in climate consciously promoted by the ministerial bureaucracy. "The time of a certain restraint on emission questions is over once and for all," Heinrich Lent proclaimed in 1956, putting the issue in a nutshell: officials were now getting serious.[185] It was a quiet, bureaucratic revolution, but one that was glowing with resolve.

It was therefore not loud fanfares or emphatic proclamations that signaled the new mood. Instead, the 1950s reforms were aimed at those inconspicuous regulating screws that are for the most part known only to insiders, but that in the end have a much greater role in determining the success or failure of policy than do warm and fuzzy words. The simple fact that specific, science-based directives were now the order of the day implied a considerable leap forward. Before, authorities had often made do with general directives that "the necessary provisions must be made so that no substantial disadvantages, dangers, or nuisances arise for the owners or residents of neighboring properties or the general public"—a formulation that officials could interpret very much at their discretion.[186] Also new was the clear primacy of emission avoidance. Previously, the response had been much more pragmatic, officials opting for financial compensation in cases where cleaning installations met with strong resistance; in smelter-smoke regions, these compensation payments all but had the status of customary law.[187] "The basic idea is to keep the air as clean as possible and not to maintain it as polluted as permitted," is how the factory inspector in Itzehoe phrased the new line of reasoning.[188] In order to have reliable information about the condition of the atmosphere, officials also initiated extensive measurement programs, something that had not been done previously.[189]

In fact, financial means on an entirely new scale were now available for air pollution control. Existing gaps in knowledge were to be filled through scientific research, and both the federal government and the states were willing to pay for it. By April 1, 1962, about 4.32 million DM from public funds had flowed into just the research projects overseen by the VDI Clean Air Commission, and by 1967 that sum had risen to 12 million DM.[190] At the same time, however, the need for further research provided no excuse for delaying concrete measures; instead, circumstantial evidence was now considered sufficient to establish threshold values. For example, while no meaningful steps against automobile exhaust were taken in the first third of the twentieth century, because researchers were unable to provide unambiguous proof that they were a threat to health, a different reasoning now prevailed.[191] As one scientist explained in the journal *Gesundheits-Ingenieur* in 1965, "In the interest of the general public, even the suspicion of a threat to health, one that has sufficient scientific justification, must not be neglected when it comes to fighting the causes of air pollution."[192] Moreover, there was now unanimous agreement that safety margins had to be included when setting threshold values; even the BDI did not question the fundamental legitimacy of such a "safety surcharge."[193] With all that, it was perfectly appropriate that the German clean air community began to talk about a "precautionary principle" in the mid-1960s, for this was essentially the operating principle that the more advanced professionals were already using in their daily work.[194] However, the phrase became part of common parlance only after the air pollution charta of the Council of Europe declared the "precautionary principle" the general guideline of air pollution control in March 1968.[195] In the case of Germany, however, this air pollution charta had more of a legitimizing than a reformist effect: it is interesting to note that Germany had already implemented virtually all the concrete steps that were mentioned in this resolution. German officials, who had played a key role in drafting the resolution, had done their homework.

Of course, these demands were not only to be formulated at the ministerial level and taken into account when setting threshold values but also carefully implemented by the subordinate agencies; for now, however, no one gave this issue much thought. The secondary importance accorded the question of implementation at the ministerial level became clear in a fall 1960 survey of factory inspectors in North Rhine–Westphalia regarding a draft of the state clean air act in the fall. The timing in itself was revealing: the Ministry of Labor and Social Affairs had announced as early as September 1957 that it was working on such a draft.[196] A carefully crafted bill was therefore already on the table in the fall of 1960, and the fact that it was only now deemed necessary to consult the subordinate agencies demonstrates the limited weight that the ministry gave to suggestions from officials in those agencies. However, the responses made it abundantly clear that implementation of the new law would be anything but smooth. For example, the factory inspection in Düsseldorf worried that it would be "inundated by a flood of complaints from neighbors," with the result that the individual offices, "given the current staffing levels, will hardly be able to get to their actual tasks."[197] What seemed to the ministry a struggle for clean air was seen by officials on the ground as a bill that "all but provokes troublemaking." "The recent press campaign encourages various citizens to complain more or less vociferously about even the most minor nuisances," warned the inspector in Duisburg.[198] The district administration of Detmold believed that there was reason to expect "considerable additional work" and that, "more so than in the past," the administration "will have to tell complainants to pursue the matter in civil court."[199] Officials in Aachen even believed that factory inspection was "at a crossroads": "Were the planned clean air act to become reality, the inevitable result, in spite of an increase in personnel, would be an utterly unsustainable overburdening of industrial inspection."[200] The disorientation of the officials on the ground was obvious.

However, the situation sounded far less dramatic in the ministry's assessment of responses. Under the heading "General," it simply stated, "The district administrations point out that the regulation of jurisdiction would create an additional burden for the state offices of trade supervision."[201] An interministerial conference about the proposed changes also gave no indication that these concerns were seen as posing a fundamental problem.[202] Thus, in the end, the only meaningful reaction on the part of the ministerial officials was that they requested an increase in staffing by a total of forty-seven positions when they drew up the next budget.[203] And this lack of attention to implementation problems was by no means unique. When concerns about "consistency in the application of the law" were raised at a board meeting of the VDI Clean Air Commission in September 1959, since implementation was being "left up to the states," officials unanimously downplayed these concerns. Heinrich Oels from the federal Ministry of Labor considered such fears plainly groundless: after all, the federal government habitually passed "the general administrative rules that the states follow." For his brother Franz, from North Rhine–Westphalia's Ministry of Labor, these problems existed, at best, "in theory."[204] The Working Committee of the States (Arbeitsgemeinschaften der Länder), which was supposed to clarify all-important issues of air pollution control, did its best to dodge the topic of implementation problems altogether.[205] At a hearing of the state assembly of Rhineland-Palatinate, an administrator from the industrial region of Ludwigshafen-Mannheim warned emphatically that officials "no longer know whether industry is doing too much, too little, or just the right thing, since they no longer have a grasp of the total complex of air pollution."[206] But this, too, did not give anyone pause.

In fact, German air pollution control was a splendid success story only if viewed from above—from the perspective of the ministries and the federal institutions. If viewed "from below," from the perspective of the day-to-day work of the bureaucracy, a different and far gloomier picture emerged: chaos ruled supreme on this level, and nobody really had a grasp of the state of affairs. "The administrative measures in the federal states show a very colorful picture," proclaimed Heinrich Oels of the federal Ministry of Health in November 1962; a "uniform line" was, in his view, "not discernible."[207] The influential German Association of Towns and Cities believed that factory inspectors were generally "not yet in a position to prepare all necessary measures, let alone implement them." In Rhineland-Palatinate, the inspection was declared to be "overextended" due to a shortage of personnel. As late as 1970, the Bavarian Ministry of Social Affairs noted internally that the view that the implementing authorities were "technically overwhelmed with questions of emission protection" was "undoubtedly correct."[208] Even the factory inspection in North Rhine–Westphalia—without a doubt the most agile executive organ of the 1960s—professed itself unable to implement existing provisions systematically. "With the exception of a few improvement programs, a systematic air pollution monitoring of existing installations was generally not possible for lack of personnel," was its terse comment in an annual report.[209] And so the bureaucratic fight against air pollution quickly evolved into an impenetrable thicket. The factory inspection in Dortmund reported at one point about a large power plant, already operating for four years, whose licensing conditions were still being negotiated by officials.[210]

Inadequate implementation became an issue only in 1974, through a report of the Council of Experts for Environmental Questions (Sachverständigenrat für Umweltfragen). "It is the general belief that the administrative agencies, but also the offices of the public prosecutor as well as the criminal courts, have so far not taken the statutory regulations of environmental law seriously enough," this circle of experts noted. In some cases, the authorities' conduct bordered "on a refusal of enforcement."[211] It might seem surprising that this discussion did not begin until almost two decades after the first reforms that were intended to move German air pollution control onto a new, more modern track. Did the officials in charge really believe that the new, tougher directives would somehow magically trickle down from the federal and state agencies to bureaucratic practice on the ground? Such a view would ignore the fact that the German age of smoke was mostly a matter of smokescreens, of secrecy and deception about what was really going on within the bureaucracy. A shortfall in implementation had always existed in German air pollution control—except that this was not seen as a problem until the postwar era. Officials had hardly ever systematically monitored the requirements laid down in licensing documents; instead, these requirements functioned more like a stopgap in case someone complained. The result was a huge contrast between the norms set at the top and implementation at the bottom: while the former was in full swing at the beginning of the 1960s, those charged with the latter followed with widely differing motivations and speed, sometimes waiting until the 1970s and beyond to initiate more than token efforts. This was the downside of a polity where ministry officials called the shots.

Progress and Consolidation

In the spring of 1967, Lufthansa German airlines came out with a special fare. American officials who wished to have a personal look at the European— especially German—accomplishments in the areas of water and air pollution control were henceforth granted favorable conditions for ticket purchases.[212] At first glance, this might seem like a minor footnote in the long history of transatlantic exchange, but in fact this special fare, offered in collaboration with the National League of Cities, fittingly reflects the general development of German and American air pollution control in the 1960s. During the early years of the postwar period, officials and experts traveled mostly in the opposite direction: German experts by the dozen had made the pilgrimage to the United States in the 1950s to glimpse the future of air pollution control. As late as May 1960, for example, Hapag-Lloyd's division of educational tours had organized a two-week excursion "to study the prevention of air pollution in the USA."[213] Thereafter, however, German interest declined

Figure 9. Residential housing in front of a coke-oven battery near Bottrop in the northern Ruhr Basin. The region became notorious for its pollution load, although changes had been under way for more than a decade when this picture was taken in 1976. Image courtesy of Ullstein Bild.

markedly; evidently, word had gotten around among German experts that the Americans were not everything they were cracked up to be. At the same time, there was growing respect for Germans among American experts: an internal report by the chemical giant Du Pont about air and water pollution control noted "a severity in Germany which has not yet been attained in this country."[214] While a growing feeling of crisis spread in the United States, the Germans were, more than anything, proud of their own success.

In retrospect, there is no doubt that the period around 1960 also marked a turning point with respect to actual emissions load. As early as 1962, a VGK survey revealed that the emission of particulate matter from its members had declined by 27 percent over the previous ten years, even though energy production had increased by about 130 percent over that same period.[215] Helmut Kettner from the federal Ministry of Health's Institute for Water, Soil, and Air Hygiene also noted in 1966 that, with a view to ten years of dust monitoring in North Rhine–Westphalia, "dust precipitation has declined noticeably during the last few years."[216] Significant progress was also reported in other sectors. According to the journal *Gesundheits-Ingenieur*, the problem of the so-called brown dust from steel converters—which was produced during oxidation in the Thomas process and, because of its fineness, could be removed only with high-capacity precipitators—occurred in the mid-1960s "only in a few installations where we encounter special technical problems."[217] BASF in Ludwigshafen reported a reduction in daily dust precipitation from 2.0 g/m2 in 1952 to 0.1 mg in 1970, a level at which the dust issue was "seen as resolved for all intents and purposes." Sulfur dioxide emissions, meanwhile, remained at least substantially constant during the 1960s, and on nitrogen oxides BASF had achieved a 60 percent reduction in spite of a rise in saltpeter production.[218] The Bayerwerk in Leverkusen, where no fewer than 153 employees worked full-time on water and air pollution control in 1970, was also able to reduce its pollution load; according to management, the plant's emissions of nitrogen oxides and organic substances were "lower than the concentrations from the traffic on the nearby Autobahn."[219] Of course, in the face of such reported successes, continuing shortcomings must not be forgotten. For instance, waste incinerators were still considered merely a dust and odor problem in the 1960s.[220] Sulfur dioxide from fuel combustion remained a marginal topic, with tall smokestacks the prime remedy of choice.[221] Regional differences remained, some of them quite dramatic in scale: as late as 1968, tests at the playground of a school near Cologne found hydrogen fluoride and hydrogen phosphide at levels equal to the maximum concentration permissible in the workplace.[222] Still, considering that the money invested in air pollution control by industry in North Rhine–Westphalia surpassed the two billion DM mark at the end of the 1960s, this much was clear: things were happening.[223]

As a result, the VDI Clean Air Commission largely sailed through calm waters in the 1960s. Even critical discussions of the technical norms produced by the commission emphasized that its critique was "in no way intended to diminish the value of the very meritorious work that had so far been done in this area."[224] The steady expansion in the commission's work is evident from the numbers alone: while Heinrich Lent had put the number of participating scientists and engineers at 130 in October 1958, an April 1964 report noted eighty-three research projects and more than 300 participating scientists—116 engineers; 72 physicists and chemists; 66 medical doctors and biologists; about 40 agronomists, forestry specialists, and veterinary scientists; and 18 meteorologists.[225]

In the legislative arena, there was little excitement in the 1960s. The most important development was probably that the clean air act the state assembly of North Rhine–Westphalia had passed unanimously on April

10, 1962, was gradually adopted by other states.[226] At the beginning of 1964, Baden-Württemberg became the second state to enact such a law, followed by Lower Saxony in 1965, Rhineland-Palatinate in 1966, and Bremen in 1970.[227] Bills for a state law were submitted in Saarland in 1964 and 1966, but they were subsequently swallowed up by the parliamentary machinery. In Bavaria, Article 18 of the State Penal and Administrative Law (Landesstraf- und Verordnungsgesetz) contained provisions that resembled in substance the North Rhine–Westphalian model, and in Schleswig-Holstein preliminary work for passage of a state clean air act was suspended in 1971 "to avoid the duplication of work," since there was reason to expect "passage of a federal clean air act in the near future." Only Hesse, Hamburg, and Berlin saw no legislative activity.[228] Compared to federal laws, however, the state clean air acts carried little weight, and they were largely uncontroversial. Until 1970, all state acts were passed unanimously by the respective state assemblies.[229] By this time, however, the broad consensus that had carried the reforms of the late 1950s and early 1960s had begun to show cracks. Since the mid-1960s, the harbingers of the environmental age were being felt in Germany, and the challenges this created for the regulatory establishment were anything but insignificant.

6   Forerunners and Pioneers

Policy decisions on air pollution problems were local and regional matters in both Germany and the United States until far into the twentieth century. As a result, it should come as no surprise that the national styles of regulation described here imply an enormous amount of regional variation. A multitude of factors accounts for sudden advances or delays, for rapid progress or stagnation. Many of these conditions are of little interest in a study of national paths toward air pollution control. For example, the peculiar development of air pollution control in New York City was due simply to the city's proximity to the nation's only large anthracite deposits. However, such particular paths of development do deserve closer attention in this context if they provide a glimpse into the future. From this perspective, two regions that could hardly be more different come into focus during the postwar years: the Ruhr Basin in Germany and Los Angeles in the United States. The Ruhr Basin, situated in the northwest, close to the Dutch border, had been Germany's industrial heartland since the mid-nineteenth century and thus subject to the pollution problems that typically went along with coal mining, steel making, and industrial chemistry. In contrast, Los Angeles had been mostly free from air pollution troubles until World War II, when a new kind of pollution—later identified as photochemical smog—came to plague the city. The one thing both regions had in common was that their pollution problems became notorious in the rest of their respective countries: mentions of Los Angeles or the Ruhr almost automatically brought to mind notions of an exceptional pollution load. And in both countries, time would show that these situations were not all that exceptional. Rather, common problems were merely more evident in Los Angeles and the Ruhr than elsewhere.

Still, these exceptional regions had the chance to become pioneers in air pollution control, blazing a path that other regions could follow—or choose to ignore. Such forerunners can become pioneers, but this requires an attentive audience willing to learn and look ahead. In this regard, Los Angeles and the Ruhr provide an interesting contrast, for only the latter region became an engine of change driving the entire country. Had it not been for the Ruhr Basin and the reform efforts it inspired in the postwar years, the reorientation of German air pollution control would surely have been delayed for a number of years or even decades. While German state governments were following the shining example of North Rhine–Westphalia, however, American policy makers followed events in Los Angeles merely out of curiosity, not because they found them relevant to their own local decisions. Americans thus missed a historic opportunity. Los Angeles was probably the best chance the United States would have to reflect critically on whether cooperative air pollution control was really the approach for the future.

The Strange Career of the Ruhr Basin

There is probably no region of Germany whose name is so closely associated with noxious fumes as the industrial region of North Rhine–Westphalia. Much as Pittsburgh is enshrined in American collective memory as the "smoky city," the name "Ruhr" evokes images of soot and dirt. And with good reason: a wealth of striking quotes documents that the Ruhr region had been, since the nineteenth century, dealing with a pollution load that was extreme even by contemporary standards. Here, for example, is a passage from a 1911 issue of the journal *Gesundheit:* "The traveler who arrives for the first time in the centers of our industry, e.g., on the route Bochum-Gelsenkirchen-Oberhausen, notices with some anxiety how much the appearance of the entire area, the vegetation, the houses, and the people is altered by chimney soot, and how oppressively the latter settles on the lung."[1] Around the same time, factory inspector Klocke in Bochum reported similar experiences in the journal *Stahl und Eisen:* "Someone passing through the industrial region on his way from Berlin will clearly notice how the train suddenly enters a cloud of haze near Hamm and remains in it until Düsseldorf and beyond, on account of which many a traveler must have asked himself how people can possibly live in such air."[2] "Especially between Ruhr and Emscher one hardly expects to catch another glimpse of fresh, happily thriving plant life," declared Gelsenkirchen's former mayor Karl von Wedelstaedt in April 1930.[3] And as late as 1959, an article in the journal *Industriekurier* opened with the statement that the silhouette of the Ruhr region "is unthinkable without dust and haze."[4]

The problem's existence was thus not seriously disputed by anyone. In many cases, even industrialists did not deny the sad state of affairs. A representative of Thyssen's pipe mill in Styrum, for example, explained with a shrug, "That the proximity of a factory, especially one as extensive as ours, brings not only conveniences for surrounding residents is par for the course."[5] But the path from recognizing that a problem exists to doing something about it was also no simple one-way street in the Ruhr region. In fact, it is readily apparent that initial conditions in the Ruhr region were in several respects significantly less favorable than in other areas of the German Reich. There was, to begin with, the broad spectrum of emission problems: blast-furnace gas; metal dusts from steel converters; smoke and soot from the use of coal by industry, commercial establishments, and households; fly ash from power plants; coke-oven gas; and manifold emissions from chemical plants. In addition, low-quality, ash-rich varieties of coal were burned on a large scale in the Ruhr region, since the transport of these fuels over longer distances would have entailed disproportionately high costs.[6] The exhaust from coking plants also presented a very tricky problem: a 1922 survey by the Chief Mining Bureau (Oberbergamt) in Dortmund revealed that only a few coking plants had equipment to capture coke-oven gases, and even most of these installations did nothing more than disperse the gases into a wider radius by means of a high chimney.[7] As a result, many contemporaries threw in the towel in the face of the enormity of the problems. For example, the popular scientific journal *Prometheus*, assessing the Ruhr region, tersely noted that "eliminating or rendering harmless merely a part of the masses of smoke and soot in this area is out of the question for the foreseeable future."[8]

However, the emergence of effective air pollution control in the Ruhr region was impeded not only by the enormity of the problems but also by the attitude of industry. To be sure, in this region, a significant number of entrepreneurs did make serious efforts to minimize the existing problems as best they could.[9] On the whole, however, Ruhr industrialists behaved much more aggressively than their counterparts in the rest of Germany, mirroring an awareness of their colossal economic might. For example, when the burning of powdered coal by the Mont Cenis coal mine caused a fly-ash nuisance, which prompted the city government to close a nearby school, the company, in a letter to the municipal authorities, declared without the least hint of concern that it simply "cannot understand this kind of action against the mine, your largest tax payer, on whose well-being the entire town depends."[10] A similar tone was struck by the Lower Rhenish Smelter in Duisburg, which felt unjustly accused in 1909: "We would find it exceedingly regrettable if you, prompted by biased accounts, should move to cause difficulties in the continued operation of our plants," the company informed the mayor.[11] At times, the factory inspector in Essen had the impression that many entrepreneurs regarded licensing requirements "as unimportant" and therefore did not adhere to them; even after he responded with stepped-up inspections, he maintained that it was "rare that all these conditions can be seen as fulfilled upon inspection."[12] In response to a complaint by a company that maintained an iron depot nearby, the metalworks Olsberg suggested "the building of sheds to protect the iron from all types of weather." As late as 1955, the management of a mine defended itself against the charge that its operations were insalubrious by simply noting that so far "no one has keeled over."[13]

While industry was thus arguing from a position of strength, the situation of those harmed by emissions was even worse in the Ruhr region than elsewhere. As in most conurbations of heavy industry, an economically independent middle class was relatively weak in the Rhenish-Westphalian industrial region, and the number of individuals who could complain about emissions without fear of losing their jobs or suffering economic reprisals was thus limited.[14] And when such individuals did speak up, their cases were oftentimes almost tragic. Walter Hoever in Duisburg, for example, was trying to run a garden restaurant downwind from the Duisburg Copper Smelter, possibly the worst air polluter in the entire western Ruhr region.[15] Moreover, complaints from bourgeois circles were not infrequently aimed at measures that benefited only a small segment of the population. Judging by the vehemence of complaints, for example, the most important polluter in Essen around 1900 was not the giant Krupp heavy-industry complex, as might be presumed, but a spirits distillery that, according to information provided by the mayor, was located in a "section of the city that is a preferred residential area of the wealthier residents."[16] And while the farmers of the Rhenish-Westphalian industrial region in many cases did feel like "victims of large industry," their demands for the most part involved less a reduction of the pollutant load than the highest compensation payments they could extract.[17] When the Rhenish-Westphalian Protective Association against Smoke and Mining Damages (Rheinisch-Westfälischer Schutzverband gegen Rauch- und Bergschäden) was founded in Bochum in February 1914, those present spoke less about existing pollutants and possible countermeasures and more about the creation of a Research Institute for Industrial Damages in the Ruhr region and about "establishing a compulsory cooperative for companies liable for damages," designed to bring about "a fair balance of the interests of both sides in an objective manner free of hyperbole."[18] Incidentally, such compensation was not rare, and it played an important part in pacifying farmers and neutralizing them as potential carriers of a protest movement. "Many questions of damage to vegetation from smoke and the like have proved amenable to being resolved in an amicable way," a representative of the Essen Mining Association declared in 1930. "Issues were only resolved in court when demands were unnaturally high."[19]

Along with the scale of the problem, the strength of the industrial lobby, and the complementary weakness of those harmed by emissions, a fourth handicap impeded air pollution control in the Ruhr region: the relative weakness of the authorities. No other German region had to wrestle with such an explosion in population and commercial activity, and no other German conurbation had such weak structures on the municipal and state levels. Officials found themselves confronted with such a panoply of complex problems that the fight against air pollution—which officials in other areas of the German Reich already preferred to put on the back burner—usually received little attention. In addition, there are indications that the mining bureaus (*Bergämter*), which were also responsible for supervising the aboveground installations of mines, took a much more permissive approach to emission problems than other agencies.[20] This became apparent in administrative proceedings regarding emissions from the coking plant of the Friedrich Ernestine mine near Essen, for which both the local factory inspector and an official from the Mining Bureau wrote an expert opinion. While the factory inspector regarded the complaint as "entirely justified," the mining official vigorously downplayed it: the emissions were "unavoidable" because of the high gas content of the coal. "Only a few scattered" private houses were found in the environs of the mine, he declared, and moreover, any danger to health could surely be "ruled out."[21]

For all these reasons, a reform debate existed in the Ruhr region before 1914 in rudimentary form, at best. While in many large cities of Imperial Germany a discussion about how to combat the smoke nuisance was well under way, an almost paralyzing lack of interest was evident in the Rhenish-Westphalian industrial region. Every official must have felt a profound sense of powerlessness when it came to the question of what effective countermeasure would look like. "In the local industrial regions, measures against the smoke and soot nuisance must be seen as entirely hopeless," wrote one municipal official with brutal honesty in 1912.[22] Revealingly, it was not an established bureaucratic organ that broke the silence, but the Ruhr Area Federation for Regional Planning, which had been set up in 1920, by a law of the state of Prussia, as the agent for systematic regional planning.[23] The Ruhr Area Federation invited representatives of bureaucracy and industry to a "meeting on questions about reducing the smoke nuisance in the industrial region" on January 13, 1925—the first round table of its kind in the region's history.[24] However, like so many other German agencies, the Ruhr Area Federation moved at a rather casual pace: although the meeting agreed to set up subcommittees that were supposed to report their initial findings "in about five weeks," in fact most reports dribbled in with considerable delay; they were finally published as part of a memorandum in February 1928.[25]

In addition to moving at a leisurely pace, the work of the Ruhr Area Federation suffered from the skepticism it encountered elsewhere. One mining official expressed his disdain for efforts to preserve the forest as early as 1921, with a bluntness that left nothing unspoken: when the Ruhr Area Federation pointed to the threat to the forest near the Franz Haniel mine in Bottrop and also mentioned the danger from "smoke and other noxious gases," the Mining Bureau official in charge responded brusquely that the establishment of a coking plant could hardly be "prevented in the long run simply to meet the intellectual interests of the Ruhr Area Federation."[26] The Düsseldorf district government was evidently also unhappy that the Ruhr Area Federation was trespassing on its jurisdiction. When the director of the federation requested in February 1925 that the Essen factory inspector be authorized to join the federation's smoke committee, the Düsseldorf government simply refused.[27] And industry, which had numerous representatives on the smoke committee, could have had no interest in legitimizing a tougher approach by the authorities.[28] Thus, the Ruhr Area Federation soon found itself surrounded by a great coalition of obstructionists on the smoke committee, a situation that was openly addressed by the director of the association. When the subcommittees set up in January 1925, with industry holding a particularly strong position, presented the results of their work after nearly a year, the director commented that "the manner of negotiating until now, where every industry presents itself as more or less uninvolved in the smoke damages, will hardly lead anywhere."[29] The hardened fronts appear to have softened a little once the Smoke Committee began to deal with individual cases of pollution that "have been raised in the press or elsewhere."[30] Even this minimal activity, however, eventually seems to have fallen victim to the Great Depression. At any rate, in May 1932, the director of the association informed the members of the Smoke Committee that "in view of the circumstances," he had refrained from calling a meeting. No further committee activities are documented in any files.[31] The first discussion about a reform of air pollution control had thus failed miserably.

The most important lesson from this episode was no doubt that a "corporatism of weakness" was doomed to failure, given the conditions of the

Ruhr region. The Ruhr Area Federation for Regional Planning had been caught in a fatal dilemma from the outset: since it had no legal authority to fight air pollution, it depended on the cooperation of representatives from the bureaucracy and industry—except that the inclusion of precisely this circle of people made reforms unlikely to begin with. The president of the Ruhr Area Federation sought a way out of this dead end by asking the former mayor of Gelsenkirchen, Karl von Wedelstaedt, to draft a memo, but the opponents of drastic change made short shrift of it: "Since the memo . . . offers nothing new, those present dispense with a reading," note the minutes of the committee meeting held on October 31, 1930.[32] "It may be said that smoke damage and smoke nuisance have so far not received the kind of attention that an effective fight against them would require," the director of the federation had declared in February 1928, and this would not change until the end of World War II.[33] The failure of the federation's initiative simultaneously put an end to any debate about possible reforms for two decades. Although complaints about the emission load continued to be voiced, they could no longer initiate a strategy debate.[34]

Therefore, only an unabashed optimist would have predicted around 1950 that the Ruhr region would soon become the driving force behind federal German air pollution control. But this is exactly what happened over the course of the following decade. By around 1960, the state of North Rhine–Westphalia, with the Ruhr region at its heart, held an uncontested position of leadership in federal German air pollution control. The establishment of the VDI Clean Air Commission, for example, would probably have been unthinkable without North Rhine–Westphalia; the 1964 financial contribution from the federal government to the commission's work was less than that from the state of North Rhine–Westphalia.[35] Individuals also reveal the Ruhr region's central role: the first chair of the VDI commission, Heinrich Lent, was a mine director, as was the chair of the BDI Emissions Committee, Karl Oberste-Brink.[36] When the States' Committee for Pollution Control was set up in 1964, North Rhine–Westphalia assumed the chairmanship, and the short-lived States' Working Committee on Clean Air, established by the conference of state ministers and senators of health, was headed by Joachim Wüstenberg, the director of the Hygiene Institute of the Ruhr Region.[37] And no other state had anything comparable to the State Agency for Air Pollution Control and Soil Protection (Landesanstalt für Immissions- und Bodennutzungsschutz) in Essen-Bredeney, founded in 1961. In a letter of March 1971, the district administration of Osnabrück compared the North Rhine–Westphalian agency, with its approximately three hundred employees, to the situation in Lower Saxony, where only a single chemist was employed in the Department of Occupational Medicine and Commercial Hygiene within the Office of State Administration.[38] Even the September 1957 "Resolution on Air Pollution Control" would hardly have come about without the Ruhr region: it went back to an initiative by the Central Duisburg Citizens' Association (Duisburger Bürgervereinigung Stadtmitte), which had raised the issue at the Conference of German Civic Associations in Kiel.[39] The impressive successes achieved in the fight against air pollution problems in the Ruhr region in the 1950s and 1960s were soon known far beyond the boundaries of North Rhine–Westphalia. Even in distant Oregon, an official brochure declared in 1968 that "Germany's heavily populated and industrialized Ruhr Valley, once a pest-hole of pollution, has been cleaned up to a large extent through the vigorous efforts of government and industry."[40] If the cleaned-up Pittsburgh was "Cinderella City," as the National Association of Manufacturers declared in a widely read brochure, then the post–World War II Ruhr Basin was "Cinderella County."[41]

What were the reasons behind this almost magical transformation? One was certainly the dissolution of the state of Prussia, to which the Ruhr region had belonged since the early nineteenth century, and the establishment of the state of North Rhine–Westphalia in 1946. This endowed the Ruhr region with new political weight: no longer merely a part of two Prussian provinces, it was now clearly the heart of a new state. In 1961, 32.3 percent of the population of North Rhine–Westphalia lived in the Ruhr region.[42] Moreover, North Rhine–Westphalia was by far the largest state of the Federal Republic, home to about one-quarter of all West German citizens and host to the capital city of Bonn. How important all this was to the reforms is evident merely from the fact that the driving force behind them was the North Rhine–Westphalian Ministry for Labor and Social Affairs. It is hard to imagine that a Prussian ministry in far-off Berlin could have played a similarly vigorous leadership role.

The importance of this state's creation was further heightened by the fact that within the boundaries of North Rhine–Westphalia lay a second area with an extreme emissions problem: the Rhenish brown-coal region.[43] As early as 1906, the Prussian Ministry of Trade had dispatched two commissioners to the Rhenish brown-coal region to investigate, in collaboration with representatives from the ministries of Culture and Agriculture, vociferous complaints about dust from the manufacture of brown coal in the communities of Bachem and Frechen.[44] In 1908, the town council of Brühl demanded "measures by the royal government, which would, after all, also be substantially in the interest of the royal court," thus alluding to a royal castle in town.[45] Shortly before World War I, the paper industry in Düren fought desperately to prevent the construction of a dust-generating briquette plant on the outskirts of the city.[46] Around 1930, Cologne's mayor, Konrad Adenauer, later the first chancellor of the Federal Republic, was planning to bring an air pollution lawsuit against the Rhenish brown-coal industry.[47] In the early 1960s, a civic association in Cologne lamented that "the southern suburbs of Cologne are not far behind the mining areas of the Ruhr region when it comes to the degree of air pollution," and this was clearly not off the mark.[48] Similar to the Ruhr region, the Rhenish brown-coal area had long been dominated by a fairly unsystematic bureaucratic approach. At the beginning of the 1950s, however, a noticeable rise in administrative interest became evident: beginning in 1954, officials and industry negotiated the requirements for flue-gas cleaning as part of a special "Brown Coal Dust Committee" (Staubausschuss Braunkohle).[49] The historical importance of these negotiations went far beyond their actual subject: the Rhenish brown-coal district here saw the first test of the corporatist mode of negotiation that was subsequently institutionalized in the VDI Clean Air Commission.

While officials were thus much more active and agile in the 1950s than in previous decades, industry was arguing during this period from a noticeably weaker position. Industry's traditional "master-of-the-house" policy came to an end in 1945, the Ruhr industrialists were compromised by their complicity with the Nazi regime, and discussions about socialization and demilitarization put industrialists further on the defensive.[50] As a consequence, they showed themselves much more open to compromise in dealing with air pollution problems. Of course, their openness was not unlimited: the establishment of an "Association for the Promotion of Air Hygiene" (Verein zur Förderung der Lufthygiene), pushed by Heinrich Lent, was thwarted not least by industrial opposition.[51] Still, it was a clear sign of the changing times that a working committee entitled "Air Hygiene" could be set up in 1954, headed by the director of a mining company, under the aegis of the Association for Combating Social Diseases in the Ruhr Coal Region.[52] A similar initiative had been launched as early as 1912, but according to a report by the district administration of Düsseldorf, it "unfortunately had to be 'postponed' because of vigorous opposition by representatives of large industry and the mining association."[53]

The most apparent difference between the interwar and the postwar periods was that substantially more time was allowed for a real reform debate in the Federal Republic. Leaving aside the discussions that took place in the Nazi state, the interwar period saw only a narrow window of opportunity for reform debates between the inflation crisis in 1923 and the beginning of the Great Depression. Still, the situation at the end of the 1920s had certainly appeared more positive than it might seem in light of the failed efforts of the Ruhr Area Federation, as tentative efforts at reform also began to emerge for the first time in the communities of the Ruhr region. In Dortmund, a special Office for Combating the Smoke and Soot Nuisance was set up in 1929 within the city's Public Health Department, and in Duisburg, a municipal Working Committee for Combating Smoke and Soot met for the first time in October 1928.[54] Public opinion was also in the process of mobilization: "It is not the money bag of industry but the health of the people as the highest good of the nation . . . that must be vigorously supported," the Social Democrat *Herner Volkszeitung* opined in August 1929, while the *Duisburger General-Anzeiger* noted that it could already detect a "Duisburg air war" between the copper smelter and the Civic Association of Hochfeld.[55] And industry's obstructionist policy had also begun to crack a little: at the end of 1930, representatives from industry signaled that future research work by WaBoLu could count on the support of Ruhr industrialists.[56] Of course, all these activities were still far from effective reforms, but at least there was some movement after decades of lethargy. The onset of the economic crisis, however, made all these first steps irrelevant overnight. By contrast, during the prosperous years of the 1950s and early 1960s a broad social debate was able to unfold—a debate for which there simply had not been enough time in the 1920s.

The background conditions for a promising reform debate were thus more favorable than ever before in the postwar period. Still, in retrospect, it must be stressed that the Ruhr region probably would never have seen vigorous air pollution control if the authorities, experts, and industry had not been under constant pressure from public opinion. To be sure, until well into the 1970s, only a few organizations were active in the fight against air pollution. The above-mentioned initiatives, by the Duisburg Civic Association Stadtmitte and by the Community of Interests against the Ash Nuisance of the Cuno Works of Elektromark, seem rather paltry in view of the fact that more than five million people lived in the region in 1961.[57] But a low level of organization does not mean that contemporaries were indifferent to the quality of Ruhr air. The true sentiment in the Ruhr region was hinted at when, for example, the *Westdeutsches Tageblatt* ventured as early as October 1959 to declare air pollution the "No. 1 Ruhr Problem"; in the Bundestag, meanwhile, the member of parliament Willy Reichstein uttered this obscure warning: "If the Ruhr region were in the Sahara, we'd all be dead already."[58] More and more, the public's discontent was looking for an outlet. In 1959, Clemens Schmeck, a physician and chair of the Civic and Transportation Association in Dellwig, filed charges against the Oberhausen Smelting

Works, and the SPD group on the Duisburg city council submitted a serious proposal "to provide two interested citizens with the financial means to pursue a model suit through all levels of the judiciary against the nuisance from industrial air pollution," a measure that would have been—in the estimation of the city manager—"an absolutely unusual procedure."[59] The pressure to do something was evidently enormous.

In this process, the public drew encouragement from a foreign model—interestingly enough, Pittsburgh. "The Americans have already demonstrated—in Pittsburgh, for example—that a substantial reduction in the smoke and gas plague is possible," declared an ophthalmologist from Bottrop in a letter to the state government in Düsseldorf.[60] "The first energetic 'Stop' was pronounced by the citizens of Pittsburgh in the USA," reported the *Westfälische Rundschau*. Sturm Kegel also praised the Americans for "proceeding much more robustly in this area," and when one speaker at a citizens' meeting in Duisburg pointed to the "American mining city of Pittsburgh," the daily *Neue Ruhr Zeitung* asserted that this reference had "stirred up a hornets' nest."[61] Of course, all of these observations referred only to the positive side of the Pittsburgh campaign; it was hardly ever mentioned that gaseous emissions were not subject to any controls in Pittsburgh, that industry dominated the key committees, and that the entire campaign had begun with the fight against the problem of domestic fires. Revealingly enough, German statements about air pollution control in Pittsburgh grew more critical the more a speaker knew about the issue. "In Pittsburgh, for example, which is always presented as a prime example, the dust conditions are no better and, if anything, worse than in Essen," Franz Oels, of the North Rhine–Westphalian Ministry of Labor, remarked somewhat tartly in 1961. In a lecture before the Health Committee of North Rhine–Westphalia's Conference of Counties, another expert went so far as to denounce Pittsburgh's restrictions on the choice of coal for domestic burning as "extraordinarily dangerous."[62] According to the surviving files, American air pollution control thus hardly played a role when it came to the drafting of crucial measures. Only the creation of a State Advisory Council for Pollution Control (Landesbeirat für Immissionsschutz) noted that this move "creates a parallel to corresponding institutions under American law," and even in this case the Advisory Council was endowed with substantially less authority than similar committees in the United States.[63] In short, when residents of the Ruhr region spoke of Pittsburgh, they were talking not about the real politics of the city in Pennsylvania, but about an imaginary city onto which they projected their own yearnings for cleaner air. Pittsburgh was not a model but a symbol—it could be done!

Photochemical Smog and the California Alternative

At a time when Los Angeles long ago became the epitome of an urban catastrophe, it is difficult to imagine that it was once considered exceptionally salubrious. In fact, its pleasant climate was critical to the development of southern California in the late nineteenth century. Tuberculosis patients, among others, provided the region's first immigration surge between 1870 and 1900, transforming Los Angeles from a provincial nest into the center of a "sanitarium belt"—meaning, as John Baur observes with a wink, that "perhaps for the first time in American history, a frontier was developed by the sickly and the invalid."[64] Moreover, Los Angeles had largely been spared the coal-smoke plague, since petroleum and gas were its chief energy sources. Since the mid-1920s, coal accounted for less than 1 percent of California's energy supply.[65] Although it had a local 1907 ordinance about combating the smoke nuisance, Los Angeles did not have the tradition of smoke inspection that dominated most American cities. "As the city is not a great manufacturing center, smoke abatement is not a subject of very live interest," noted a 1924 survey, according to which the local ordinance was for all intents and purposes not enforced.[66] While the rest of the country followed Raymond Tucker's antismoke campaign with bated breath, it aroused at most humorous interest in Los Angeles: a former resident of St. Louis reported that fellow Californians constantly teased him about the notorious smoke problems of his native city—"and unfortunately I haven't much of a defense to offer."[67] Only this tradition of an almost complete lack of concern about urban air quality explains why the photochemical smog that made its first acute appearance in 1943 was immediately recognized in Los Angeles as a threat to the very foundations of the city's existence.[68] "Los Angeles County has always enjoyed an enviable reputation as a healthful land and a tourist paradise"—thus began in March 1944 the first scientific study of what was at that time still an entirely mystifying phenomenon.[69] At a conference two years later, a participant read a letter from an asthmatic who had come to Los Angeles from Seattle for the clear, healthful air, only to experience the first smog episode three months after his arrival.[70]

To a far greater degree than elsewhere, air pollution here was a burning problem—quite literally so: one of the consequences of smog was irritation of the eyes. The famous physicist Richard Feynman, who came to the California Institute of Technology in Pasadena in 1951, describes in his memoirs how the effects of the smog almost drove him out of the city: "I was standing on a corner, and my eyes were watering, and I thought to myself, 'This is crazy! This is absolutely INSANE! . . . I'm getting out of here.'"[71] "Upon arrival in Pasadena, I was compelled to put cold compresses over my eyes for ten to fifteen minutes before I could pay attention to close work," recounted a motorist's complaint in April 1950.[72] "The people of Los Angeles, while objecting to eye-irritation per se, are really concerned about it as indicating the possibility of more permanently damaging effects on their health," a leading official opined in 1956.[73] A sense of alarm pervaded even circles of medical experts. When the Los Angeles County Medical Association conducted a 1950 survey among its members on whether smog should be seen as a real threat to health or more as a nuisance, the response rate was the biggest ever in the history of the organization. Nearly 70 percent of members responded, and more than 95 percent affirmed the existence of a health threat—a result so shocking to the Los Angeles County Medical Association that it decided to keep the data under wraps. "We have not released these figures for fear that public panic might be created," one of those involved explained in a private letter.[74]

The smog of Los Angeles was thus from the very beginning understood to pose a health problem—unlike in the rest of the country, where the health argument came to prevail only around 1960. However, this observation alone does not suffice: the health argument acquired its real potency only from the fact that it joined hands with material interests.[75] These twinned motives were perhaps most starkly evident in Stephen W. Royce, owner of the Huntington Hotel, a winter meeting place for wealthy men from the East Coast. After several guests informed him that they would not be returning the following winter because of the smog, Royce turned into a passionate champion of air pollution control.[76] Harold Silbert, the president of the Los Angeles Coat and Suit Manufacturers Association, also had obvious economic reasons when he called for a rigorous fight against the problem in 1949. "During the past week when the smog was particularly irritating and was perhaps at its worst, about one-half of our factories were required to shut down," Silbert reported. If his employees had to wrestle with irritated and watering eyes, they were hardly in a position to perform their work with the requisite precision.[77] At times during World War II, even the output of war-essential enterprises had suffered from the smog situation.[78] Damage to agricultural plants was estimated at more than 4 million dollars a year at the end of the 1950s.[79] Pessimism was also quite justified with a view toward the city's attractiveness as a tourist destination: "Smog . . . will surely kill the tourist business of Southern California," was the terse assessment in a 1953 letter.[80] Finally, those who owned real estate in the city were generally at risk of a substantial drop in the value of their assets: in 1955, the Citizens Anti Smog Action Committee estimated that this loss amounted to at least 50 million dollars in Pasadena alone.[81] This final motive, in particular, likely played a significant role in mobilizing the broader public. Revealingly, the managing director of the women's group Stamp Out Smog, probably the most important organization dedicated to fighting smog in greater Los Angeles, also sat on the board of the Cahuenga Property Owners Association around 1970.[82]

This union between health-related and economic arguments infused air pollution control in Los Angeles with an impetus that other places could not even approximate. Already the first protests in 1943, which focused chiefly on a malodorous butadiene plant of the Southern California Gas Company, led to the temporary shutdown of the plant, even though it was producing for the war economy.[83] October 1943 saw the establishment of the Los Angeles County Smoke and Fumes Commission. Composed of three scientists and two technical laypersons, it carried out the first, still very rudimentary, studies of the problem.[84] In November 1944, the city of Los Angeles passed an ordinance that created a Bureau of Air Pollution Control, the county following suit in February 1945 with an almost identical law.[85] However, the county law applied only outside of incorporated towns and cities, and the chances of prevailing upon all the other cities of the county to enact similar laws were quite poor, given the chronically fragmented Los Angeles cityscape.[86] Comprehensive control became possible only in June 1947, when a state law created the possibility of setting up special Air Pollution Control Districts. The Los Angeles Air Pollution Control District (APCD), the establishment of which was set in motion immediately after passage of the law, began work in October 1947 and would become the central agency in Los Angeles for decades.[87]

From the moment of its foundation, the APCD was operating under a kind of pressure that was unheard of in other cities. "We need to tell the plant he is a violator and let him go ahead and find out his own methods of stopping it. If he doesn't, shut him up," one member of the Los Angeles County Board of Supervisors, the APCD's political supervisory body, declared as early as May 1946.[88] "There is no other problem facing our county that is more important than to restore [a] healthy and clear atmosphere to our citizens," proclaimed Kenneth Hahn, also a member of the Board of Supervisors.[89] A ten-year plan modeled after Pittsburgh's was thus out of the question in Los Angeles from the very beginning. When, in 1953, the APCD gave the county's refineries between eighteen and twenty-four months to make certain technical changes, Hahn pushed for an even faster pace.[90] Since such tough regulations could hardly be implemented with the personnel levels that were normal in other cities, the APCD quickly became by far the largest agency of its kind: at the height of activities in fiscal year 1957–58, the APCD had 467 employees and a budget of nearly 4 million dollars.[91] As late as 1961, when the APCD's budget was down to around 3.4 million dollars, this represented more than 40 percent of the expenses for local air pollution control programs in the entire country.[92]

To fully appreciate the enormous speed of the APCD, one must bear in mind that the causes of the smog were initially entirely unknown. The officials of the APCD were in the uncomfortable position of having to take vigorous measures against a problem that they did not understand.[93] As a result, the agency's work during the first few months was focused entirely on dust emissions and sulfur dioxide. Nobody knew whether this was in fact the root cause of the problem, but dust and sulfur dioxide were by far the best-known pollutants, and taking steps against them at least made it possible to demonstrate to an impatient public that vigorous action was being taken.[94] The general fog of ignorance began to lift a little only in November 1950, when the native Dutchman Arie Jan Haagen-Smit, a professor of biochemistry at the California Institute of Technology, speculated that smog was produced by a reaction of nitrogen oxides and hydrocarbons under the influence of sunlight.[95] It took some time, however, for Haagen-Smit's theory to be accepted by the scientific community; only toward the mid-1950s was it generally accepted, at least in its basic principles. "We believe the state of our knowledge in this community is just beginning to reach the point where we can realistically assess the magnitude of our problem," wrote Lauren Hitchcock, the president of the independent Air Pollution Foundation, in November 1954.[96] But once again, the APCD did not wait for a final clarification of the questions. As early as 1953, it passed regulations aimed at getting emissions of hydrocarbons under control.[97] Of course, such quick action bore the risk of errors, something that the APCD conceded with remarkable candor. "We do believe that we certainly have made mistakes," declared Gordon Larson, head of the APCD, in March 1953, before a committee of the State Assembly.[98] But precisely because the APCD did not shrink from issuing stringent directives on the basis of fragmentary scientific knowledge, it was always operating at the very edge of what was technically and scientifically possible. "There are no basic smog control regulations which could be adopted by this District at this time, other than those already on the books," said S. Smith Griswold, Larson's successor, at a hearing in May 1956.[99] No other American official could have claimed as much at the time.

Externally, officials in Los Angeles cultivated the policy of cooperation between authorities and industry that was also evident in other parts of the country. For example, one professor at Caltech, who as a member of the Hearing Board had considerable insight into the ongoing activities, declared, "I am greatly impressed by the efforts that industry is continually making to reduce harmful emissions to the atmosphere and have been aware of the very considerable sums of money that are being spent for this purpose."[100] But this was in fact a different kind of cooperation, one where APCD officials usually had greater clout and did not shrink from putting massive pressure on industrial enterprises by threatening to revoke their operating licenses.[101] Further, it was a cooperation to which industrialists agreed only after it had become apparent that opposition to air pollution control in Los Angeles could not be sustained. Specifically, the oil industry had initially fought against the 1947 law that made the establishment of the APCD possible; eventually, however, it was forced to capitulate to the media power of the *Los Angeles Times*, which had put all its clout as the West Coast's leading daily behind the law.[102] The oil industry thus yielded to the inevitable. "Regardless of who or what is to blame, the oil industry is determined to do everything in its power to help solve the smog problem," the Western Oil and Gas Association proclaimed only a year after its failed attempt to block the law.[103] Just how little leeway remained for measures taken on personal initiative in the face of the tough approach of the authorities is revealed by the failure of an APCD plan to dispense with formal directives in cases of smog alarms and to go instead with voluntary restrictions on production. Several large companies considered this plan simply impracticable and explicitly asked for clear legal directives.[104] While industrialists in other cities pointed with honest pride to their willingness to cooperate, Los Angeles entrepreneurs tended to clench their teeth when they heard such words. For example, in 1951, one industrialist in the oil business spoke of an "anti-refinery campaign."[105] The formulation by one APCD official put matters in a nutshell: "Implementation of our source surveillance program is based on compulsory rather than voluntary cooperation."[106]

The work of the APCD was, of course, not above criticism. For example, in the early 1960s, the *Los Angeles Times* found that the "bureaucratic process . . . is patently cumbersome, unwieldy and frequently reluctant and agonizingly slow."[107] Moreover, in retrospect, it is apparent that the APCD always focused on technological measures and never made any serious effort to raise fundamental questions about energy consumption at the time. In particular, residents' transportation habits were long considered sacrosanct by the APCD, even though it was becoming clear after only a few years that it would be impossible to combat smog effectively without a reduction in automobile emissions. But in the end, this simply reflected the attitude of city dwellers: while the vast majority of the population was certainly in favor of drastic measures against smog, they also took for granted that the car would remain the chief means of urban transportation. Contemporaries were usually unaware of the contradiction that later generations would see in this posture: fighting smoke by giving up one's own car—this was at most a hypothetical possibility. Nothing reflects this inconsistency better than the plan of California governor Ronald Reagan to finance the fight against air pollution through the sale of special license plates. "The plan will provide every motorist with the opportunity to help solve the problem he has helped to create. The personalized plates will serve as a symbol of his concern for improving our environment," Reagan bluntly declared.[108] Evidently, it had not occurred to anyone in the governor's office that this proposal could end up looking like a parody of Reagan's proclaimed "all-out war against the debauching of our environment."[109]

For all the legitimate criticism of the work of the APCD, however, there is no doubt that it was in an uncontested position of leadership compared to all other agencies in the country. "The first city to come to grips with air pollution in all its complexity was Los Angeles, which still holds the lead in the battle for clean air," declared a much-read brochure in 1967.[110] Arthur Stern took a similar view: "The greatest reservoir of air pollution control talent lies in the organization of the Los Angeles County Air Pollution District," he counseled a Pittsburgh professor who had sought his help in looking for a new head for the Allegheny County Bureau of Air Pollution Control.[111] "The world's most smog-controlled city is Los Angeles," declared an essay in *Reader's Digest,* while a study conducted by the Public Health Service came to this sweeping conclusion: "Los Angeles County . . . is known to have the most restrictive air pollution control requirements in the world."[112] Los Angelenos themselves laid claim to a leadership position with some confidence: "More advanced control measures have been adopted here than anywhere else in the world at any time," boasted Griswold in May 1956. Five years later, the *Los Angeles Times* declared the city the "smog capital of the world."[113]

But all the bureaucratic accomplishments notwithstanding, the public remained unhappy with the regulatory style in Los Angeles. Public discourse increasingly followed a logic all its own, one that had little to do with technical and administrative possibilities. It called for tough, relentless action against those who had robbed Los Angeles of its good air—a demand that was shaped more by an angry public's desire for revenge than by a sober assessment of ways and means. Time and again, what comes through in public statements is the hope that the problem could be resolved once and for all with a single, vigorous action—a hope that naturally made the day-to-day work of the APCD seem rather pallid. As one typical critique put it, "Their approach has been academic, abstract and theoretical rather than direct and purposeful."[114] Large technological fixes that promised to solve the problem in one fell swoop seemed to hold an almost irresistible fascination for smog-plagued Los Angelenos. There were plans for gigantic tunnels in the surrounding mountains or huge ventilators to disperse the smog, designs for underground canals to collect the exhaust, and suggestions to use airplanes to break up the inversion layer—while experts, who could unmask such plans as unrealistic with even rough calculations, simply shook their heads.[115] The examples of Pittsburgh and St. Louis were repeatedly invoked, two cities that had supposedly defeated the smoke nuisance in a very short period of time through a radical campaign—a view that was, first, historically inaccurate, and, second, had little to do with the problem in Los Angeles. "Pittsburgh and St. Louis are usually brought into the smog picture here by uninformed people," wrote a clearly exasperated Gordon Larson in December 1952.[116] An added factor was that Los Angelenos had initially tacitly assumed that the smog would remain a brief episode in the history of the city. When this hope proved illusory, frustration was targeted at the agencies in charge.[117] Thus, completely independent of the quality of the work done by officials, a popular way of talking about the smog problem developed that allowed public unhappiness to escalate to the point at which Gordon Larson eventually had to step down as the head of the APCD at the end of 1954—even though experts were unanimous in their opinion that from a purely professional point of view it would have been hardly possible to accomplish more than he had.[118]

The sequence of events during the mid-1950s cannot but look prodigious in retrospect. This was the environmental revolution of 1970 in miniature: public sentiment became so agitated that it finally dominated policy, sweeping aside sober calculations about ways and means. The outcome was also quite similar to that of 1970: the successor chose to get tough, irrespective of costs. Henceforth the slogan was, "Maximum enforcement against business and industrial plants."[119] Within a very short period of time, the personnel of the Enforcement Division were doubled, and it was no coincidence that more lawsuits were filed in the first half of 1955 than in the previous seven years combined.[120] However, the actual effectiveness of the APCD is unlikely to have kept pace with this surge in lawsuits: the majority of the reported violations were presumably cars with visible exhaust fumes, which were supposed to be relentlessly punished in accordance with the enforcement policy introduced at the end of 1954. Internally, therefore, officials made no secret of the fact that the real motivation behind the new approach was less its immediate effectiveness and more its symbolic effect. "An undefined, indefinite, or unrecognized enforcement policy breeds indecision and ineffectiveness in the minds of the public and the enforcement agency," explained an APCD memorandum.[121]

It is unclear whether this approach had any notable impact on the extent of the problem. However, it was quite effective from a public relations perspective, certainly contributing to the decline of public discontent after 1955. It was only in the 1960s that renewed public interest in the fight against photochemical smog was apparent, and this was strongly linked to the rise of the modern environmental movement.[122] Of course, this approach did not win the public's unlimited trust in the APCD; given the heated atmosphere regularly generated by discussions of the smog problem, this would no doubt have been too much to expect. Revealingly, employees of the APCD made a habit of concealing as much as possible what they did: "Whenever their friends or neighbors learn they are smog men they immediately become subjects of impassioned oratory, much of it scornfully derogatory," the *Los Angeles Times* reported in 1961.[123] And a complaint in November 1958 went so far as to claim that "the public attitude seems to be that of a resigned determination to suffer."[124]

Thus, even after the dismissal of Gordon Larson, the APCD had to reckon with the unhappiness of an irate public. To be sure, this did not only have a downside: the knowledge of where public opinion stood allowed for the kind of rigorous action through which Los Angeles set itself apart from the rest of the country. Stamp Out Smog—an organization founded in October 1958 by a woman from Beverly Hills that had 62,500 members two years later—in particular represented a potent lobby for air pollution control.[125] But there were also citizens' groups that sought to make up with stridency what they lacked in professional competence. As late as the end of the 1950s, the Citizens Anti Smog Action Committee was still loudly and vehemently proclaiming that the source of the problem was chiefly sulfur emissions, even though the emission of sulfur had by this time been below the 1940 level for several years.[126] "In matters as ubiquitous as air pollution, the thoughtful editorial greatly outweighs the controversial headline," the Air Pollution Foundation admonished on the occasion of its dissolution in May 1961.[127] The same could have been said with equal justification a decade later, when the environmental revolution had delegitimized existing ways of doing things across the country.

Obviously, the rest of the country could have learned a lot from Los Angeles. The city's fight against photochemical smog provided a window into the future, a preview of things to come. It showed that there was an alternative to the pseudocorporatism dominating the rest of the country, and it demonstrated that growing public discontent could overwhelm even successful policies and officials. Even more, there was so much news coverage of the Los Angeles situation that one observer noted as early as 1949 that the Los Angeles "sky haze" was "perhaps the most publicized in America."[128] In point of fact, other cities watched the work of the APCD with an equal measure of fascination and distance. There was a clear and undeniable hesitation in other cities to set up control organs modeled after the APCD. In the first few years, this was excusable, since smog's causes were still undetermined, and nobody therefore knew whether Los Angeles was perhaps simply a special case. Beginning in the mid-1950s, however, this excuse became increasingly dubious, as other cities were now showing symptoms similar to those in Los Angeles. San Francisco in 1955 and San Diego in 1956 witnessed their first serious smog episodes, and evidence was also accumulating outside of California.[129] "The problem we face here in Los Angeles is one that in varying degrees could happen, and is happening, in every other major metropolitan area of this nation," Griswold declared in 1956. In subsequent years, the vast majority of experts came to agree with the assessment published in the *Engineering News-Record* as early as October 1955: "The Los Angeles phenomenon is a phenomenon of our particular motorized industrial society, and it is showing up all over the world."[130]

Since the beginning of the 1960s, therefore, there was little doubt that Los Angeles was a portent of things to come, rather than a special case. Still, the interest in other cities remained muted. Some cities even made desperate efforts to highlight the great differences between the local problems and the situation in Los Angeles. "We have an air pollution problem in Michigan but it is not the same as that in Los Angeles County," Michigan's Department of Public Health emphasized as late as 1967, going so far as to claim that there was no photochemical smog in the state.[131] Pittsburgh, too, drew a reassuring conclusion from a comparison of the two cities: "The climatology in the Pittsburgh area has never been known to create conditions which result in smog similar to the Los Angeles area," declared the head of the Allegheny County Bureau of Air Pollution Control.[132] "In Los Angeles . . . the problem was evidenced by frequent public complaints," the Illinois Manufacturers' Association explained, averring that it was "unaware of any appreciable volume of public complaints in this area."[133] Subjectively, such defensive reactions might be understandable: any official who proclaimed Los Angeles the shining example ultimately had to concede that all previous efforts by his own city or state had been hopelessly inadequate—and most officials were simply not ready for this kind of radical self-criticism. But what was subjectively understandable was also objectively disastrous: Los Angeles revealed the future of air pollution control, and nobody bothered to learn from it.

Up until 1970, the only major impact that Los Angeles had on air pollution control in the rest of the country was that it was spearheading the fight over automobile exhaust. In some sense, this was the result of its own success: after the APCD had gained control of all other sources of emissions by 1957, automobiles were the only unregulated air polluters in the city.[134] At the same time, however, automobile exhaust became by far the most frustrating aspect of the city's fight against photochemical smog. For while the people of Los Angeles were waiting impatiently for a solution, Detroit showed a pronounced lack of interest in the problem. This spawned a controversy between Los Angeles and the American automobile industry—a controversy fought with ever-increasing bitterness, especially on California's part, and a controversy whose consequences would radiate far beyond the atmosphere of the city on the Pacific.

7   Environmental Revolutions and Evolutions

An American Nemesis

In the 1960s, the American regulatory tradition, born during the Progressive Era and modified after World War II, faced a double crisis: an internal crisis arising from strategic problems (see chapter 5) and an external crisis produced by several changes in the general context in which local officials were operating. First, it was discovered that automobiles were major contributors to the country's air pollution problem. Needless to say, municipal authorities were in no position to get a grip on an inherently mobile pollution problem; furthermore, the American automobile industry showed only lukewarm interest in solving the problem. Second, new federal policy departed considerably from the municipal tradition and represented a tacit break with the pseudocorporatism of the 1940s and 1950s. Third, the general public was slowly beginning to embrace a new way of looking at the problem, which I will label here, for lack of a better word, the "ecological" perspective. These developments, especially the third, amounted to a fundamental challenge to the regulatory establishment—a challenge, to be sure, that this establishment in many cases failed to recognize, let alone accept. Indeed, while the spread of ecological ideas gave birth to new political demands, the work of the authorities was characterized by a strange attitude of "business as usual."

Needless to say, this merely postponed the conflict and only increased the rift over time. Two different ways of talking about air pollution problems established themselves and took on a life of their own. While questions of what was technically and economically feasible continued to guide "insiders" in industry and regulatory agencies, the effects of air pollution and popular fears about them were dominating the ecological discourse of "outsiders." With both discourses gathering momentum throughout the 1960s, the repercussions were enormous when the outsiders' perspective suddenly became politically potent around 1970. The regulatory establishment, oriented toward consensus and cooperation for decades, not only now saw itself confronted with harsh accusations but also had to struggle with the fact that these two discourses on the same problem were largely incompatible. In order to understand the depth of the regulatory crisis of 1970, it is crucial to note that the clash was not simply about divergent political goals—it was also a disaster of communication, with escalating attacks on what insiders and outsiders all saw as the "other side," attacks that often betrayed a stunning degree of mutual ignorance. It is an open question whether this debacle could have been avoided: the conclusion of this book will offer some thoughts in that regard. Its consequences, however, were abundantly clear: the pseudocorporatist regulatory tradition came to an abrupt and inglorious end in the late 1960s. The new environmental policy, codified in the federal Clean Air Act of 1970 and in many ways in place to this day, would leave behind the smoking ruins of the previous approaches without so much as a second look.

Automobile Exhaust

When Los Angeles took up the fight against the phenomenon that would later be identified as photochemical smog in 1943, suspicion also fell on automobile exhaust with remarkable speed. "The large number of automobiles in the Los Angeles area contributes greatly to the total volume of obnoxious gases," the Smoke and Fumes Commission noted as early as March 1944, and laypeople had similar suspicions.[1] Of course, at the time this was little more than speculation, which is why the APCD initially focused on other sources of pollution; after all, automobiles were emitters difficult to control by their sheer number alone, and they were, moreover, not seen as classic polluters.[2] It was only when Haagen-Smit developed his smog theory that the APCD took on automobile exhaust, thus starting a debate that lasts to the present day. "The development and enforcement of effective steps to control pollution from automobile exhausts appear to be among the most serious problems of the air pollution control program," noted a December 1953 report.[3] A month later, supervisor Kenneth Hahn told the Southwest Kiwanis Club, "The automobile is the key to complete smog control, and if the automotive industry cooperates, Los Angeles skies can be clear by 1956."[4]

At first, officials in Los Angeles were quite happy with the automobile manufacturers. "Detroit is interested and is doing something about our smog problem," declared a report from the APCD in November 1954, and in March 1955 the Los Angeles County Board of Supervisors sent a letter of thanks to the automobile industry "for their continued interest in the matter."[5] Soon thereafter, however, disillusionment set in. As early as August 1956, the Air Pollution Foundation voiced doubts about the commitment of the automobile industry, which set off alarm bells for the first time with Griswold, the head of the APCD: "Your Air Pollution Control Officer . . . cannot understand why the industry appeared to have developed a current 'do nothing' attitude," he reported to the County Board of Supervisors.[6] Two years later, little had changed, in Griswold's view. "Little technical progress has been realized . . . in the development of practical control remedies," he testified before the National Advisory Committee on Community Air Pollution.[7] "It is time, and more than time, for the automotive industry, to increase greatly its effort to control air pollution from automotive exhausts," admonished the Smog Committee of the Grand Jury of Los Angeles County in July 1958.[8] Skepticism was also on the rise outside of Los Angeles: "We are coming more and more to believe that the automobile industry cannot be counted upon to take any drastic step toward solution to this problem," a 1959 PHS memorandum noted.[9] And when Kenneth Hahn published his correspondence with the automobile industry in 1962, he wrote with evident bitterness in the foreword: "The automobile industry has failed to meet its responsibilities to the people of Los Angeles County, California, and the nation."[10]

And these impressions were by no means unjustified, for a strange inertia characterized the efforts of the American automobile industry. When a representative of the Automobile Manufacturers Association listed, in a letter of February 1954, what he believed to be the most urgent desiderata of research, there was no mention at all of possible countermeasures; instead, he declared the first order of business to determine the exact ratio of automobile emissions in the total pollution load, to identify the medical and other consequences, and to develop precise procedures for measurement and investigation.[11] On a different occasion, this same person invoked the enormous difficulties that allegedly confronted a solution. "The situation is analogous to finding a cure for cancer," he asserted, reiterating his call for thorough scientific study of the problem.[12] At a hearing before the California State Assembly in February 1955, one representative of the automobile industry claimed, without further explanations, that brand-new cars caused virtually no emissions. He therefore called for systematic inspection of the engine timing in all vehicles on the road; only if this did not produce any significant success should the question of special cleaning devices again be taken up.[13] And in 1963, a PHS official expressed his puzzlement over the apathetic behavior of automobile manufacturers in the Technical Task Group on Automobile Engineering Research Studies: "The automobile industry representative has never been too communicative or outspoken at these meetings, and has more or less adopted the attitude of observer with no particular comments for or against anything."[14]

All this even though there was no lack of suggested approaches that promised to get control of the problem. The general principle of catalytic cleaning was already known among experts before World War I.[15] "The possibilities of such a method do not appear to have been developed, possibly . . . because of the lack of any demand for improvements to be made," noted a 1947 British publication about the technical potential of a catalytic cleaning of exhaust in automobiles.[16] From the mid-1950s, at the latest, the catalytic converter was thus generally considered one of two basic technical approaches to the cleaning of exhaust gases that deserved closer investigation; the other technical option was afterburning through an open flame in the tailpipe.[17] However, it was clear early on that a catalytic converter would be rendered ineffective if the exhaust contained lead; the use of tetraethyl lead as a gasoline additive, already questionable on health grounds, therefore also impeded the fight against the automobile's other emission problems.[18] And the prospect that a proven antiknock agent might have to be dispensed with must have put an additional dampener on the already less than overflowing enthusiasm of the automobile industry.

There is thus hardly any doubt that the American automobile industry initially took a very lackluster approach to dealing with the exhaust problems of its products.[19] The only lingering question concerns the motives behind the automobile industry's stance. And this question is of interest not least because some contemporaries developed nothing short of a conspiracy theory to explain this behavior: the automobile industry, under this theory, secretly agreed to dispense with effective action against automobile emissions; the industry's joint research program and the agreement that patents concerning exhaust-gas cleaning would be valid for all manufacturers were presented as key evidence for this allegation.[20] "It is more than merely possible that this interlocking arrangement between the large automobile companies has served to deter the development of an exhaust control device rather than to accelerate it," Griswold declared as early as 1958, arguing that the agreement had "many of the characteristics of a trust."[21] Similar suspicions attracted growing attention in the following years, eventually prompting the Los Angeles County Board of Supervisors in January 1965 to demand that the U.S. attorney general investigate these allegations.[22] The investigation, however, yielded no clear result. The process dragged on for several years and eventually ended in October 1969 with an out-of-court settlement in which the automobile industry did not acknowledge any wrongdoing—a settlement that drew massive criticism from the nascent environmental movement.[23]

This conspiracy theory has also exerted a certain fascination over environmental historians.[24] However, two sets of circumstantial evidence in fact render the existence of such a conspiracy unlikely. First, the papers of Kenneth Hahn, a key person in this story thanks to his role on the Board of Supervisors, raise considerable doubts. The situation was in fact much more complicated than Hahn made it seem to the outside world. For example, in January 1966, Hahn published his correspondence with the automobile industry, clearly intended as an indictment. "There will be no relief from smog and air pollution for the people of Los Angeles County and California until the automobile industry turns its attention from higher horsepower and fancier styling to meeting its responsibility to the public health," Hahn declared in his foreword.[25] However, absent from this documentation was a letter that did not at all fit with the picture of general failure that the book otherwise seemed to reflect. While Hahn on all other occasions called for effective measures, Chrysler received a very different letter from him in the summer of 1963. "I want to express to you my sincere appreciation for your research, your work and your accomplishments," Hahn had written appreciatively, even including copies of the rather harsh letters he had addressed to Ford and General Motors at the same time. A few months later, the Board of Supervisors decided to equip the cars of agencies under its authority with a retrofit developed by Chrysler.[26] In yet another instance, the wording itself is revealing: in a letter to Attorney General John Mitchell, Hahn spoke of "negligence" and "conspiracy" as if the two were synonymous.[27] Finally, Hahn's credibility is not enhanced by the fact that he responded to the termination of the lawsuit in October 1969 with an even grander conspiracy theory, now casting suspicion not only on the automobile companies but also on the oil industry. On February 16, 1970, he wrote to the attorney general, in all seriousness: "It appears a giant conspiracy exists between the automobile industry and the petroleum industry not to manufacture automobiles with adequate smog control devices or to produce smog-free gasoline—a conspiracy which has existed during the last 15 years."[28]

Second, the very behavior of the automobile industry itself casts doubt on the conspiracy theory. Indeed, its behavior often evidenced such breathtaking incompetence that it is hard to imagine any long-term calculation on the industry's part. A prime example is the testimony of automobile manufacturers at a State Assembly hearing, at which automobile representatives quite matter-of-factly presented the smog of Los Angeles as the special problem of a single city, a thesis that would have required at least justification four years after the first appearance of similar symptoms in San Francisco.[29] The representatives then declared that the question of a threat to health was still unresolved and required further study, which also did not sound very convincing to contemporaries, who had long viewed smog as a health problem.[30] Afterward, they went on to present a peculiarly convoluted statement about Haagen-Smit's smog theory, which by this time was generally seen as definitive: "The theory of photochemical smog is not an automotive industry theory.... We have accepted this as a working basis, and have based our work on it, but it is not our theory, nor are we putting it forward."[31] Finally, one representative declared that the automobile industry's research program had insufficient funds, only to assert in the same breath that its work was, of course, not suffering from this situation.[32] It thus seemed like a diversionary maneuver when the presentation concluded with a plan to introduce an extensive inspection system as a precursor to installing cleaning devices. No one, save the industry representatives themselves, could have been surprised when an assemblyman declared, "I don't like that kind of approach to the problem." The other listeners were also less than enthusiastic: "Its net effect was zero, a fiasco," an observer from the PHS reported.[33] Given all this, the fact that the plan was anything but persuasive on its own merits—it had very weak experimental underpinning, industry representatives could say nothing concrete about its costs, and its technical feasibility rested entirely on the apodictic, highly dubious assertion of a lobbyist—was of only incidental importance.[34]

The behavior of industry representatives was also remarkably inept in other situations. At a meeting between Arthur Fleming, the head of the Department of Health, Education, and Welfare, and representatives from the Automobile Manufacturers Association, the latter demanded that the department back their demands for further research, without any indication that they would, in turn, help the department secure the necessary funds.[35] In addition, the manufacturers repeated the previous year's faux pas, presenting photochemical smog as the special problem of Los Angeles. When it was pointed out to them that San Francisco now had a similar problem, they agreed only with evident reluctance.[36] "The meeting adjourned... without reaching any specific conclusions or agreements," the minutes eventually recorded, the wording itself hinting at the puzzlement of the government representatives.[37] In fact, the auto industry was never able to organize a potent coalition around its concerns. Matthew Zafonte and Paul Sabatier, who have studied automotive pollution control policy with a view to the

Advocacy Coalition Framework, describe the 1960s anti–pollution control cluster as consisting of "a rather bizarre set of actors."[38] At times, the industry's assertions seemed to come from another planet. When Kenneth Hahn in 1961 lamented auto manufacturers' inadequate efforts, the Automobile Manufacturers Association responded indignantly: "We do not understand how the industry's long record of cooperation with the officials of Los Angeles and the state of California could lead to such a conclusion"—a formulation that reveals either incredible mendacity or abysmal naiveté.[39] In short, in a host of cases, the industry's conduct was so clumsy and incompetent that the question of far-reaching strategic considerations can be addressed on principle: by no stretch of the imagination is it possible to detect any kind of planned strategizing behind the manufacturers' short-term tactical behavior—unless they were systematically working to discredit themselves. To put it another way, if there was indeed an industry conspiracy, these were some of the most incompetent conspirators in world history.

Rather, there are plenty of indications that the auto industry for a long time did not take the problems of Los Angeles seriously. After all, manufacturers in the 1950s were poorly sensitized to the fact that their products were producing exhaust problems; the last discussions in which the automobile had figured as a potential polluter had taken place twenty-five years in the past.[40] And since the patent agreement ruled out competition among manufacturers, the incentive for car makers to act essentially consisted of merely a few complaints from a large city on the Pacific—and from the perspective of distant Detroit, this provided awfully weak motivation to undertake resolute measures.[41] Reading through the correspondence Hahn published, it is striking that manufacturers' answers were often written by technicians within the company's hierarchy, even though Hahn's letters were always addressed to the companies' presidents.[42] Moreover, it would appear that the automobile industry was initially still hoping that it could resolve the problem by improving the burning within the motor. At any rate, one advisor to the APCD reported in November 1954 that "any type of afterburner will have to be very good or very cheap (or both) to change the basic faith of the automotive industry in the 'cure at the source' approach."[43] Finally, until passage of California's Motor Vehicle Pollution Control Act in 1960, the automobile industry had no legal obligation to tackle its emission problems.[44] The patent agreement in fact had only a single precedent: the development of a uniform standard for the sealed-beam headlight—a fact that throws a revealing spotlight on how little significance the U.S. auto industry initially accorded the exhaust problem.[45] Thus, industry attention was focused on other issues, while its representatives, with their loose talk about the emission problem, lost all public credit.

Incidentally, it should be emphasized that Kenneth Hahn's motives in propagating a conspiracy theory were certainly self-serving: after all, this put enormous pressure on the foot-dragging auto industry to demonstrate the seriousness of its efforts.[46] Moreover, the cartel suit brought by the Board of Supervisors reflected precisely the mood in greater Los Angeles: since the frustration over photochemical smog had long since boiled over, there was a clear need to find a scapegoat—and a desire by elected officials to move out of the line of fire. Interestingly, at a Board of Supervisors meeting on January 26, 1965, Kenneth Hahn emphasized the effect of the planned suit outside of the courtroom: "I think what you are going to have to bring out is publicity," he bluntly stated. In contrast, the question of whether there was in fact any good evidence for the alleged conspiracy became secondary.[47] It is no exaggeration to say that conspiracy theories were literally in the air in Los Angeles. For example, several individuals, including the mayor of Los Angeles, poured forth hints that the PHS's research was somehow being controlled by the automobile companies.[48] And California's Motor Vehicle Pollution Board—the first institution of its kind in the country, and probably the world—also found itself the target of such suspicions within a short period of time.[49]

State regulation of automobile emissions thus confronted an exceedingly difficult task: not only was it supposed to impel a lethargic industry into taking resolute measures, but at the same time it had to stand up to the critical scrutiny of an irate public. With respect to the former, regulation had some success. In the course of the 1960s, engineers were in fact able to develop the first cleaning devices for automobile exhausts, and their use was made mandatory; the 1970s saw the mass use of catalytic converters, which eventually became the universally accepted approach in automobile exhaust cleaning technology. In fact, this prompted some observers to issue optimistic pronouncements that seem premature from today's perspective: "The day is fast approaching when the automobile will no longer be regarded as a major polluter," the magazine *Nation's Business* declared in August 1968.[50] Yet these measures could not even begin to keep pace with continuously rising public discontent. For example, in 1969, a few California environmental groups, including Stamp Out Smog, called for a boycott against the automobile and petroleum industries to force them to develop vehicles with low exhaust.[51] And it was not long before even this initiative seemed almost conservative. "Ban the internal combustion engine" was the

Los Angeles League of Women Voters' blanket demand at the beginning of 1971, and at the time there was in fact a serious debate in California about just such a ban.[52] The 1970 Clean Air Act's provision that emissions of new cars had to be reduced by 90 percent within five years looked almost retrograde against this backdrop—even though it represented a radical break with earlier, much less ambitious directives: in 1966, a representative from the Department of Health, Education, and Welfare had still been aiming for a reduction of 40 percent by 1985.[53]

The immense importance of these debates becomes clear only if it is borne in mind that no other conflict over emissions was more intensely publicized in the 1960s than the controversies swirling around automobile exhaust. Publications from the nascent environmental movement were scathing in their attacks: assessing the attitude of car makers, Donald Carr spoke of a "national scandal of major proportion," and the report by Ralph Nader's "Task Force" on air pollution contained a chapter entitled, "The Automobile Industry: Twenty Years in Low Gear."[54] The extraordinarily uncooperative stance of the American automobile industry thus had consequences that reached far beyond the issue itself, discrediting the efforts of American industry as a whole. "So much for how one industry has shown how unreliable voluntary cooperation can be," Howard Lewis concluded his account of the controversy, in his book entitled *With Every Breath You Take*.[55] Against this backdrop, it is easy to forget that the automobile industry and its behavior were really more of an exception.

Feds on the Move

Federal agencies had repeatedly dealt with air pollution problems since the turn of the century. During the Progressive Era, the Bureau of Mines had published several studies on the coal-smoke issue and had contributed to the scientific study of the smelter-smoke problem in the Selby Smelter Commission, and the Bureau of Mines and the PHS had jointly taken on the automobile-emissions problem in the 1920s.[56] However, these were always isolated projects rather than part of a more comprehensive environmental agenda, and they were never truly at the center of the agencies' work. This pattern of sporadic action initially continued after World War II. The Donora catastrophe became the subject of an investigation by PHS experts, as did air pollution from riverboats in international waters between Detroit and Windsor, Canada, and the U.S. Technical Conference on Air Pollution in May 1950 was organized and run by the Bureau of Mines.[57] Still, it was not until Congress passed the Air Pollution Control Act of 1955 that the federal government launched a sustained policy on air pollution issues, which continues to the present day.[58] However, it took some time before this policy was anchored institutionally: a Division of Air Pollution was not set up within the PHS until the beginning of 1960, but it then established itself quickly as the federal government's central organ on questions of air pollution control.[59]

The Air Pollution Control Act of 1955 focused on creating a research program, education and training measures, and technical support for other agencies.[60] As these were all largely uncontroversial areas, the federal program was largely free of rancor in its early years.[61] "Federal activities in air pollution are considered so necessary to the eventual conquest of the problem that virtually every agency concerned, public or private, favors their continuation and expansion," a 1959 article declared.[62] On the other hand, the federal government was originally not involved in any way with the practical fight against air pollution problems, and this too was apparently in accord with the wishes of a broad majority.[63] The fate of a bill by Congressman Paul Schenck was revealing: it called for specific emission values for vehicles, but in the course of deliberations, the bill was watered down into a resolution that amounted merely to a PHS study on the health effects of automobile exhaust.[64]

Within the PHS, as well, there were serious reservations about taking a more active role. At this time, the PHS saw its tasks generally in the field of scientific research and was therefore opposed to taking on executive and enforcement authority; at the beginning of the 1960s, the only meaningful enforcement authority it had was in the area of quarantine measures.[65] "In general the principal role of the Federal government is to support research, to provide for technical assistance and training and to participate in the solution of international air pollution problems.... The setting and enforcement of regulations is primarily a local matter," a PHS memorandum declared, in no uncertain terms.[66] At times, this position even placed it in open conflict with the Department of Health, Education, and Welfare, to which the PHS was formally subordinated.[67] The PHS only ceased to resist further authority after President John F. Kennedy intervened in early 1963.[68] The PHS acquired certain enforcement authority for the first time with the Clean Air Act, passed that same year.[69]

The main focus of the Clean Air Act of 1963 was still the strengthening of local and state authorities through consultation and especially money. The success of this financial support was in fact sweeping: the PHS estimated that funds for air pollution–control programs doubled within the space of about two years.[70] Thus, the federal government approached the actual fight against air pollution problems only indirectly, which was probably a good thing in hindsight, for it meant that the PHS did not become entangled in a host of individual conflicts right away. Instead, the PHS maintained a broad and fairly independent perspective, which allowed it to nurture something that other American bureaucrats were rarely able to develop: an overarching vision. The PHS was one of very few institutions that considered what an ideal air pollution–control strategy might look like during the 1960s, and the result, roughly, was that the starting point of an effective control program was to be the definition of certain target values for local air quality. After that, the current emission load, the type of pollutants, the critical polluters, the geographic conditions and their influence on the distribution of emissions, and the expected growth of a given large city were to be determined in as much detail as possible so as to derive specific threshold values for individual emitters—and, needless to say, the control program was to encompass the entire conurbation. In other words, all factors that influenced air quality were to be thoroughly analyzed in the hope that certain technical standards could be derived from the results. Clearly, this approach drew on classic concepts of conservation—the scientific management of scarce resources. In fact, this intellectual tradition was already evident in the strategy's programmatic epithet: the "air resource management approach."[71]

To be sure, this approach came with its own set of problems. For example, it is doubtful whether it would in fact have been possible to deduce—with compelling logic—usable standards from a dispersed set of findings, and even champions of this approach conceded shortcomings in some respects.[72] Yet this kind of criticism fails to appreciate the value that the "air resource management" approach possessed within the regulatory context of the 1960s: it was the first model that proceeded not from given administrative and technical possibilities, but from desired air quality. Unlike pseudocorporatist air pollution control, which started from the existing state of technology and the wishes of industry, the "air resource management approach" directed its attention first at the extent of the problem. Above all, however, the PHS now had a standard against which to measure the results of air pollution–control efforts—and the results were generally unambiguous. "The pragmatic campaigns that have been conducted in the past will no longer suffice," noted Vernon MacKenzie, the head of the PHS Division of Air Pollution, at a September 1965 conference in New England.[73] Compared to the self-praise of pseudocorporatists, this was a welcome dose of sober realism.

The PHS, however, lacked the means for direct implementation of the "air resource management approach." The enforcement mechanisms provided in the 1963 Clean Air Act were so complicated that they were of only limited use; moreover, regarding problems that concerned only a single state, the PHS could intervene only with the consent of the agencies of the state in question, which practically ruled out such interventions altogether.[74] And so the PHS essentially had to rely on good counsel in setting up and reforming control programs to help the "air resource management approach" achieve success—counsel, however, that many contemporaries were also actively seeking. "We are frequently called upon to assist State and local agencies in drafting, reviewing and revising air pollution laws and regulations," a member of the Division of Air Pollution reported in April 1964.[75] A survey conducted at the same time revealed that PHS reports were more highly respected by state and municipal officials than all other official reports.[76]

And when the PHS was asked for advice, its recommendations made it abundantly clear that it pursued very ambitious goals. In the end, expert PHS opinions broke down some basic rules of pseudocorporatist air pollution control. Above all, the PHS was convinced that only legal coercion would propel industry to take meaningful measures. "It is not practical to assume that industry will voluntarily control air pollution to a significant degree," noted an internal memo in 1960.[77] Consequently, the PHS rejected in principle the membership of industry representatives on control boards. And officials did not find industry's objection that only its participation would prevent arbitrary bureaucratic action very convincing. "The great concern is not that the control program will be too restrictive or arbitrary, but that it may not act in a sufficiently effective and timely manner to protect the public health and welfare," PHS officials noted in an expert opinion.[78] In addition, the PHS was in favor of stiff punishment: "Violations should carry substantial penalties in order to discourage any violator from incurring the penalty as a cost of doing business," noted a comment appended to a Montana bill.[79] And when ASME and APCA together planned a "guide to industry for the control of air pollution"—a project, incidentally, inspired by the work of the VDI—the PHS took a noticeably reserved stance.[80] "I gave no promises, commitments, nor even undue encouragement to ASME and APCA," said one file note, the very wording suggesting the official's skepticism—indeed, his distrust of the whole enterprise.[81] The PHS time and again emphasized the shortcomings of all previous efforts compared to the problem's true dimensions—in open contrast to the champions of pseudocorporatist air pollution control, who liked to boast about their own successes. "We have only begun to scratch the surface," Vernon MacKenzie declared in a speech to the American Petroleum Institute, while Arthur Stern stated apodictically in a hearing, "The problem of air pollution continues to grow faster than the combined Federal, State, and local efforts to deal with it."[82]

To be sure, expert PHS opinions and financial support for local control programs did lead to a considerable surge in bureaucratic activity in many parts of the country in the 1960s. At the same time, however, mere words were unsatisfactory for an institution that was committed to a broad vision of a new, more aggressive approach to air pollution control. And so the PHS threw itself with great zeal at the only direct contribution it could make to the diffusion of the "air resource management approach": the development of threshold values for the concentration of outdoor air pollutants. At the time, this was pioneering work: "We are only beginning to approach a full understanding of the connection between man's well-being and the quality of his environment," MacKenzie said in a 1965 lecture.[83] At the same time, however, there was a clear demand for such threshold values: in a 1965 survey, researchers, officials, and other experts declared standards for outdoor air one of the most important desiderata of research.[84] And in fact, the results published by the PHS after years of investigation would assume considerable—though entirely unintended—importance: in the public hearings in the wake of the Air Quality Act of 1967, the PHS's stringent threshold values served as almost indispensable points of reference for environmentalists.[85] Moreover, without the years of research by the Division of Air Pollution, the federal government most likely would not have been able to fulfill the directive of the 1970 Clean Air Act to publish specific "ambient air quality standards" within thirty days.[86]

It is surely not surprising that industrial circles were watching all these activities with a good deal of suspicion. Throughout the 1960s, industry categorically rejected a federal policy that went beyond a mere research program.[87] Only the automobile industry favored federal emission standards since 1965, hoping to avoid a myriad of divergent state laws.[88] To be sure, there were not yet the hardened fronts that often characterized environmental politics in the United States in later decades. "We cannot clear the air without close cooperation—among all levels of government, industry, and the public," MacKenzie declared in a lecture. As late as 1969, a PHS brochure pointed out that many company presidents were making serious efforts to control their air pollution problems without being forced to do so by law.[89] Still, the united front of those representing industry interests drove federal officials to seek a lobby of their own. And so the PHS invested a substantial portion of its energies in intensive public relations work that went far beyond brochures and lofty words. For example, after a hearing in New Jersey dominated by industry lobbyists, the PHS observer decided to cultivate future contacts with the few representatives of the affected public who were present, "with the idea that their opinions might bear valuable weight in subsequent conference action."[90] The PHS also supported the National Tuberculosis Association in its air pollution work, especially during the early years, and the Montana Clean Air Association, which had done nothing more than borrow a film about air pollution problems from a special office of the PHS, received a letter from the Division of Air Pollution that virtually overwhelmed the association with offers of help.[91] Even the report from Ralph Nader's study group, above suspicion of whitewashing bureaucratic activities, complimented the public relations work of the PHS.[92] In the end, all this served a higher goal: "An important effect of raising the degree of public awareness of contemporary air pollution problems will be to strengthen the hand of the control official," MacKenzie explained in May 1965 before the Committee for Nuclear Information in St. Louis, one of the few environmental groups that existed at the time.[93] That the federal government, in its efforts to get the lethargic public to protest loudly, would soon be swamped by its success was something that MacKenzie surely would not have dared to dream about in 1965.

Ecology Arising

In March 1964, the Public Health Service received the results of a survey it had commissioned, in which about one thousand residents of greater St. Louis spoke about their attitudes toward air pollution. Many of the findings were certainly not surprising for experts in the field. "Nearly all residents of the St. Louis area feel that governmental action is desirable," noted the poll, and most individuals also considered it acceptable if administrative programs increased their own living costs by five dollars a year. At the same time, those polled showed a marked preference for control programs on the local and regional levels, and the majority also felt that the current strategy of cooperation was right. The answers also mirrored some limitations and shortcomings in problem awareness: air pollution was not yet a truly burning issue, and industry continued to be seen as the most important polluter; all other sources of pollution were mentioned far less frequently. Only in one respect did the survey show a result that was completely new—the full significance of which the authors of the study did not yet appreciate: "Many persons who do not feel the air pollution problem in their own residence neighborhood, are nonetheless 'bothered' by air pollution."[94]

It takes the wisdom of hindsight to recognize the fundamental importance of this new state of mind. For there was one thing that the countless public statements about air pollution problems since the late nineteenth century had in common: the speakers were personally, directly affected by air pollution. Of course, that is not to say that public protest was an adequate reflection of the actual emissions load: the previous narrative offers ample evidence for the social imbalance of the antismoke movement. But whether

Figure 10. During the 1960s, an alarmist tone began to surface in air pollution rhetoric. This cartoon is from a book that the Public Health Service produced in preparation for the 1966 National Conference on Air Pollution. Image courtesy of the *Chicago Sun-Times*.

the issue was smoke in Pittsburgh or smog in Los Angeles, air pollution was always a problem that occurred in a person's immediate environment. And the fact that in the 1960s air pollution also became a problem for those not directly affected could mean only one thing: a major change in the perception of the problem was under way. Air pollution was no longer merely a social problem, but a social construct.

Therefore, the clearest sign of the change in public opinion in the 1960s was a new kind of rhetoric. And this rhetoric revealed something that air pollution—except in Los Angeles—had never before evoked in the United States: fear. Air pollution was no longer merely a phenomenon with a few unpleasant consequences, but a reason to be afraid. In many cases, the names of the new organizations mirrored this increased urgency: at least four chose, independently, the acronym "GASP."⁹⁵ This could indicate both a response to the enormity of the problem and a basic physiological reaction, and this was clearly no coincidence: the threats to health were now generally at the center of popular fears. "The ultimate objectives of the City Air Pollution Division should be aimed at reducing health aspects of air pollution with economic considerations playing a secondary role," the Ohio Pure Air Association declared in December 1960, for example.⁹⁶ The women of the Cleaner Air Committee of Hyde Park–Kenwood, who had initially approached air pollution very traditionally as an issue of cleanliness, soon changed their view, as well. "Five years ago we began our fight against air pollution because we were weary of the dirt," reported one of these women in 1964. "Today we are continuing because we feel that the evidence is becoming overwhelming that our health and that of our children demand it."⁹⁷ The popular media, especially, painted the health threat in glaring colors: "If you live in any city with busy factories or numerous cars, you have filthy and possibly deadly air in your lungs," *Harper's Magazine* warned as early as 1959.⁹⁸ The impact that such reporting had on contemporaries is evident in a 1966 complaint from a Detroit-area resident who had just seen a TV report entitled "The Poisoned Air": "While watching the program, I could almost feel myself being choked by the pollutants, and I wonder how long it will be before we will all choke as a result of dirty air," he wrote to the governor of Michigan.⁹⁹ "Public concern over air pollution is centered mainly on its effects on health," an article composed at the end of the 1950s asserted.¹⁰⁰ And an "Air Pollution Seminar" in Minnesota in October 1969 proclaimed somewhat flippantly, "Health effects are the name of the game."¹⁰¹

However, the new prominence of the health argument in complaints and petitions was by no means the only change in public discourse. Especially in comparison with earlier statements, the tone became noticeably more abstract in the course of the 1960s. For example, the Coalition for the Environment in St. Louis declared in its statement of purpose, "If human life is to continue to flourish, the people in every corner of the world must learn to protect and support the environment that in turn supports them; to guide the pace and the nature of the changes man makes in the natural world; to take care that we do not sacrifice the basis of life for future generations to some temporary advantage for this generation."¹⁰² Moreover, statements increasingly stressed the global dimension of the problem. "Air pollution is causing increasing concern not only nationally but on a worldwide level," the environmental group Clear Air Clear Water Unlimited, based in greater Minneapolis, declared as early as 1958, pointing among other things to the radioactive fallout from nuclear testing.¹⁰³ A 1961 article in the otherwise staid magazine

House Beautiful had this to say: "Even if you live in the remote vastness of Alaska, or on the sunny shores of Waikiki, you can't escape breathing polluted air."[104] With striking frequency, contemporaries now also invoked the fragility of the earth's atmosphere, which earlier generations had simply not cared about. "'Air' is that thin band of mixed gases that envelops the planet on which we live"—so began, for example, a widely read 1967 brochure.[105] In general, air pollution was increasingly being seen as a truly urgent problem, and the practical fight against it was tracked with great impatience. "Progress is significant, albeit too slow to satisfy the more hotheaded members of the Air Pollution Committee like its chairman," reported the chair of the Air Pollution Committee of the Chicago-based Hyde Park–Kenwood Community Conference, regarding the work of local control agencies.[106] Finally, the fact that air pollution was now understood as part of a broader ecological crisis of society was also a sign of the shift in popular discourse. A 1966 publication by the Conservation Foundation began, "This is about air pollution, one of the many ways in which industrialized, urbanized—and sometimes civilized—man is devastating his environment."[107]

Thus, the debate over air pollution mirrored a more general sentiment within American society. It was no coincidence that "Eve of Destruction" rose to the top of the sales charts in August 1965, becoming "the fastest-rising song in rock history," according to some disk jockeys.[108] The song obviously captured the mood of the time, pessimism over the course of American society and apocalyptic scenarios gaining a prominence rarely found in U.S. history. Indeed, turmoil was the dominant state of political debates. Maurice Isserman and Michael Kazin even go so far as to speak of a "Civil War of the 1960s."[109] The escalating war in Vietnam was dividing the country, as were the civil rights struggle and the war on poverty—and all along, the American business community was losing much of its former legitimacy. It was not surprising, then, that the debate over air pollution also grew more divisive. For the first time in the history of American environmental politics, the concern over pollution grew into full-scale horror scenarios, and apocalyptic descriptions of the hazards of airborne pollutants gained currency politically. Even the official journal of the National Parks Association warned in 1966, "The world appears to be moving toward dangerous and perhaps catastrophic environmental conditions with a relatively unknown end result because of the rapidly increasing rate of pollution."[110]

How novel this perspective was is perhaps best revealed by the fact that it was precisely the venerable organizations from around the turn of the century that remained strangely untouched by this development. One example was the Civic Club of Allegheny County, whose struggle against the Pittsburgh smoke nuisance reached back to the creation of a special committee in April 1897.[111] Together with the Chamber of Commerce, the Civic Club had secured the passage of a local ordinance in 1906. It was a founding member of the Smoke and Dust Abatement League; the United Smoke Council, which supported the campaign against the problem of domestic fires in the early 1940s, was basically an outgrowth of the Civic Club.[112] But when the demands of the nascent environmental movement dominated a hearing in the state of Pennsylvania in 1969, the Civic Club stood up squarely against the spirit of the times. "I have been personally disturbed at the apparent emotional anti-industry stand of many of the participants at the public hearing," the club's secretary declared, parts of his statement reading like the remarks of an industry lobbyist.[113] Even more extreme was the case of the Air Pollution Control League of Greater Cincinnati, which had been founded in 1906 as the Smoke Abatement League of Cincinnati.[114] While other organizations had long since been emphasizing the global nature of the problem, the Air Pollution Control League was still describing air pollution as a local problem in 1969, going so far as to add truculently, "It will always be a local problem."[115] While other groups were discussing Rachel Carson's warnings against the abuse of pesticides, the league was blithely investing a substantial part of its energy in a weed-eradication program to provide relief to asthmatics and allergy sufferers.[116] When the ecological discourse became a strong, organized movement in 1970, the Air Pollution Control League watched this development with obvious unhappiness. "Finger pointing is the order of the day among many interests in air conservation," the league complained in October 1970, pleading stubbornly for cooperation by all parties involved.[117] In the 1970s, the League also came out in favor of the construction of nuclear power plants—against the will, incidentally, of Robert Kehoe, a professor of medicine at the University of Cincinnati and a staunch defender of leaded gasoline, who had been affiliated with the league through many years of active work.[118] As late as 1981, the Air Pollution Control League was still expressing regret over the politicization of air pollution control and calling for the collaboration of all parties—at a time when a bitter conflict raged across the country over Ronald Reagan's antienvironmental revolution.[119]

To what extent this change in mentality went hand in hand with a change in the social basis of air pollution control is difficult to judge, given the lack of systematic studies. However, there are indications that the long process of broadening the movement's social basis, already evident in the immediate postwar period, continued in the 1960s. To be sure, the emerging environmental movement was not a representative mirror of American society.

In 1966, a study of air pollution control in Chicago found that substantially more complaints came from the wealthy North Side than from all other sections of the city.[120] But it is striking that unions were finally joining the camp in the late 1960s. "What was once viewed as a threat to the health and lives of industrial workers has increasingly become an issue of survival for the entire nation and the whole world," the United Automobile, Aerospace and Agricultural Implement Workers of America noted in March 1970; the previous year, the United Steelworkers of America had organized the first conference on air pollution by an American union.[121] Occasionally African Americans also became active, although Andrew Hurley's thesis that the steel town of Gary had a "multiclass and multiracial environmental coalition" at the end of the 1960s is surely an exaggeration.[122] Freddie Mae Brown, the president of the African American environmental organization Black Survival Inc., candidly conceded at the beginning of the 1970s, "Blacks feel the ecological movement isn't their thing."[123] And Cleveland's mayor Carl Stokes—the first black mayor of a major U.S. city—made no bones about his reservations when he was asked for a statement about environmental pollution in the summer of 1970: "Attention to ecological problems should not become a cop-out. The war on poverty, on hunger, and on inadequate health care should be America's first priority."[124]

To some extent, the new awareness was an ironic by-product of the successes that postwar pseudocorporatist air pollution control could claim regarding pollutants perceivable by the senses. Needless to say, the United States continued to have air pollution problems that the senses could detect: "The dust at our house is so bad that sometimes I have to use a razor blade to scrape it from the windows," stated a 1966 complaint from the city of Petoskey in northern Michigan. As late as 1964, the Cleaner Air Committee of Hyde Park–Kenwood presented a dirty handkerchief as evidence of the local emissions load, a reenactment of Progressive Era tactics.[125] On the whole, however, many places witnessed a noticeable reduction, especially of visible and smellable pollutants. However, this did not lead to gratitude toward and confidence in the authorities, as officials and industrialists had hoped; instead, the discontent came to focus on invisible pollutants, in many cases subject to no or only loose controls. Common clichés invoked air pollutants as an invisible but dangerous threat to life and health. This had a lot to do with medical research of the postwar period, which was slowly able to paint an increasingly precise picture of the consequences of air pollution.[126] Moreover, society seems to have been highly receptive to these new health threats, since many of the diseases that had once spread fear and terror were now under control.[127] But the situation surely also had a lot to do with the fact that greater threat could be projected onto dangers that escaped immediate sensory perception.

It goes without saying that it is difficult to pinpoint this change in mentality with any precision. The rise of the ecological perspective unfolded in the kind of nonuniform, discontinuous way that should be expected in a prolonged and geographically dispersed process of social change. Still, as a rough periodization, it might be said that the first signs of a change in public opinion date from the second half of the 1950s; that these signs appeared so frequently after about 1960 that a "movement" can be alluded in the narrower sense of the word; and that from around the middle of the 1960s a growing intensification of efforts can be observed, as well as increasing supraregional interconnections. However, there is no apparent pattern in the formation of new organizations. Some organizations were still entirely within the tradition of neighborhood organizations. For example, the Buhrer Air Pollution and Civic Association, set up at the end of the 1950s, focused entirely on air pollution in the heavily industrialized valley of the Cuyahoga River in Cleveland.[128] Other organizations took a comprehensive view of the problems in a given city: in 1960, the Ohio Pure Air Association had been formed in Cleveland, and Clear Air Clear Water Unlimited had been founded even earlier, in 1956.[129] Women's organizations were also active in several cases: the League of Women Voters of Salem, Oregon, proclaimed the local air pollution problems the study topic of the year in 1959, and the Federation of Women's Clubs of Missouri did the same two years later.[130] The National Tuberculosis Association played an especially prominent role in the formation of new organizations. Because of the sharp decline in tuberculosis cases in the postwar period, the organization was looking for new tasks and therefore set up a National Air Conservation Commission in 1966. The founding of the Delaware Valley Citizens' Council for Clean Air in Philadelphia, the Metro Clean Air Committee in Minneapolis, and the Air Pollution Control League of Rhode Island is hard to imagine without the massive support from the local branches of the National Tuberculosis Association.[131] Finally, the new interest in air pollution was also evident in organizations that had previously devoted themselves to classic nature protection: 1959 marked the first time that emission problems were discussed at the annual meeting of the Hudson River Conservation Society, the breathtaking river landscape of the Hudson, which formed the meeting's backdrop, blighted by the clearly visible emissions from a nearby power plant.[132]

The ecological cause received a boost in 1962 from Rachel Carson's *Silent Spring*, even though the book was discussed less within the various organizations and agencies than might be expected.[133] Evidently, the heated conflict between Carson and the chemical industry was initially seen as an exception, without broader significance. At any rate, a general anti-industrialism cannot be detected within the early environmental movement. Seemingly, most efforts were still characterized by the constructive cooperation that had been established everywhere after 1945. "We agree with a policy of peaceful cooperation with industry if it is not at the expense of our home owning community," declared the Southeast Air Pollution Committee in Cleveland.[134] When talking to industrialists, Clear Air Clear Water Unlimited regularly underscored its intent "to help industry to help itself," and Gals Against Smog and Pollution, from Missoula, Montana, following a demonstration in front of a paper factory, wrote a letter of thanks for the friendly reception they received from factory representatives.[135] As a rule, there was no general reluctance in dealing with representatives of industry. As late as December 1968, the Metro Clean Air Committee voted unanimously in favor of cooperation with the Minnesota Association of Commerce and Industry, and the Delaware Valley Citizens' Council for Clean Air even had a representative from the chemical giant Du Pont on its board of directors.[136] But under the surface, a certain change could be detected even here: the almost boundless trust that industry had enjoyed in the immediate postwar period had largely vanished. The Ohio Pure Air Association provides a prime example of the ambivalent attitude of emerging environmental groups. While it gave a "Best Neighbor Award" to the steel company Jones & Laughlin for its outstanding contributions to air pollution control, it also supported a "taxpayer's lawsuit" against Republic Steel.[137] It was characteristic of the attitude of the nascent environmental movement that while a brochure from the National Tuberculosis Association recommended doing everything to gain the understanding and support of local industry, it simultaneously warned against giving industrial associations a seat on the board since if activists did so, their "air conservation program may be handicapped, perhaps beyond salvation."[138] The general rule was that resolute determination was seen as crucial to achieving the desired cooperation from industry. "Regulations should have teeth," Howard Lewis wrote in *With Every Breath You Take,* and a message from the Buhrer Air Pollution and Civic Association in the spring of 1962 put this opinion in plain terms: "The fact that industry knows that 678 persons are aroused, and organized, and sitting at the top of Jennings R[oa]d Hill, has a constructive effect in strengthening industry's 'Social Conscience.'"[139]

A similar change took place in the emerging environmental movement's relationship to municipal and state authorities. There still was active collaboration between citizens and bureaucrats—but the friendly agreement that had prevailed in earlier times was no longer there in quite the same way. In fact, it seems public organizations in the 1960s cooperated with authorities more or less only because working against the regulatory establishment seemed even less promising. Occasionally, a fundamental critique of the administrative approach could be heard. "It seems to me personally that certain powerful leaders in the townships squash, one way or another, complaints of air pollution in order to make it appear that the townships are happy to have industry move in, come what may," one representative of the Sharon Civic Association in Pennsauken, New Jersey, declared.[140] In general, however, the nascent environmental movement was much more gentle in its dealings with existing institutions.

A larger vision of a new kind of air pollution control, however, was not discernible behind the statements from environmental organizations. Moreover, it would appear that there was no real interest in developing such a vision within the emerging environmental movement. It is interesting to note that the "air resource management approach" advocated by the Public Health Service had very little resonance in these organizations, even though in principle it would have been quite compatible with their goals. At best, one can detect in their statements fragments of a new strategy: for example, there were repeated calls for high fines, which by the very amounts involved would have had a deterrent effect.[141] In essence, however, the organizations were only interested in one thing: the toughest possible directives and their rapid implementation. "In general, efforts to combat air pollution have moved at an all too leisurely pace," the Conservation Foundation criticized, and the San Francisco–based Citizens Against Air Pollution simply advocated "realistic industrial emission standards that will stamp out smog."[142] For most organizations, the question of how these general demands were to be translated into concrete regulations seemed in the final analysis a technical one.

Incidentally, when it comes to assessing the nascent environmental movement, it is important to remember that vociferous complaints about the apathy of the public were common in the 1960s. Environmentalists voiced "scattered cries of protest," as the *Science News Letter* observed in 1961, but they were far from constituting a popular movement. The Committee for Clean Air Now, headquartered in Palo Alto, California, acknowledged with a sigh that the fight for clean air was "not the most glamorous thing in the world." As late as September 1969, the members of the Metro Clean Air Committee racked their brains about how to arouse the public from its disastrous lethargy.[143] In fact, in retrospect, it is hard to deny that these complaints had some substance. To be sure, existing organizations could not complain about a lack of interest: as early as 1958, Clear Air Clear Water Unlimited had more than one thousand members.[144] But if one compares the dramatic press reports from the 1960s with the decade's rather small number of organizations—in 1961, the Cleaner Air Committee of Hyde Park–Kenwood was evidently the only organization of its kind in all of Chicago—it is hard to escape the impression that the organizations documented merely a fraction of the public discontent.[145] In part, this discrepancy between actual awareness of the problem and its organizational representation can be explained with the help of Mancur Olson's logic of collective action, which in this case constituted a brake—as subtle as it was effective—on all organized activities. However, it was of at least equal importance that the ecological discourse of the 1960s was at heart a discourse far removed from the everyday qualms of regulation. For if the emerging environmental movement failed to develop a new vision of regulatory policy, this was, in a way, perfectly consistent behavior: at the center of the ecological discourse stood the dangerous effects of pollutants and personal fears about them, and the question of strategies for regulatory action was in some sense merely a by-product—indeed, almost an appendix—of these popular fears: "The concept that the air is a dumping ground for waste products is unacceptable to us. Air is to breathe, not a depository for wastes, and we want to keep it for that purpose."[146] Statements like these document merely a vague sense of unease about bureaucratic agencies, one that was difficult to translate into concrete political demands. And so an inner logic was certainly apparent, when, in spite of all the cooperation between the authorities and environmental organizations, two largely distinct and separate discourses came increasingly to the fore: the insider discourse of the regulatory establishment, focused on the technical and economic aspects of air pollution control, and the outsider discourse, overwhelmingly focused on the effects of pollution. And since most insiders did not quite realize that their approach was losing public credibility, the practice of air pollution control in the 1960s increasingly resembled a dance on a volcano.

"Business as Usual"

The ecological discourse presented industry with three problems at once. First, representatives of industry were increasingly pushed onto the defensive in the 1960s. After air pollution control had been for decades a comparatively uncontroversial issue from the perspective of business, manufacturers were now feeling the wind of opposition in their faces. Second, there was every reason to believe that the heightened public demands would entail increasing costs for industry. "There is little doubt that the tab for pollution control comes high," noted *Dun's Review and Modern Industry* in March 1963.[147] However, industry would probably have been able to master these two problems with little effort, had there not been a third problem: massive communication difficulties. Since industry and environmental organizations cultivated two completely different forms of discourse, only fragmented communication between the two parties was possible. If representatives of the emerging environmental movement spoke about danger and fear, industry saw this at best as the demand for stricter threshold values—which was obviously only part of what was being said. Industry did not come up with an effective response in the 1960s because it failed for the most part to understand what was happening.

As a result, industry's reaction was often simply a verbose defense of cooperative air pollution control. For example, in September 1966, the Greater Detroit Board of Commerce declared, "Industry is as interested in developing an adequate and useful program of air pollution control for Michigan as any other segment of the population."[148] "Industry in general has been cooperative with the controlling agencies," the California Manufacturers Association asserted, labeling itself "part of the total family."[149] The American Petroleum Institute claimed in 1964 that "the oil industry has never hesitated to, or failed to, face up to its air pollution problems," while the National Association of Manufacturers published an in-depth account of air pollution control in Pittsburgh as an example "of the truly remarkable results that cooperative community action against air pollutants will produce."[150] Individual entrepreneurs also emphasized their constructive attitude: "Cooperation and teamwork . . . are the only sure paths to continued progress in the air and water conservation programs of the future," a brochure from Minnesota's largest oil refinery declared. Standard Oil's in-house magazine for employees and stockholders suggested that "the battle for clean air can be won, but the winning will require the intelligent cooperation of all segments of society—businessmen, voters, the scientific community, and local, state and Federal governments."[151] "If a certain newspaperman is authoring unfair articles about industry, it is helpful to invite him to the plant to acquaint him with the size and complexity of your problem, and what you are doing about it," a representative of Granite City Steel explained in 1967 before a committee of the Illinois Manufacturers' Association.[152] And Kerryn King, head of Texaco's PR department, seemed almost delighted: "The current 'crisis of concern' over air and water pollution presents a classic example of the need for effective public relations."[153]

If industry's response was thus largely intensified PR work, this did not mean, of course, that industry believed that everything was in fact fine. Praise

Figure 11. By the 1960s, air pollution control had evolved into an established field of engineering with complex machinery, as this switchboard for a number of filters amply demonstrates. Little wonder that industrialists, as "insiders," thought they had done their duty! Image courtesy of the Hagley Museum and Library.

for cooperation always entailed the obligation to take effective measures. For example, the Interlake Steel Corporation in Chicago declared, "We recognize the problem as national in scope, and we feel that it is our corporate responsibility to do everything within social and economic reason to help correct it."[154] Industrialists in fact took pains to back up this claim with concrete measures. After the American Petroleum Institute had been accused of negligence during a congressional hearing, it approached the Public Health Service directly and asked for information "relative to what the API could do to indicate that they are doing their full share in the battle against smog."[155] The Manufacturing Chemists' Association organized a workshop in Denver in 1966 "to encourage industry to fully meet its responsibilities for air pollution control." At the end of 1968, the Illinois Manufacturers' Association went so far as to call for a substantial increase in the budget of the Illinois Air Polluting Control Board, as well as a reform of the rather cumbersome enforcement procedures.[156] New York saw the founding of the New York State Action for Clean Air Committee, which brought together Associated Industries of New York State, the state Air Pollution Control Board, and the

State Tuberculosis and Respiratory Disease Association. A similar organization was set up in the heavily industrialized Ohio River Valley, between East Liverpool, Ohio, and Parkersburg, West Virginia, where creating an administrative program was especially difficult because of the interstate nature of the problem. New York City even saw the establishment of a New York Business Council for Clean Air "to stimulate citizen action."[157]

In fact, industrialists did not even gloss over the fact that there were huge differences among the contributions of individual businesses. Texaco's Kerryn King noted tersely that a minority of "extremists" was found not only among the public but also in industrial circles, and the "coordinator of air and water conservation" of the Humble Oil & Refining Company put it bluntly: "A small percentage of industry is irresponsible."[158] On the whole, industry's reaction was thus not simply a narrow-minded battle against tougher requirements, but a constructive attempt to preserve the well-practiced procedures of pseudocorporatist air pollution control even as background conditions were changing. In fact, many industrialists believed that they were responding honestly and honorably to the public's growing discontent—except that their approach always remained bound by the criteria of the insider perspective. From industry's point of view, the ecological discourse certainly presented a good reason to make a greater effort, but it was by no means a fundamental challenge.

A perfect example of how industrial circles insulated themselves against ecological protest is a decision the Allegheny County Air Pollution Control Advisory Committee made in June 1959. Under consideration was the Allegheny County Citizens against Air Pollution's application for membership on this Advisory Committee, which, as mentioned earlier, was tilted in favor of industry. Not only did the Advisory Committee unanimously vote down its petition, but it also saw the attempt to join the committee not as a substantive challenge to its actual work but as evidence of a PR problem. "The need for increased publicity of the progress made by industry in helping to reduce air pollution was mentioned," the meeting minutes revealingly declare.[159] Representatives of the nascent environmental movement encountered similar problems in Cleveland. In 1964, the Ohio Pure Air Association called for the creation of an Air Pollution Advisory Committee, whose existence was in principle even mandated by law. This demand evidently fell on deaf ears: the 1965 annual report of the Cleveland Division of Air and Stream Pollution makes no mention of such an Advisory Committee.[160] And decisions like these were by no means random: they document the self-confidence of the regulatory establishment in general and industry in particular, the notion that they stood for the only rational approach to air pollution control. It thus seems that the chemical industry's arrogant response to Rachel Carson's *Silent Spring* was not a simple slipup but perfectly rational behavior from the insider's perspective.[161]

In the absence of a meaningful strategy debate, industrialists also failed to grasp that the crisis within pseudocorporatist air pollution control was growing increasingly acute in the 1960s. The élan of the early period, when considerable success was possible with little effort and expense, had by now largely evaporated. Especially in regions with well-developed control programs, tough negotiations now prevailed where there had formerly been a lively and uncomplicated working relationship. The case of southern Detroit's Great Lakes Steel Corporation offers a prime example of just how much the prevalence of industrial interests could shape these negotiations. In 1966, after lengthy preliminary talks, the company proposed a two-million-dollar control program that promised to eliminate all of the plant's emission problems by 1971.[162] However, the government representatives were anything but thrilled: Michigan's Department of Public Health objected that the plan provided for no meaningful reduction in the emissions load in the first two years, while county representatives were calling for comprehensive control within three years. Only the town of Ecorse, where the plant was located, could muster any enthusiasm for the company's offer. That the proposal was nevertheless accepted in the end was mostly because of the lack of alternatives. Since the company categorically rejected speeding up its program in any way, this left only recourse to the courts—with the outcome uncertain. "If this matter were brought to court it could possibly be delayed for a period longer than five years," warned Michigan's Department of Public Health, and its position eventually carried the day.[163] It is obvious that such stories were not well suited to winning new supporters for much-vaunted cooperation.

At times, the internal crisis of pseudocorporatist air pollution control in the 1960s had disadvantages even from industrialists' perspective. Above all, the failure to develop a regular process of setting threshold values in the immediate postwar period now came back with a vengeance. Industrial lobby groups had to constantly solicit the desired cooperation, and quarrels over the modalities of the consultations were often the order of the day.[164] And since technical standards were often negotiated by ad hoc committees that were immediately dissolved once their work was done, industrial lobbyists had little reason to restrain themselves. One of the more turbulent negotiations took place at a meeting of the Illinois Air Pollution Control Board in September 1965. When the representative of the Illinois Manufacturers' Association vigorously attacked the draft of an ordinance and suggested setting up a special subcommittee to revise it, the representatives of the public authorities candidly declared that the existing standards were "idealistic in their approach to the problem": the fact was that they had so far ignored technical and economic considerations, and the draft was actually "primarily a trial balloon in order to get the ball rolling and have something concrete to start with"—a statement that all but guaranteed freewheeling haggling in subsequent negotiations.[165] And in Pittsburgh, the conflicts over the development of new threshold values at the end of the 1960s even became public, a clear alarm signal in a city that was associated—as perhaps no other—with cooperation and consensus. Evidently, the cohesive force of pseudocorporatist air pollution control had weakened to the point where even members of the Advisory Board of Allegheny County on Air Pollution no longer felt obliged to defend their joint decisions. While a representative of U.S. Steel offered lavish praise for the result of the negotiations ("as a member of the Board, I am very proud of its effort"), another member of the Advisory Board made no bones about his disappointment: "These Regulations . . . represent the typical compromise that normally comes out of committee action."[166] Morton Corn, another member of the Advisory Board, went a step further and offered his critique of the agreement in a public hearing.[167] The contrast to the VDI Clean Air Commission could hardly have been more striking.

Threshold values for sulfur dioxide, in particular, became hotly contested in the 1960s. In principle, such threshold values were certainly long overdue. "The electric utility industry . . . must install adequate equipment to remove sulfur from its flue gas," the *Harvard Business Review* noted in September 1966.[168] But since sulfur dioxide was invisible, industry had no interest from a PR point of view in resolving the problem; moreover, it was clear that any solution would become a very costly undertaking.[169] Consequently, industry pushed hard for the most permissive threshold values possible or even demanded that the introduction of such standards be postponed entirely. At the end of the 1960s, the National Coal Policy Conference, a coal-industry lobbying group, published a booklet that sought to demonstrate the overwhelming lack of secure scientific insights and urged a "cautious approach."[170] And in fact, draft laws frequently teetered on the edge of meaninglessness: one draft by New Jersey's Air Pollution Control Commission from the outset excluded all sulfur dioxide emissions produced by the burning of fossil fuels, while Morton Corn demonstrated that the standards planned for Allegheny County in practice merely codified the status quo.[171]

To be sure, the crisis of air pollution control in the 1960s should not be exaggerated. The mere fact that a debate about sulfur dioxide was getting under way documents considerable progress vis-à-vis the 1940s and 1950s.

Whereas earlier all of the more difficult problems had simply been put off, now the totality of all air pollutants was increasingly coming into view. For example, Michigan's Department of Public Health was now also pushing for effective measures with smaller foundries, even though these measures were often more expensive than the plants themselves: the necessary costs amounted to between four hundred thousand and one million dollars for each cupola furnace. In justification of its position, the Department of Public Health declared, "It is impossible to allow complete exemption for one class of industry without undermining the whole air pollution control program."[172] Considerable sums were still being invested in air pollution control.[173] The authorities, too, were generally satisfied with industry's attitude: "In many cases, reductions in pollution levels have been gained which are not covered explicitly by 'Code' provisions but through voluntary efforts by industry," the Division of Air Pollution Control in Cleveland reported, for example.[174] At times, industry's successes could even strike a curious note: *Nation's Business* gleefully related the story of a photojournalist who in 1961 photographed a seemingly emissions-free steel plant to illustrate a story about the crisis in the steel industry—while the plant was actually running at maximum capacity.[175]

Of course, there were black sheep among the entrepreneurs who were critical toward air pollution control, as well as various regions that refused to follow the general trend.[176] On the whole, however, many industrialists were convinced that they were responding constructively to the public's mounting unhappiness. For example, the journal *Steel* proclaimed in May 1966, "There's little room for doubt about industry's good intentions."[177] And so many industrialists could hardly conceal their astonishment that public protest just would not quiet down in spite of all their efforts: "The paradox is that public concern has increased during the last few years when much more was done to reduce pollution than in any preceding period," Kerryn King noted.[178] It was not unusual for industry's self-confidence to go even further: oftentimes the rhetoric of industrialists reveals that they regarded their own approach as the only useful—indeed, the only rational—approach possible. Time and again, one senses the faith in a golden middle way that would emerge virtually by itself in calm and reasonable negotiations. "The proper and reasonable industry approach to air pollution control planning lies somewhere between the extreme position of a small number of zealous but not well-informed air conservationists and a small but vocal contingent of inflexible but not well-informed industrialists," a manager proclaimed at a conference at Tufts University, while the Standard Oil Company in California complained about "emotion and ignorance" in the debates over air pollution and declared pompously, "What we must do is establish rational standards for the purity of air and water—standards that are scientifically proper and which protect the health and well-being of the community—and which are economically attainable within these requirements."[179] One section of a statement from the Illinois Manufacturers' Association carried the revealing title "Emotion vs. Reason."[180] Industry failed to notice that champions of the ecological discourse could all too easily misunderstand such pronouncements as arrogant defensive reactions.[181]

Therefore, it is likely that many industrialists had a clean conscience when it came to their own pollution problems. And thus, they saw no reason why they should not make their willingness to cooperate contingent on certain conditions: If there was only one rational approach to this problem, was not meeting these conditions a compelling necessity? "We have been opposing proposed regulations and laws with constructive suggestions that we believe responsible corporate citizens have an obligation to make," a Du Pont manager blithely noted in an internal memo.[182] The editor of *Steel* also presented cooperation on the part of industry representatives matter-of-factly as a substantively necessary consultation to ensure that the work was appropriate: "Participation and leadership by industrial leaders will make problem solving programs more realistic, more efficient, more effective, and will tend to keep the costs within the limits of our total resources."[183] A speaker at a meeting of the National Association of Manufacturers in November 1968 was even more blunt: "I think we in industry are rapidly becoming victims of a tyranny of pollution control decisions made by many people who know so much 'that ain't necessarily so.'"[184] This being the case, industrialists believed they simply had to be proactive to ensure that reasonable work was done in the future. For example, the Illinois Manufacturers' Association demanded in March 1966 that the five-member Technical Advisory Committee of the Illinois Air Pollution Control Board have at least three engineers from industry—very much in line with industry's classic goal of attaining veto rights.[185] And it was not rare for lobbyists to bring out the big club: the warning that factories might close and industry leave. For example, the Greater Detroit Board of Commerce observed to the governor of Michigan that "if the regulations are adopted as they now stand, they may discourage the growth and development of Michigan industry."[186] And here, too, misunderstandings were almost inevitable: what seemed to insiders a struggle for the only reasonable approach often struck outsiders as mere obstructionism.

Industrialists were not always able to have it their way. Contrary to the wishes of the Illinois Manufacturers' Association, only two representatives from industry were placed on the Technical Advisory Committee.[187] In general, however, industry could argue from a position of strength. "I would like to emphasize the difficulty that a local, unpaid board, limited in time and money, finds when it has to compete against the hordes of lawyers and lobbyists hired by industry," declared a member of the Granite City Air Pollution Control Board in Illinois in January 1968.[188] The result was that industry was able at least to put its clear stamp on most laws and ordinances in the 1960s. For example, the Air Quality Act passed by Congress in 1967 was so profoundly shaped by industry's demands that it seemed to some nothing but "coal's law."[189] Among other things, industry pushed through its proposal that ambient air-quality standards, which were to be developed by the states in a complicated process, would become subject to public hearings.[190] For an industry that had a clean conscience when it came to emissions, this was an almost ideal starting point for exerting effective control: through well-rehearsed appearances at these hearings, industry lobbyists would be able to guide air pollution control onto sensible tracks. That other individuals might also get a chance to speak at these hearings was evidently a possibility that never occurred to industry. This may have been the greatest error in the history of air pollution control.

Two Worlds Collide

The gap between public awareness of the problem and organizational representation of that awareness attained almost grotesque dimensions toward the end of the 1960s. A mood bordering on panic had long since spread among broad segments of the population, and a series of polls by the Opinion Research Corporation in Princeton brought it to light: the percentage of those surveyed who believed that air pollution problems were "very serious" or "somewhat serious" had risen from 28 percent in May 1965, to 48 percent in November 1966, to 55 percent in November 1968. The numbers were even more dramatic for residents of large cities: by 1968, no less than 84 percent were convinced of the seriousness of the problem.[191] However, in most cities, this view continued to be represented by only a handful of organizations, which, needless to say, were hardly in a position to keep up with industry's well-organized lobbying machine. The Olson paradox was still blocking the formation of a potent alliance that could give political weight to the demands of the ecological discourse. The public's unhappiness was beyond doubt—the only thing missing was an occasion for citizens to voice their anger.

In this context, the hearings held under the Air Quality Act became the central point around which public protest crystallized. Throughout the country, these hearings were dominated by numerous representatives of an irate public, who did not mince words. For example, some four hundred people attended the Pittsburgh hearings, and the threshold values under review were roundly rejected by most of the forty-five statements, the first speaker describing them succinctly as "legalized murder."[192] In Atlanta, the hearings were moved to a nearby church because the conference room at the State Health Building could not accommodate the large public turnout, and here, too, there was no lack of passionate voices. "We are talking about life and death, and survival," warned a professor from renowned Emory University, who was speaking for an organization called Citizens for Clean Air, while the "information director" of the Georgia Tuberculosis and Respiratory Diseases Association warned that this would not be the first civilization to become extinct "as a result of man's greed."[193] Similar gloomy tones were also heard in Chicago in August 1969: "We must question if we Americans, in our astounding affluence, have somehow deluded ourselves into believing that anything could be more important than our health," one speaker said.[194] Health was also the chief argument in Minnesota. "We should be very concerned by the rising incidence of respiratory disease including emphysema, asthma, chronic bronchitis, and lung cancer," said a representative of the Metro Clean Air Committee.[195] "Let us ... guard the life-sustaining atmosphere of earth to keep it a good place to live," one speaker from the American Association of University Women demanded at a hearing near St. Louis.[196] And at a hearing in Fairfax, Virginia, about fifteen Girl Scouts presented an immortal slogan: "Make Love, Don't Pollute the Air."[197]

In this heated atmosphere, hardly anyone could still muster much sympathy for industry. After all, the battle lines had been clearly drawn. "We are engaged in a struggle to preserve the quality of life versus the right to sell or make a product which tends to destroy it," proclaimed the California-based Clean Air Council.[198] Even the Metro Clean Air Committee, which shortly before had sought cooperation with the Minnesota Association of Commerce and Industry, brushed off a reference to possible costs with a laconic remark: "Everything is relative."[199] And by the standards of the day, this was still rather moderate criticism. For example, the Minnesota Emergency Conservation Committee declared to Bethlehem Steel Corporation on the first "Earth Day" (April 22, 1970), "You have trampled on people and created a situation whereby America has become a nation forbidding existence for future generations."[200] "Industry has contributed greatly to the pollution of our environment, and now it is their moral responsibility to pay for it.... Let the polluters know that to expect us to sacrifice human lives on their behalf is unrealistic," one congressman in Chicago thundered.[201] Others professed utter incomprehension: "I like so many others do not understand the logic of Industry," a woman wrote to the mayor of Cleveland. "Do they think that the air that surrounds them is different than the air we are trying to protect?"[202] Even Pittsburgh's Group Against Smog and Pollution, which stressed constructive, science-based work like few other groups, occasionally indulged in anti-industry tirades. "The company's callous disregard of human life and health is equaled only by its dishonesty," it noted of U.S. Steel, for example.[203] And an editorial in the *Cleveland Press* provided the history lesson to go along with this state of mind: "The history of pollution abatement is clear: Polluters just don't seem to get around to cleaning up their dirty discharges until they have to."[204]

On the outside, the hearings seemed like the spontaneous outcry of a tormented public. In fact, however, at least one institution had been systematically working toward such a result: the Division of Air Pollution within the Public Health Service. Time and again, John Middleton, the commissioner of the National Air Pollution Control Administration, pointed emphatically to the significance of these hearings. For example, at a conference at Colby College in Maine, he proclaimed, "I cannot emphasize too strongly the importance of broad participation in these public hearings."[205] And the officials did not end with high-minded words: in close collaboration with the Conservation Foundation and the National Tuberculosis Association, the National Air Pollution Control Administration supported special workshops all over the country in which participants were given the knowledge they would need to make an effective presentation.[206] Federal officials were also helpful on details: the above-mentioned Girl Scouts were able to demonstrate in Fairfax only because an employee of the National Air Pollution Control Administration had previously established the necessary contact.[207] Of course, this is not to say that public discontent was merely the result of deliberate manipulation. The work of federal officials was consciously limited to supportive and encouraging measures, and it was only too obvious that the statements of the environmentalists mirrored an authentic feeling of the times. Still, it is at least doubtful whether the protests would have been quite as extensive and sophisticated without the determined work of the federal government.

At the same time, the ecological cause was also benefiting from developments in other areas of politics. In the conflict over the Vietnam War, exhaustion set in after the escalation during the presidential campaign in 1968. The same was true for the civil rights struggle: after the assassination of Martin Luther King Jr. on April 4, 1968, it passed "into another stage characterized by bickering and shouting."[208] Against the background of the divisive politics of the late 1960s, a topic that united all Americans—in fact, all human beings—obviously had a special charm: the environment was a topic that finally brought everyone together under one big roof, perhaps not healing the wounds of the previous disputes but at least allowing Americans to forget about them for some time. The Earth Day celebration, not coincidentally, was intended as a strictly nonpartisan event, and contrary to popular belief, students were not prominent in the hearings pursuant to the Air Quality Act. Indeed, a representative of the Conservation Foundation, listing, in August 1969, the various kinds of organizations whose members had participated in the foundation's workshops, did not once mention student organizations.[209]

For industry representatives, all this was deeply unsettling. Greatly vexed, they discovered that they had lost all public credibility overnight. At the Pittsburgh hearing, the sole defender of the proposed threshold values was one representative from the Pennsylvania Chamber of Commerce, while the Georgia Business and Industry Association did not even bother to attend the hearing in Atlanta, submitting its comments in writing instead.[210] The statements of industrialists commonly reflected their utter incomprehension about unfolding events. "Radicals Kidnap Pollution Baby," the journal *Iron Age* reported in January 1970, an assessment that was remarkably ignorant given the actual makeup of the protest movement.[211] Hardly less peculiar was an industrialist's remark at the end of a hearing in Wisconsin: "I would like to strongly suggest that in our zeal to correct air pollution problems as promptly as possible, we not let ourselves get carried away with impractical, unrealistic and highly emotional approaches, which can only lead us further down the path of socialization."[212] An internal memo from the Illinois Manufacturers' Association stated, "Pollution control is now such an emotional issue that housewives and hippies were joining forces in the recent Board meetings." And as though this were not already abstruse enough, the author added this gloomy rumination: "There is increasing indication that radical militants are working in this area to discredit industry and government."[213]

Such statements should not be summarily dismissed as clumsy scaremongering. After all, the irritation that made industrialists bluster about an alliance between hippies and housewives was quite understandable given the entrepreneurs' realm of experiences: to anyone who was used to decades of cooperation between authorities and industry, the arguments of environmentalists looked irritating indeed. For decades, industry had been emphasizing that air pollution control was one of the natural preconditions of industrial production—and now the environmental movement was accusing it of irresponsibility. For as long as anyone could remember, industrialists had routinely been in touch with officials and the public regarding air pollution—and now the environmental movement was asserting that it was impossible to talk to industry. As recently as November 1965, the business magazine *Fortune* had published an essay with the programmatic title "We Can Afford Clean Air"—which, as a "Fortune Proposition," had the explicit backing of the publisher, its content praised even by critical officials of the Public Health Service.[214] And now all sides were accusing industry of hiding behind the cost argument. Environmentalists must have struck industrialists like people from another world. And in a certain sense, that is indeed what they were.

To be sure, there were also industrialists who expressed understanding for the public's ire. For example, after an executive memo from the Illinois Manufacturers' Association reported on environmentalists' alleged exaggerations, Chicago's Quaker Oats Company responded by saying, "Play more the role of the devil's advocate, less that of the apologist."[215] In general, however, the situation around 1970 seems like a classic case of miscommunication: while environmentalists and industrialists believed they were quarreling, they were in fact talking right past each other. Of course, the differences between the two sides did not amount merely to different manners of discourse: the two sides were pursuing genuinely contrary goals, and this certainly justified an open conflict. But what made the issue truly explosive was the fact that the antagonists were talking about completely different things: one side was talking about technology and costs, the other about dangers and fears. While industrialists placed questions of technological and economic feasibility front and center, environmentalists dismissed these issues with few words. As New York's Citizens for Clean Air put it, "Human health and well being must be considered before financial ease and corporate convenience."[216] "Any society that can send men to the moon can also solve such pollution problems as these we have here on earth," observed Montana's Gals Against Smog and Pollution, and in Atlanta the representative of an environmental organization declared, "We hear much about our technology today, but the assertion that technology will solve our problems evidences a blind faith that is the antithesis of technology itself."[217] The *Environmental Handbook* published by Friends of the Earth, founded in 1969 by David Brower, even included a recommendation from the environmental initiative Boston Area Ecology Action, directing activists not to enter into any discussion about threshold values but to demand instead "an immediate end to air pollution."[218] Many a technician must have heard such words with a sense of desperation: anyone who was planning an industrial plant around 1970 was virtually flying blind when it came to standards and threshold values. As the editor of the journal *Power Engineering* wrote in September 1969, "At the moment, our concern is not so much that the standards are too strict or too loose, but that the target is constantly moving."[219] Among environmentalists, such problems were not even discussed.

The explosive power of this miscommunication is especially apparent when the debate is examined from the perspective of regulatory strategies. The only thing that was clear was that certain approaches would not work any longer: in the wake of the ecological revolution, pseudocorporatist air pollution control had clearly been discredited. Around 1970, hardly anyone could still imagine a trusting collaboration among industry, officials, and the public. But the champions of the ecological discourse offered at best only vague outlines of a new vision of regulatory policy, the environmental movement continuing to evince a pronounced lack of interest in such questions. To be sure, the call for tough punishments was repeatedly heard, "Fines for large industries . . . should be such that they would prefer to install the necessary control equipment rather than pay fines," the Cleaner Air Committee of Hyde Park and Kenwood maintained. But such demands were more like the legal equivalent of accumulated popular anger than the elements of a comprehensive strategy.[220] In the final analysis, environmentalists' demands were aimed at only one thing: the most stringent possible approach, which would eliminate the existing problems as quickly and comprehensively as possible. "Our health and our long-range economic welfare require firm action based upon strict controls," was the word from Pittsburgh.[221]

In political terms, this created a huge vacuum: the old policy was undoubtedly finished—but it was largely unclear what should take its place. And thus arose an odd situation: at a time that marks a historical highpoint in general ecological awareness, the impulse toward reform came not from the environmental movement itself but from two other developments only indirectly related to the environmental revolution. First, implementation of the Air Quality Act of 1967 had by now proved so exceedingly difficult that another change to the law seemed entirely sensible.[222] Second, Edmund Muskie, who as the longtime chair of the Senate subcommittee in charge of air pollution control was identified with this issue perhaps more than anyone else, was the frontrunner for the Democratic nomination in the 1972 presidential campaign, offering President Richard Nixon the chance to steal his likely opponent's thunder with a bold bill.[223] This was made all the easier for Nixon since industry had decided on short notice to give up its traditional opposition to a stronger federal role. For unless the federal government took the issue off the agenda with a comprehensive law, industry ran the danger of an endless series of battles over new laws at the state and municipal levels—a horrifying vision after the traumatic experience of the hearings.[224] The result was the passage of the Clean Air Act at the end of 1970, large parts of which have remained relevant to American air pollution control to this day.

Even if the Clean Air Act was thus a rather indirect result of the ecological revolution of 1970, it was clearly a reaction to broad public protest. The act called on the newly founded Environmental Protection Agency (EPA) to establish "National Ambient Air Quality Standards," which would define the maximum permissible ambient air concentration for a number of important pollutants. The EPA was explicitly instructed to develop these standards with an eye only toward health effects and to ensure an "adequate margin of safety." Moreover, specific technical standards were to be put in place for new installations; for existing polluters, guidelines would have to be developed to reduce their emissions. Implementation of these directives would be achieved by agencies of the various states, which had to demonstrate to the federal government in special "State Implementation Plans" how they planned to reach these goals. These implementation plans had to meet certain criteria and required the approval of the EPA. For automobiles, a 90 percent reduction in emissions was required by 1975, a mandate that was not yet achievable with contemporary technology. With such stringent directives, a cooperative approach could hardly be expected anymore. As though to remove the last doubts about the future course of air pollution control, William Ruckelshaus, the first head of the EPA, told the *New York Times* in December 1970 that, on principle, industrialists should not be members of control agencies: "It's not a good idea to have the regulated controlling the regulators."[225] In an essay he penned for the *Annals of the American Academy of Political and Social Science,* Ruckelshaus went so far as to call his own work "the beginning of the new American revolution."[226] Indeed, anything that promised less than an ecological revolution was considered hopelessly inadequate around 1970. The Clean Air Act of 1970 was, as Richard Lazarus notes, "dramatic, sweeping, and uncompromising," setting a precedent for the environmental statutes of the early 1970s.[227] The pseudocorporatism of the previous decades was now merely an unfortunate piece of history that people were glad to have put behind them.

Afterthoughts on the Environmental Revolution

The environmental revolution of 1970 was something of an American nemesis. After all, not only did it mean the collapse of a proud and long-standing regulatory tradition and thus the final end of the age of smoke. It also meant the end of the arrogant attitude that a quarter-century of pseudocorporatist air pollution control had produced within the American business community. For decades, industry had entertained the belief that it was doing everything "within reason"—and now it finally learned that "reason" was not necessarily defined by industrialists. Environmentalists won a resounding victory, one that they would recall for a generation and more. However, it was a victory that turned sour with amazing speed. Environmentalists quickly realized that clean air was not merely a matter of tough legislation. Even more, it gradually dawned on the environmental community that the approach defined in the 1970 Clean Air Act was exceedingly bureaucratic and cumbersome. And so the summer of 1971 saw the establishment of a Coalition to Tax Pollution, which advocated the use of free-market instruments to solve the sulfur dioxide problem.[228] This approach was not entirely new: as early as 1959, the Rand Corporation had developed the concept of a "smog tax."[229] But only now did the idea of taxing sulfur dioxide emissions garner meaningful support among environmentalists. The director of the Sierra Club candidly admitted that with this the environmental movement was bidding farewell to a self-imposed taboo: "Traditionally conservationists have been reluctant to consider economic factors in relation to controlling industrial pollution. They have been fearful that shifting the focus to these factors tends to move the discussion into an arena which is basically more favorable to the pleas of polluters."[230]

The Coalition to Tax Pollution could certainly not complain about lack of support. Its January 1972 membership roster reads like a veritable "Who's Who" of the environmental movement at the time: the National Audubon Society, the Federation of American Scientists, Environmental Action, Friends of the Earth, Zero Population Growth, the Wilderness Society, and the Sierra Club were corporate members, in addition to thirty-five local organizations from twenty-three different states.[231] The concept of a "sulfur tax" also drew lively backing from economists: the positive voices that the Coalition to Tax Pollution gathered in a special brochure ranged from Milton Friedman and Wassily Leontief to Paul Samuelson and James Tobin.[232] The problems with implementation of the Clean Air Act, which became ever more obvious after 1970, also underscored the merits of such a concept. "Studies have shown that State Clean Air Act implementation plans are, for the most part, extremely ineffectual in dealing with this problem," the director of the Oregon Environmental Council complained in May 1973.[233] Still, the concept was not put into practice: the special situation of 1970, when the momentum of public opinion presumably would have carried the "sulfur tax" through Congress, no longer existed.[234]

Thus, the failure of the Coalition to Tax Pollution reveals that the environmental movement squandered an opportunity in 1970—the opportunity not only to influence politics through sentiments but to shape it through concrete concepts. The "sulfur tax" in fact amounted to the first draft of a regulatory strategy against air pollution that was actively championed by the environmental movement. And it was no coincidence that the movement was a few months late with this draft: the thrust of ecological discourse had left environmentalists unable to present a comprehensive concept of reform. In 1970, environmental policy wasn't made—it simply happened.

Numerous studies have chronicled developments since 1970, and there is no need to recount subsequent events here: the energetic start and the subsequent disillusionment once the administrative challenges of air pollution control became clear; the reinvigoration under the Carter Administration; the backlash under Reagan, and so on. However, few have noted the peculiar atmosphere in which policy was made in 1970: this was a strangely ahistoric decision-making process, with traditions and previous experience playing almost no role. The goal was not to reform air pollution control but to create it from scratch, without attention to what already existed. Even technical standards, usually subject to intensive scientific debate, were now defined with unprecedented carelessness, "We just picked that number because it sounded like a good goal," an advisor to the Senate Environment and Public Works Committee later explained, regarding how the 90 percent reduction mandate for automobiles had come into being.[235] Everyone worked under the same tacit assumption: that existing laws and institutions, and the regulatory tradition they represented, had become discredited and obsolete and that the new policy would have to work with a new set of guidelines, rules, and regulations—a stance that made the federal program hugely unpopular with state officials, who bitterly resented the EPA's involvement, rather than seeing the agency as a welcome ally in the fight against pollution.[236] The age of smoke had ended without fanfare or glamour; in fact, in the heat of the environmental revolution, few people noted—or cared—that there had been an age of smoke. Within the nascent environmental movement, there was no place for a regulatory tradition that required time, or cooperation, or trust in industry. Some sixty years after the reforms in Chicago, and thirty years after the movement's apogee in St. Louis and Pittsburgh, the smoke inspectors' reign was finally over.

The Environmental Revolution, German-Style

During the federal election campaign in 1961, Willy Brandt, the candidate of the opposition Social Democrats (SPD), put forward a bold demand in his speech at the party convention: "The sky above the Ruhr region must become blue again!"[237] This sentence was not given any particular prominence within the speech: clean air was simply one of the policies with broad appeal, like the nation's health and urban renewal, that Brandt put at the center of his domestic political agenda, hoping to make the SPD viable to voters beyond its classic working-class circle.[238] But this sentence achieved enormous resonance with contemporaries and eventually entered into the Federal Republic's collective memory. The "blue sky above the Ruhr" became one of the most frequently quoted campaign slogans in the history of the Federal Republic; for many Germans, Brandt's sentence is pretty much the only thing they know about the history of air pollution control. Far less well known, however, is the fact that Brandt's foray did not come as a surprise to the governing CDU. Interestingly, the Chancellor's Office, under Konrad Adenauer, had already proposed an inquiry into further measures as early as the summer of 1960, "also with a view toward the coming elections."[239] The politicization of air pollution control had, so to speak, been in the air for some time.

This episode reflects the growing importance that air pollution was assuming in the postwar period. Already around 1960, clean air was considered so important that leading politicians addressed it in high-profile, public ways. To be sure, heads of state had occasionally taken up air pollution problems even before this time. In the 1880s, Reich Chancellor Otto von Bismarck had tried in vain to pass a smoke ordinance for Berlin, and in 1905, Theodore Roosevelt, in his "State of the Union" address, had called for a tough fight against smoke in the District of Columbia. But these remarks were altogether different from a catchy sentence couched in a party convention speech.[240] Brandt's wording also contained just the right dose of utopianism, which has always been the guarantor of successful symbolic policy: to promise a "blue sky" in a region that was synonymous with smoke and dust—this was indeed the articulation of the collective yearnings of a population that was less and less willing to endure massive pollution loads with the traditional thick-skinned stoicism. Even though Brandt ended up losing the 1961 elections and the CDU remained in power, his brilliant rhetoric was a real blow to the CDU. Years later, Helmut Kohl, a CDU assemblyman from Rhineland-Palatinate and later German chancellor from 1982 to 1998, was still fuming, during a committee meeting, that Brandt's making air pollution control a topic in the campaign had paved the way for "a general political hysteria."[241]

What was Brandt after in the spring of 1961? First of all, he was concerned about symbolic policy. This observation is value-neutral, as the phrase "symbolic policy" has a dual meaning from a political-science perspective. First, it can describe policies considered inadequate in substance; second, it can refer generally to policies that use the power of symbols.[242] While the first usage has an invariably negative connotation, the second is neutral: political symbolism can be open to criticism if the substance of its reference is misleading, but it is not inherently empty. As the political scientist Volker von Prittwitz emphasizes, "Symbols are a common 'currency' of policy, which can activate resources that can be given a positive or negative valuation."[243] In this sense, symbolic policy can certainly also be successful in a substantive way. In fact, the editors of a collection of essays on the topic even consider successful symbolization "an essential prerequisite for successful politics, and all the more so, the less an individual person can grasp the problems and the consequences of certain policies."[244] From such a point of view, it is clear that Brandt's emphatic remark played a crucial part in giving air pollution control a new visibility and thus also a new significance in the political realm, obviously a boon to ongoing efforts. Of course, Brandt's primary intent was to present himself as a progressive candidate for all Germans, but the cause of air pollution control also benefited, indirectly, to a considerable extent.[245]

Brandt's staff had also examined the issue thoroughly beforehand and had asked the German Association of Towns and Cities for information no fewer than three times in the run-up to the speech.[246] At the same time, however, these requests demonstrate that Brandt's interest was rather limited. As the mayor of Berlin, Brandt had not launched any meaningful initiatives in matters of air pollution control, just as the SPD as a whole had fewer accomplishments to show than the CDU-led governments on the federal level and in the state of North Rhine–Westphalia. Brandt had tersely claimed in front of his party delegates that air pollution control had been "almost completely neglected until now"; this, however, truly amounted to nothing more than campaign bluster.[247] In fact, a Brandt-led government would have looked rather helpless had it attempted to implement the candidate's election promises politically. After all, since the establishment of the VDI Commission and the changes to the Industrial Code in 1959, the future of air pollution control had depended largely on the application of these new instruments—and this was exclusively a matter of state policy. In November 1962, the Consultative Committee of the Federal Government on Questions of Air Pollution Control, which had been created by the 1959 changes to the Industrial Code, stated that further legal initiatives were "unnecessary for the time being," since "the federal regulations provide an adequate foundation for the fight against air pollution."[248]

Brandt's statement is thus a perfect example of the possibilities and limitations of symbolic policy. It gave air pollution control an unprecedented political virulence—but it also stands at the beginning of a political tradition of cheap words that are not backed by political substance. This was even more evident in the next legislative session, when the Bundestag discussed (and in 1965 eventually passed) a "Law on Preventive Measures on Air Pollution Control": in retrospect, this law seems like a parody of the good intentions of those who championed it.[249] Behind its promising name, the law was nothing more than a call for "measurements about the nature and extent of the particulate and gaseous air pollution in the atmosphere," which were to generally form "a basis for measures to reduce them."[250] However, since the federal government had no authority, according to constitutional law, to either take these measurements or implement the resultant measures, the law's wording was so vague and pliant that the minutes of an interministerial meeting in North Rhine–Westphalia opened with the remark that "neither the text of the proposed bill nor its justification make it entirely clear what purpose is to be accomplished with the law."[251] In practical terms, the 1965 law was meaningless: by 1970, its implementation had not progressed beyond the operation of an experimental station in Frankfurt.[252] But for all these shortcomings, nobody wanted to give the impression of being *against* "preventive measures for air pollution control," and this, in itself, was sufficient for the law's passage. Even Bavaria, which at first had resolutely attacked federal interference in the states' prerogative, ended up voting for the law in the State Chamber (Bundesrat), because "among the public the failure to vote in favor could be taken as a position against air pollution control."[253]

Comparing these German events with events in the United States highlights their peculiar nature. In Germany, the age of environmentalism began "from above," with politicians trying to capitalize on the growing discontent of wide sections of the public. In contrast, the American environmental revolution began as a grassroots protest, with civic leagues being formed and gradually becoming more aggressive in their demands. In a way, Brandt had a counterpart in Senator Edmund Muskie, who worked extensively on pollution issues in the Senate Public Works Committee. Ultimately, however, Muskie was riding a wave of discontent, rather than creating that wave with aggressive rhetoric.[254] One might thus ask what was going on at the grassroots level in Germany, among the concerned citizens whose anger journalists had invoked forcefully since the 1950s. The answer, in a nutshell: less than in the United States but more than at any previous time in German history.

Civic Discontent and Energetic Politicians

From a transatlantic perspective, one of the key questions is whether German officials had developed an insider mentality similar to that of American officials and industrialists in the postwar years. Like all administrators in complex fields, German officials faced a great temptation to develop such a mentality, in some sense perhaps even greater than for their American counterparts. Only slowly did the tradition of active civil engagement develop in Germany—a tradition that, in the United States, reached back to the Progressive Era—and then there was always the pride of the German official. Some protagonists' responses evidenced this pride above all else. For example, when two national civic associations accused the VDI Commission of partiality, Heinrich Lent, at an October 1957 committee meeting, virtually erupted with indignation. "It is an insult for all of us who are volunteering our work and time that we are being put down in this way. I've now finally had it with taking all this accusation without fighting back," he fumed, demanding that his comments be "incorporated verbatim into the minutes."[255] But Lent soon calmed down again, the announced counterattack never came, and the VDI Commission focused on its scientific work. When all was said and done, the VDI Commission was content as long as "no nonobjective and hateful attacks against the commission appear[ed]" in the press.[256]

In Germany, too, there was alienation between the bureaucracy and the public; however, it stayed within limits, and at least until around 1970 it did not escalate as it did in the United States. The situation remained at the level of mutual declarations of unhappiness. "It is undesirable for environmental protection to become the object of a creed," a 1972 essay about the rise of ecological rhetoric declared.[257] Conversely, the period around 1970 saw the first protests intimating the full sting of ecological criticism. In December 1970, for instance, the Aktion Lebensschutz collected signatures for a class-action lawsuit against the Bavarian State Ministry of the Interior and Munich City Hall, purporting that there was sufficient evidence of "negligent bodily harm."[258] But this was still a far cry from the American debate in the 1960s, in which discussions ultimately split into two self-contained discursive worlds. At any rate, there was no hermetical separation between insiders and outsiders: on the whole, a conciliatory approach by the German officials predominated, one that took the public's discontent seriously and accepted it as legitimate. For example, the factory inspection in Hesse saw its task overtly as one of acting "as a mediator between the one who is causing a problem and the one inconvenienced by it."[259] And for an employee of the TÜV in Essen, who was lamenting the "atmosphere of mistrust, prejudices, and passions" that permeated discussion sessions, the issue was not simply complaining about the public's emotionalism but arriving at "mutual understanding in the area of the economy, the public, and the administration": "Not confrontation, but mutually complementing collaboration . . . is the solution of the future."[260]

In spite of all the conflicts, officials succeeded in a remarkable number of cases in allowing open communication between the various parties involved. A prime example of this tendency is the establishment of a State Advisory Board for Clean Air Issues in North Rhine–Westphalia, linked to the desire "to somehow bring together the polluters and the sufferers."[261] Mutual understanding also characterized events organized at seminaries. As early as 1958, for example, the Protestant Seminary of Westphalia in Bochum had brought together, under the slogan "Love your neighbor," Heinrich Lent and other experts with representatives of civic organizations, with the goal of "awakening understanding for the situation of the other side." And the Hermann Ehlers Seminary in Kiel, Schleswig-Holstein, held a similar conference in 1971, where a local pastor, who was also the chair of the municipal working group Environmental Protection, forcefully expressed the "demands of the citizen on environmental protection."[262] Only civic protest against atomic energy, which rapidly developed into a mass movement in the mid-1979s, evidenced sharp lines between opposing camps—a situation that resonates in the German debate over atomic energy to this day. But the rift remained mostly confined to this specific topic, especially since the conflict's bitterness also quickly stirred doubts within the antinuclear camp. Indeed, two demonstrations, in late 1976 and early 1977, came to be known as the "battles" of Brokdorf and Grohnde because of the exceeding violence on both sides, arousing doubts among many protesters about whether the movement was on the right track. If legitimate protests took on the character of military campaigns, something was clearly wrong.[263]

But these were developments of the 1970s. In the 1960s, the agenda was still clearly set by politicians, not by protesting citizens, and these politicians did not let the less-than-impressive results of the 1965 federal law discourage them. Soon after passage of this bill, parliamentarians began to consider yet another change in the law, namely a uniform federal clean air act that would combine and consolidate the regulations scattered among various state and federal laws.[264] In fact, such a demand made sense: after all, there was no logical explanation for why federal law applied to automobiles and industrial installations, while state law applied to all other emitters—only a historical one. Moreover, North Rhine–Westphalia's clean air act had established itself since the mid-1960s as the dominant model, which meant that creating a uniform law for all of Germany seemed much more reasonable than simply waiting for every state to pass similar regulations.[265] In practice, however, the creation of a federal clean air act proved a very tricky business. One crucial problem was the uncertain position of the states, which would have to approve such a law in the State Chamber. Passage of the 1965 law had already shown that North Rhine–Westphalia and Bavaria, especially, were not willing to simply go along with Bonn's ideas.[266] Moreover, from the states' perspective, the creation of a uniform clean air act was by no means an urgent matter. As the minutes of a conference of state officials in September 1968 put it, "From a legal standpoint there is no 'crisis' [*Notstand*]."[267] This foot-dragging on the part of the states took on special weight from the still-unresolved constitutional situation: federal legislative authority could be deduced from the Grundgesetz (West Germany's equivalent of an actual constitution) only indirectly—from its attribution of responsibility for the "laws relating to the economy," for example—and therefore rested on shaky ground. The attempt to introduce concurrent legislation for air pollution control in accordance with Article 74 of the Grundgesetz was voted down in the State Chamber in early 1969.[268] Moreover, it soon became clear that most of industry rejected a new law and could at best warm to an amendment to the Industrial Code—by no means an insignificant problem, because the previous reforms had been enacted so smoothly not least because industrial interest groups had approved them in principle.[269] Finally, the Ministry of Commerce also balked at times because the planned federal clean air act was supposed to also encompass traditional licensing procedures and would therefore erode the substance of the Industrial Code.[270]

Given this broad spectrum of opponents, officials in the Federal Ministry of Health, unsurprisingly, followed the project with lackluster interest. Although a first "rough draft" of a federal clean air act dates from May 1966, it was not until the summer of 1968 that the work of ministry officials started to pick up speed. However, the attempt to rush the bill through parliament before the federal elections in September 1969 failed miserably.[271] The federal minister of health, Käte Strobel, thus had little to go on when she promised in March 1969 to present a federal clean air act in the next legislative session.[272] On the whole, the project to create a uniform federal law looked like quite a mess when responsibility for air pollution control was shifted to the Federal Ministry of the Interior with the formation of a social-liberal government in the fall of 1969.[273] But there was no need for a countrywide law to take effective steps against air pollution—or was there?

Symbolic Policy and Realpolitik

Five years on, Germany had a national air pollution law, passed by the German parliament without a dissenting vote in 1974.[274] Even more, Germany now had an ambitious environmental policy and a minister who was not afraid to use big words in describing the peril confronting the natural environment. It was absolutely urgent that the nation "grasp the environmental crisis at its roots," Interior Minister Hans Dietrich Genscher declared when he presented the federal government's environmental program on December 3, 1971. Effective environmental protection was possible "only through a new environmental law, one that makes the protection and development of the natural foundations on which all our lives and survival depend one of the paramount tasks of the state in securing the future and taking precautions."[275] This promise was fulfilled through a series of bills that were unprecedented in German history in such concentration and with such a programmatic claim. The legislative offensive began in 1971 with the Law on Aircraft Noise and the Law on Leaded Gasoline; moved on to the DDT Law and the Waste Law (both 1972), the Federal Forest Law, the Detergent Law, and the Law on the Transportation of Hazardous Materials (all 1975); and then shifted to the Federal Conservation Law (1976) and eventually to the Chemicals Law (1980).[276] At the same time, Genscher created the previously mentioned Council of Experts for Environmental Questions and, in 1974, the Federal Environmental Agency (Umweltbundesamt), which has since, together with the older Federal Agency for Nature Conservation (Bundesamt für Naturschutz), been the highest scientific authority on questions of environmental protection in Germany.[277] And all this ran under the broad banner of *Umweltpolitik,* a literal translation of the English term *environmental policy* that Hans-Dietrich Genscher, more than anyone else, helped to popularize. After more than a decade of doing environmental policy in spirit, West Germany was now doing environmental policy in name.[278]

As the motives behind this political offensive have been exhaustively analyzed by Kai Hünemörder, a brief summary will suffice here.[279] To begin with, there was the international context: the Council of Europe had declared 1970 the International Year of Nature Protection, and the international environmental conference was coming up in Stockholm in 1972, which would remain the most important global conference of its kind until the famous environmental summit of Rio de Janeiro in 1992. In addition, Genscher was motivated by party tactics: his liberal party was the junior partner in a coalition with the Social Democrats, and with Chancellor Willy Brandt pursuing his widely publicized Ostpolitik, which made him a key figure of 1970s détente, Genscher was seeking to earn credentials for his party and himself by launching a bold policy offensive on the environment. Finally, the U.S. environmental revolution was seen as a shining example in Germany, not surprising for a country with a long-standing fascination

Figure 12. This scene is from the 1973 film *Smog,* one of the first television programs dealing with environmental issues in Germany. The director, Wolfgang Petersen, later directed Hollywood action movies such as *Air Force One.* Image courtesy of Ullstein Bild.

with all things American. Genscher was following American news coverage closely, and even though there was nothing resembling the 1970 Earth Day celebrations in Germany, he saw a need to develop some forceful policy initiatives. In fact, Genscher sought to encourage citizen action much like the PHS had in the late 1960s. Further, just like federal officials in the United States, he found himself overwhelmed by civic unrest once environmentalists' debates had gained momentum. His advisors made several attempts to turn the environmental protesters into supporters of the liberal party, but these efforts were quickly stalled. Once aroused and organized, the nascent environmental movement proved unwilling to act as a proxy.

But what exactly was Genscher trying to reform with his environmental policy? He boldly proclaimed at the beginning of 1970 that he would "replace pollution control based on neighbor law and dating back to the preindustrial age with a modern concept of air pollution control," but this was more of a rhetorical flourish than a real program.[280] Further, this proclamation invariably angered those who until then had seen themselves as the avant-garde of air pollution control. For example, in April 1970, the minister of labor of

North Rhine–Westphalia, Werner Figgen, wrote a letter to Genscher that has few parallels in official correspondence. On the surface, Figgen was asking for "clarification about how the new law relates to our administrative work to date," but the sardonic-ironic undertone that pervades the letter hints that more was at stake. Figgen noted emphatically that the authorities responsible for pollution control in urban areas—"e.g., in North Rhine–Westphalia"—had "already achieved notable successes in preventing environmental contamination and environmental damage" on the basis of the very same regulations that Genscher was consigning to the preindustrial era. At times, the letter made Genscher's ambitious environmental policy seem like the work of a veritable dilettante, as when Figgen emphasized that he, as a state minister, had to be exceedingly careful "that the continuity of our work is not endangered by ambiguous formulations or immature blueprints." This letter was the revenge of a state ministry that had done much for environmental protection and now had to watch, stunned, as a newcomer to the field was trying to exploit the issue politically with a lot of rhetorical bluster.[281]

Genscher responded characteristically, namely with a conciliatory letter that could not have been any more different from his no-holds-barred public rhetoric. It was, he said, "certainly not my intent, as you seem to assume, to criticize the law governing installations requiring permits that is enshrined in the Industrial Code." What is more, he informed Figgen, "I agree with you that continuity in the area of air pollution control should be preserved as much as possible." Genscher went on to heap exuberant praise on North Rhine–Westphalia's policy, lauded the State Clean Air Act of 1962 as a "pioneering achievement," and even expressed the hope that "the advances that North Rhine–Westphalia and other states have achieved can be put to use for the entire country for the benefit of all."[282] For a politician who shortly before had publicly announced a comprehensive modernization of the German approach to environmental policy, this was a remarkably defensive posture. In essence, it was tantamount to surrender.

Thus, one thing was clear: the comprehensive reform that Genscher was advocating was really more a reform in spirit than in substance. His goal was to draw attention to the topic, to arouse interest among concerned citizens willing to organize pressure groups, and to increase the political weight of the topic—but he did not want to redesign the German regulatory system from top to bottom. In fact, if the air pollution control law of 1974 is compared with the legal status quo of the more advanced states around 1970, any claim at institutional reform vanishes. For all those states that already had a clean air act of their own, the law essentially amounted to very little that was new.²⁸³ Genuine innovations concerned only procedural issues: §47 obligated the authorities in heavily polluted areas to come up with comprehensive "clean air plans"; §53 obligated the operators of installations subject to licensing to appoint, under certain conditions, a company officer for pollution control; and §61 called for regular reporting to the federal government.²⁸⁴ The substance of regulations, on the other hand, remained largely unchanged; revealingly, the only meaningful modifications were measures instituting deregulation through the introduction of a simplified licensing process (§19) and the possibility of partial permits and preliminary decisions in (§§8 and 9).²⁸⁵ In the final analysis, the Federal Clean Air Act of 1974 amounted to yet another ratification of the political decisions of the 1950s, and the decision was not even a dubious one. How the achievements since the mid-1950s should be assessed is a matter of debate, but they were clearly not a complete failure crying out for change. In fact, arguably, the key obstacle to Genscher's reform efforts was that he could not dismiss the previous policies outright. Faced with more than a decade of fairly stringent air pollution control, Genscher confronted a fateful choice: he could improve the regulatory setting through an ambitious program of institutional change, but the first effect of such a program would have been to break the momentum of existing practices with widely recognized merits. This was a tough choice that environmental reformers have faced many times since, and Genscher was the first to balk at the prospect of radical change with uncertain results.

Genscher's rhetoric promised a comprehensive approach to environmental policy that sought a universal tightening of regulation, but the actual approach was far more selective. Specifically, Genscher failed to push the issue of sulfur dioxide emissions even though the problem was ripe for a forceful initiative. Everyone had known for some time that sulfur dioxide was harmful to plants. German scientists had demonstrated as much in the nineteenth century, and the renown of their findings had spread as far as the American West: in 1949, a Stanford professor had declared that the turn-of-the-century German studies still held "a place in the reference file of every worker in this field."²⁸⁶ However, research and practical efforts had long focused on smelter smoke with its particularly high concentration of pollutants, while the sulfur emissions produced by the burning of coal, which had low concentrations but added up to an enormous total amount, had been marginalized for decades. As a standard work on smelter smoke put it in 1883, "Surely no one has yet thought of an absorption of the acidic gases of pit coal smoke, and for the moment there is no option but to simply let these gases escape into the air."²⁸⁷ But this ignorance had begun to break down in the later 1950s. "The danger stems not from the dust emission, but from the emission of sulfur dioxide," explained the responsible ministerial expert in the Federal Ministry of Labor in 1958. And Heinrich Lent announced at a conference in October 1962, "Every participant is aware that the task of controlling the sulfur dioxide problem must be resolved in the coming decades with technically sensible and economically acceptable means."[288] Beginning in 1964, the "Technical Directive on Air Pollution Control" recommended that when licensing installations, once a certain level of sulfur dioxide emissions was exceeded, "one should examine whether the licensee should be required to provide space for the installation of equipment for the desulphurization of flue gas."[289]

In terms of the problem at hand, Genscher would thus have stood a good chance of achieving a satisfactory solution with a little effort and commitment. However, this would have led to conflicts with the strong German coal lobby, which was evidently not something Genscher wanted to risk. At any rate, there are no indications of a forceful federal initiative, and the matter moved ahead at a glacial pace. According to expert opinions, the first flue-gas desulfurization plants were ready for practical use around 1970.[290] And if there was need for even more evidence that the desulphurization of flue gas was technically feasible, an article about a study trip engineers undertook to Japan and the United States in 1974 supplied it, declaring that "flue gas desulphurization equipment has been successfully tested in plants at least in Japan and is therefore . . . in line with the current state of technology."[291] Still, it would be another three years before Germany's first commercial flue-gas desulfurization plant went into operation in Wilhelmshaven in 1977.[292] And even thereafter, the pace continued to be slow: at the end of 1982, only seven out of some ninety large coal-fired power plants in the Federal Republic were equipped with desulfurization equipment.[293] Only warnings in the early 1980s about the large-scale dying of forests created the political pressure necessary to find a quick solution, and in 1983, a landmark decree compelled large German power plants to install desulfurization equipment within five years.[294] Because of the enormous costs associated with this law—energy producers estimated them at 14.2 billion DM by the end of the 1980s—contemporaries considered it a spectacular success, but in retrospect it is clear that a similar policy would have been conceivable a decade earlier, under Genscher.[295]

On the issue of automobile exhaust, as well, the federal government moved at a leisurely pace, while the topic was very high on the agenda in the United States. This was the case even though the prime minister of Baden-Württemberg, Kurt Georg Kiesinger, had stated as early as May 1966, in a letter to Hans Filbinger, his interior minister and eventual successor, that "the German automobile manufacturers [should] be prevailed upon to provide for domestic sales the same equipment for exhaust purification that they do for exports to the USA."[296] A few months later, Kiesinger was chancellor, but there was no sign of any action along these lines during his three years in office, and Genscher also avoided taking on the mighty German automobile lobby. Upon closer inspection, Genscher's environmental policy therefore seems more like a curious blend of symbolic policy and realpolitik. There can be little doubt about Genscher's constructive—indeed visionary—ambitions, and passage of the federal Clean Air Act is unanimously considered a milestone in the history of air pollution control: one contemporary lawyer described the law as a "vigorous consolidation."[297] Moreover, symbolic gestures, of which Genscher was a master, were especially important in an area in which implementation was chaotic and impenetrable. In discussions with polluters, a nimble official could thus invoke an ally at the very top, and this surely played a role in the fact that pollution loads tended on the whole to decline rather than increase in the 1970s. Still, Genscher's environmental policy is notable not only for what it achieved but also for what it left out.

Hans-Dietrich Genscher quit his job as minister of the interior in 1974 to become the West German secretary of state, a post that he would fill to great international acclaim until 1992.[298] One year after Genscher's change of office, German chancellor Helmut Schmidt lashed out against excessive environmental regulation in a meeting with industrial lobbyists at Gymnich Castle near Bonn, an event that environmentalists still recalled with a shudder even decades later.[299] But in the end, the German lag in environmental policies during the late 1970s was weak compared with the antiregulatory backlash of the Reagan administration in the early 1980s. In fact, Reagan's antienvironmental policies coincided with a surge of environmental activism in Germany in the wake of the aforementioned forest debate, environmentalism entering the German mainstream to a remarkable extent. Born out of this broad sentiment, the German Green Party continues to be one of the most successful parties of its kind worldwide. Further, since the 1980s, the European Union has increasingly sponsored ambitious environmental policy initiatives; some of the more important German policies, like those addressing fine dust or the greenhouse effect, in fact resulted from European-wide efforts directed from Brussels. German environmental policy has now evolved to a point at which some environmentalists are starting to wonder whether deregulation might be good for the environment—certainly an improbable situation from an American perspective.

After all, the rise of the European Union as a policy maker has merely underscored the general pattern of the environmental polity: high-ranking officials and expert commissions define ambitious environmental standards and then hope that they will somehow trickle down into decrees and licenses. To some extent, they do, but few officials have any deeper knowledge of the morass that is implementation, and environmentalists have proven loath to discuss these problems. Discussions about policy alternatives have popped up at times, but the general impression is that the German bureaucracy has shown remarkable resilience against alternative approaches, specifically approaches that make administrative and industrial performance more transparent. Readers will probably not be surprised by this outcome: much has been said about the momentum of the German bureaucracy, and there is no need to repeat it here. National styles of regulation are not like clothes that can be put on and off at pleasure; to stick with the metaphor, more often than not they have the feel of a straitjacket. And yet a country's regulatory style can be subject to change, thanks to skillful policy brokers who take stock of the potential and the limits of national approaches and adjust them judiciously. National styles of regulation can be transformed, sometimes with surprising speed; but they have roots that can only be appreciated fully if one steps back from the hustle and bustle of daily politics and takes a long view. And this is why history is so important for making wise environmental decisions.

8   Conclusion
WAS THE ENVIRONMENTAL REVOLUTION NECESSARY?

The 1970 Clean Air Act defined national ambient air-quality standards for six pollutants: carbon monoxide, sulfur dioxide, nitrogen dioxide, ozone, lead, and particulate matter. Three decades later, aggregate emissions had declined by 29 percent, and air-quality levels showed noticeable improvement. Progress was greatest for lead and particulate matter, where emissions declined by 98 and 88 percent, respectively. Sulfur dioxide emissions had declined by 44 percent, much of that due to the Acid Rain Program, which seeks to cut sulfur dioxide emissions in half between 1980 and 2010. Carbon monoxide emissions were down by 25 percent, a remarkable achievement against the background of a 143 percent increase in vehicle miles traveled over the same period. The trend was more mixed for ground-level ozone, where the situation had actually worsened in some areas. Ozone is not emitted directly into the air but forms through a reaction of organic compounds and nitrogen oxides in the presence of heat and sunlight, and the two key pollutants followed a different trajectory. While the emission of volatile organic compounds had declined by 43 percent, national nitrogen oxide emissions had actually increased by 20 percent from 1970 to 2000, though most of that increase occurred during the 1970s. However, the general picture is clearly an encouraging one: in spite of increases in many key factors influencing pollution loads—gross domestic product, energy consumption, population—emissions have declined notably or have at least increased to a far smaller extent than might have been expected in the absence of environmental regulation.[1] Thus, it might seem futile to ponder questions about the necessity of the environmental revolution. Isn't it obvious that the air is cleaner now than it was a generation ago?

Of course it is, but this should not amount to the end of all questions. After all, the decline of pollution loads since 1970 was by no means unprecedented: throughout the age of smoke, people were fighting against airborne pollutants, and the inroads they made were often impressive. The air was clearly cleaner in American cities in 2000 than in 1970, but the same holds true for 1970 compared with 1950, or 1950 compared with 1900—not for every city and every pollutant, but for enough of both to raise doubts about declensionist narratives. It is time to abandon the old myth that nothing was happening in air pollution control until the environmentalists got things going: in this study, the modern environmental movement is merely the latest chapter in a long struggle dating back to the Progressive Era. To be sure, environmentalism is arguably the most popular and successful of several successive waves of antipollution protests, but it is neither unprecedented nor the only effective attempt. Marginalizing earlier efforts underestimates the range of activities during the age of smoke: the accomplishments of hundreds of air pollution control programs on the city, county, and state levels; the work of countless engineers and other experts; the willingness of industry to cooperate with control programs once they had gained a certain momentum; and, not least, the activities of countless citizens who deserve a place in the collective memory of the environmental movement. Denying the accomplishments of the age of smoke would mean ignoring almost a century of civic activism.

Obviously, the environmental revolution was a more complex event than conventional narratives have it. A popular cliché, sometimes nourished by environmental historians, sees the environmental revolution as a strangely monolithic event, a kind of general awakening that allowed people to see hazards that they had not recognized before, but against the background of this study, such a reading looks exceedingly simplistic and in some respects utterly false. In the history of air pollution control, it seems, there was not one but two environmental revolutions, one societal, one institutional—a transformation of social values that gave new urgency to air pollution problems and a transformation of policies and institutional settings for pollution abatement. It seems crucial that the two not be conflated and that the latter not be assumed basically a consequence of the former. After all, the two revolutions look quite autonomous in this study, with distinct sets of actors, distinct rationales, and distinct ways of monitoring progress. Of course, they were not completely independent, but their relationship looks anything but linear.

Contrary to popular myth, it was not environmentalism that inspired the first reforms of air pollution control after World War II. In the United

States, a first wave of protests arose almost immediately after 1945, often pointing to the spectacular campaigns in St. Louis and Pittsburgh as proof that a more vigorous approach was possible. In fact, public pressure was so strong that even the established expert community gave in, the Smoke Prevention Association of 1906 reluctantly changing its name to the Air Pollution and Smoke Prevention Association in 1950 and to the Air Pollution Control Association in 1952. A similar trend held sway in Germany, though it lagged a few years behind the United States, for obvious reasons. But with postwar exigencies left behind, the call for clean air emerged swiftly in the West German public, and more energetic measures against air pollution had become a mainstream concern by the mid-1950s, with the highly industrialized Ruhr Area in the vanguard. Somehow, these early efforts have not entered the collective memory in either Germany or the United States, and the myth persists that people were simply too preoccupied with economic growth during the postwar years to bother about pollution. But in the 1950s, Germans and Americans alike saw no reason to choose between economic growth and environmental protection. On the contrary, the two goals meshed neatly: if the economy was healthy, industry could certainly afford clean air.

In response, air pollution policies changed significantly in both Germany and the United States. In fact, this was the most fundamental reformation of air pollution control since the early 1900s, and the extent of these reforms would influence the course of environmental policy in both countries far beyond the 1950s. In the United States, the common path was upgrading municipal smoke inspection departments to bureaus of air pollution control, responsible for all urban pollutants. Substantial support came from the industrial community, but this was a decidedly mixed blessing: on the one hand, plant managers were willing to invest in control equipment on an unprecedented scale, but on the other hand, industrialists sought, and usually achieved, a defining influence over the pace and extent of control policies. German reforms were narrower geographically, with the state of North Rhine–Westphalia the undisputed leader and the other states following its path with differing speed until the West German government federalized environmental policy in the early 1970s. But the German campaign against pollution was no less vigorous than its American counterpart, especially when viewed against the long tradition of bureaucratic inaction in the highly industrialized Ruhr Area, the economic heart of North Rhine–Westphalia. And all this, to be sure, happened long before the civic environmental revolution had reached full swing.

In other words, environmentalism did not lead to the discovery of problems that had been neglected before. Rather, environmentalism was a new way to talk about problems that city dwellers had been complaining about for a century and more. It was, at its core, a shift of rhetoric, though one with profound implications: it emphasized health, as opposed to traditional concerns about cleanliness and property; it was broad in focus, ultimately depicting pollution as a global issue, while smoke had been perceived as a completely local problem; it tied air pollution with other environmental problems into one great challenge to humanity; it was impatient and suspicious toward authorities, two hallmarks of Progressive Era antismoke protests that had been mostly lost over the following decades; and it revealed a growing sense of fear about invisible pollutants that had been foreign to the age of smoke. All in all, environmentalism implied a new kind of challenge for regulators and a hope that the revolution in civic awareness would be matched by an equally staunch institutional revolution.

To some extent, this revolution was already under way in both countries by 1960. Ever since the postwar reforms, the prevailing view saw air pollution control as a progressive enterprise, an idea foreign to antismoke activists of the Progressive Era. Smoke inspection had focused on one pollutant only, coal smoke, with the tacit assumption that the movement's work would be finished with the end of the smoke nuisance. But since 1945, a broad consensus saw a need for constant advances in order to keep air pollution in check: in both countries, investments in air pollution control programs increased enormously, and so did investments in control equipment; from a narrow focus on smoke and dust, policies expanded to include heretofore unregulated pollutants. The institutional revolution had taken on a momentum of its own: environmental protesters could challenge the speed of the transition, but there was no way to deny the reforms' general direction. And yet this institutional revolution proved sustainable only in Germany, not in the United States. Whereas German environmental policy built on features from the age of smoke, America's regulatory tradition ended with the environmental protests of 1970. Since then, everything that smacked of cooperation with industrialists was thoroughly discredited, and a new, more adversarial environmental policy emerged from the smoking ruins of the former approach.

Thus, the question of the necessity of the environmental revolution might be rephrased: Why did the two regulatory traditions show such different abilities to adjust to the challenge of environmental sentiments? Why did the German institutional revolution keep pace with civic discontent while the American one did not? A good part of the explanation may lie in the different trajectories of environmental activism in the two countries. In

West Germany, environmental protest was gathering steam over more than a decade and did not culminate until the early 1980s, when the fear of forest death was galvanizing everyone's attention. Even more importantly, the peak of Germany's environmental protests took place in the wake of an exceedingly bitter confrontation over nuclear power, instilling doubts among environmentalists about overtly divisive politics. In contrast, the American environmental revolution erupted almost overnight, to the huge surprise of regulators and industrialists alike, producing a sudden pressure to adjust policies that was unknown in Germany.

But the different paths of institutional reform probably had more to do with events before 1970 and the respective national styles of regulation. By the mid-1960s, the American style of regulation was clearly in trouble, torn between growing public discontent and the limited potential of ostensibly "cooperative" air pollution control. In contrast, the West German regulatory tradition looked generally quite promising around that time, even in the eyes of dedicated reformers like Genscher and his associates. In fact, there had been a notable difference in the long-term prospects of each country's regulatory style as early as the mid-1950s. The German approach was notable for its openness, allowing input from experts, industrialists, and the general public. In contrast, the American approach stood out for its stark imbalance of power: industrialists were defining the limits of acceptable policies almost autonomously, and outsiders had a hard time even entering regulatory commissions, let alone influencing their decisions. Fearful of public hysteria even in the widespread absence of aggressive protests, industrialists set out to dominate the American approach to air pollution control, making the "cooperative approach" look like pseudocorporatism in the vast majority of cases. And once in power, industry held on to its prevailing authority until its power base eroded in the heat of 1970.

The culprit would thus seem clear: America's industry. By seeking dominance over air pollution control policy, it pushed critics into a realm outside existing institutions, where anger built up until it was too late. In retrospect, it might have benefited industrialists to attend more closely to the shifting preferences of the general public since the 1950s and to leave more room for criticism within the regulatory system, for this might have prevented the total delegitimation of cooperative approaches. But putting all the blame on American industry misses one crucial point: German industrialists were no less eager to gain control over air pollution control. They fought for the routine consultation of their lobbyists within the VDI Clean Air Commission, they put their biased experts onto working committees, and they opposed

Genscher's environmental policy. The difference is that German industrialists—unlike their American counterparts—failed in all these attempts and thus generally failed to dominate air pollution control. In other words, the difference was not between cooperative German industrialists and short-sighted American ones; the difference was that German industrialists failed where American ones succeeded.

The key issue, then, is not so much the stance of industry as the existence of countervailing forces. Interestingly, the bias of pseudocorporatist air pollution control stemmed from the same feature of the American political system that allowed the astounding success of the American smoke-abatement movement during the Progressive Era: the openness of the U.S. political process. Conversely, the German postwar debate profited from attitudes that had hindered the smoke debate of the early 1900s: the strength of the German bureaucracy and the autonomy of the engineering community. German officials were far more independent politically than municipal air pollution–control engineers in the United States, and this was especially true of ministry officials in North Rhine–Westphalia, who were calling the shots. The German engineering community was still skeptical of bureaucratic regimentation, but with the Committee on Dust Technology founded in 1928, it had an institutional nucleus for a different approach to rule-setting that gave engineers and other experts more autonomy and influence than a state-centered approach would have. Without the stance of officials and engineers as a counterweight against industrial interests, the German style of regulation would likely have been just as biased as the American one.

If there was a single reason for this notable about-face, it was open communication. The birth of the VDI Clean Air Commission provides a case in point: it came into being because Heinrich Grünewald, the director of the VDI, sat down with Wolfgang Burhenne of the IPA and talked things over. The same kind of candor was a typical feature of discussions within the Clean Air Commission: the guiding principle for the working committees was to include experts from diverse backgrounds and to keep on discussing and researching until everyone, or at least a large majority, could agree on a certain set of rules and regulations. In contrast, the American style of regulation did not include a similar "middle ground," notwithstanding the pervasive talk about "cooperation" and "good neighbors." Committees were not open to the public, or were open only to a very limited extent, and PR activities, such as the annual Cleaner Air Week, presumed communication to flow in only one direction, from the nice regulators and their friends in industry to the grateful public. Thus, it is no surprise that environmentalists became suspicious of backroom deals and emphatic pledges of "cooperation." Why should they have confidence in proceedings from which they were excluded?

What all of this makes clear is that the most precious resource in environmental policy is probably neither money nor expertise: it is trust. There is no scarcity of difficult problems in air pollution control, but they become infinitesimally more complicated if there is no sense of trust between the parties involved. Discussing the myriad details of air pollution control in an atmosphere of distrust is almost nightmarish, guaranteed to produce solutions only in the long term. To be sure, trust is not necessarily a matter of personal relations, and it probably cannot be, as environmental policy has ceased to be a purely local affair. In fact, even the local cooperation between smoke inspectors and civic leagues was probably not a matter of personal acquaintance in the first place: civic reformers usually supported smoke inspectors only after they had thoroughly checked their performance; charismatic people like Raymond Tucker, who enjoyed public confidence as soon as they took office, were clearly the exception to the rule. In a complex modern society, trust is usually a question of institutions.[2]

In retrospect, the age of smoke looks almost like a golden era for building trust. Never thereafter has it been so easy to monitor the performance of an air pollution control official: all it took were a few stack observations and a number of interviews. In fact, the key instrument for stack observations, the Ringelmann Chart, came into use primarily because even laypeople could use it easily; in terms of precision, the Ringelmann Chart was a notoriously crude device.[3] But with visible pollutants on the decline and invisible and otherwise imperceptible pollutants moving onto the political scene, it has become much more complicated to assess environmental policies. It usually takes a good deal of expertise and insider knowledge to understand proceedings, and even then the issue remains exceedingly complicated.

The tricky thing is that it is far easier to destroy trust than to build it up. It takes time to develop a middle ground where the different parties can meet, learn about each other's ideas and interests, and ultimately develop a sense of trust. In spite of the cooperative tradition of the Committee on Dust Technology, the VDI Clean Air Commission did not work as smoothly as envisioned until several years had passed. Further, successful cooperation is always in the details, and it is usually difficult for outsiders and laypeople to evaluate those. Finally, cooperative bodies seem to need a good dose of cozy rhetoric about "cooperation" and "common interests" in order to function properly, but an enraged public easily mistakes this same rhetoric for a smokescreen. So was erosion of trust inevitable during the transition from the age of smoke to the environmental era? Is the preservation of certain spheres of trust, like the VDI Commission, the best that can be hoped for? It is perhaps no coincidence that the decline of cooperation went along with the rise of the modern mass media, for cooperative policies are arguably incompatible with their demands. Negotiations in smoke-filled rooms will never have the TV appeal of a dramatic clash between noble environmental activists and evil captains of industry. For all its merits, the VDI Clean Air Commission will never be as popular as Erin Brockovich.

All this, to be sure, in no way denies the accomplishments of civic activism. After all, the history of the age of smoke presents a clear-cut lesson in this respect: nothing gets done without an active citizenry. If anything, this book demonstrates the endurance that civic activism needs to demonstrate and the enormous sacrifices involved in a civic drive against smoke and pollution. "To organize a movement you're going to have to first find someone who has the time to give up at least six months of his life," Douglas LaFollette, director of a Wisconsin-based environmental pressure group, sighed in 1971. No one who has read this book will dare to challenge his words.[4] And yet these civic energies may remain futile, or at least below their potential, if they are not backed up by the skills of "policy brokers": people like Lester Breckenridge, Paul Bird, and Raymond Tucker in the United States, or the Oels brothers and Wolfgang Burhenne in Germany, who combined different inputs and negotiated deals that all parties could live with and then implemented this approach forcefully and energetically. It takes both kinds of people, the vociferous crusader and the skillful manager, to start an effective drive against environmental problems. After all, the times are past when it sufficed for an environmentalist to simply point to a problem and demand abatement. In fact, as this book has shown, that time never existed.

ABBREVIATIONS

AHC	Atlanta History Center
AIS	Archives of Industrial Society, University of Pittsburgh
APCD	Los Angeles County Air Pollution Control District
ASME	American Society of Mechanical Engineers
BadGLA	Badisches Generallandesarchiv, Karlsruhe
BArch	Bundesarchiv
BayHStA	Bayerisches Hauptstaatsarchiv, Munich
BCA	Baltimore City Archives
BDI	Bundesverband der Deutschen Industrie (German Association of Industry)
BGB	Bürgerliches Gesetzbuch (Civil Code)
BML	Becker Medical Library, Washington University School of Medicine, St. Louis
BrLHA	Brandenburgisches Landeshauptarchiv, Potsdam
BYUL	Special Collections and Manuscripts Department, Brigham Young University Library, Provo
CACR	Cleaner Air Committee of Hyde Park–Kenwood Records, Department of Special Collections, University of Chicago Library
CCC	City Club of Chicago Records, Chicago Historical Society
CDU	Christlich-Demokratische Union (Christian Democratic Party)
CGC	Charles Gruber Collection, Cincinnati Historical Society
CHS	Cincinnati Historical Society
CSA	California State Archives, Sacramento
CWC	Chicago Woman's Club Records, Chicago Historical Society
DBC	Dorothy Boberg Collection, Urban Archives Center, California State University, Northridge

DÜV	Dampfkesselüberwachungsverein (Association for Steam Boiler Surveillance)
DVRPCR	Delaware Valley Regional Planning Commission Records, Urban Archives, Temple University, Philadelphia
DWHC	Daniel Webster Hoan Collection, Milwaukee County Historical Society, Milwaukee
ERWP	Edwin Richard Weinerman Papers, Manuscripts and Archives, Sterling Memorial Library, Yale University, New Haven
FWTC	Frederick Winslow Taylor Collection, Stevens Institute of Technology, Hoboken
GDAH	Georgia Department of Archives and History, Atlanta
GPMR	Greater Philadelphia Movement Records, Urban Archives, Temple University, Philadelphia
GStA	Geheimes Staatsarchiv, Berlin
GWRGP	George W. Romney Gubernatorial Papers, Bentley Historical Library, University of Michigan, Ann Arbor
HAK	Historisches Archiv der Stadt Köln, Cologne
HCAP	Henry Carter Adams Papers, Bentley Historical Library, University of Michigan, Ann Arbor
HessHStA	Hessisches Hauptstaatsarchiv, Wiesbaden
HML	Hagley Museum and Library, Wilmington
HPKCCP	Hyde Park–Kenwood Community Conference Papers, Department of Special Collections, University of Chicago Library
HRCSP	Hudson River Conservation Society Papers, Franklin D. Roosevelt Presidential Library, Hyde Park
HStAD	Hauptstaatsarchiv Düsseldorf
HStAS	Hauptstaatsarchiv Stuttgart
HSWPA	Historical Society of Western Pennsylvania Archives, Pittsburgh
HU	Widener Library, Harvard University, Cambridge
IMAR	Illinois Manufacturers' Association Records, Chicago Historical Society
IPA	Interparlamentarische Arbeitsgemeinschaft für naturgemäße Wirtschaft (Working Group of Federal and State Representatives)
JAFC	John Anson Ford Collection, Huntington Library, San Marino
JRP	James Roosevelt Papers, Franklin D. Roosevelt Presidential Library, Hyde Park
JTWP	Jay T. Williams Papers, Western History/Genealogy Department, Denver Public Library, Denver
KHC	Kenneth Hahn Collection, Huntington Library, San Marino
LAB	Landesarchiv Berlin
LASb	Landesarchiv Saarbrücken
LASH	Landesarchiv Schleswig-Holstein, Schleswig

LHAK	Landeshauptarchiv Koblenz
LHAM	Landeshauptarchiv Magdeburg
LMDP	Leon Mathis Despres Papers, Chicago Historical Society
LWVLAC	League of Women Voters of Los Angeles Collection, Urban Archives Center, California State University, Northridge
LWVMR	League of Women Voters of Michigan Records, Bentley Historical Library, University of Michigan, Ann Arbor
MCACR	Metro Clean Air Committee Records, Minnesota Historical Society, St. Paul
MLCP	Morris L. Cooke Papers, Franklin D. Roosevelt Presidential Library, Hyde Park
MnHS	Minnesota Historical Society, St. Paul
MoHS	Missouri Historical Society, St. Louis
MUA	Milwaukee Urban Archives, University of Wisconsin–Milwaukee
NA	National Archives of the United States of America, College Park
NieHStA	Niedersächsisches Hauptstaatsarchiv, Hanover
NIPCR	Northeastern Illinois Planning Commission Records, Chicago Historical Society
OHS	Oregon Historical Society, Portland
OSA	Oregon State Archives, Salem
PCCA	Pittsburgh City Council Archives
PCCR	Portland Chamber of Commerce Records, Special Collections, Knight Library, University of Oregon, Eugene
PHS	Public Health Service
PNJDMPR	Pennsylvania–New Jersey–Delaware Metropolitan Project Records, Urban Archives, Temple University, Philadelphia
RAKA	Robert A. Kehoe Archives, Cincinnati Medical Heritage Center, Academic Information Technology and Libraries, University of Cincinnati Medical Center
RRTSAC	Raymond R. Tucker Smoke Abatement Collection, Washington University Archives, St. Louis
SHStA	Sächsisches Hauptstaatsarchiv, Dresden
SHSW	State Historical Society of Wisconsin, Madison
SJGP	Steve J. Gadler Papers, Minnesota Historical Society, St. Paul
SPD	Sozialdemokratische Partei Deutschlands (Social Democratic Party)
StAA	Staatsarchiv Augsburg
StAAc	Stadtarchiv Aachen
StAB	Staatsarchiv Bremen
StABi	Stadtarchiv Bielefeld
StABo	Stadtarchiv Bochum
StABr	Stadtarchiv Braunschweig

StAC	Stadtarchiv Chemnitz
StAD	Stadtarchiv Düsseldorf
StADo	Stadtarchiv Dortmund
StADr	Stadtarchiv Dresden
StADu	Stadtarchiv Duisburg
StAE	Stadtarchiv Essen
StAF	Stadtarchiv Frankfurt
StAFr	Stadtarchiv Freiburg
StAH	Stadtarchiv Hanover
StAHe	Stadtarchiv Herne
StAHei	Stadtarchiv Heidelberg
StAHH	Staatsarchiv Hamburg
StAL	Stadtarchiv Leipzig
StALh	Stadtarchiv Ludwigshafen
StALu	Staatsarchiv Ludwigsburg
StAM	Staatsarchiv Munich
StAMs	Staatsarchiv Münster
StAMü	Stadtarchiv Mülheim an der Ruhr
StAN	Staatsarchiv Nürnberg
StAS	Stadtarchiv Stuttgart
TÜV	Technischer Überwachungsverein (Association for Technological Inspections)
UCB	Bancroft Library, University of California, Berkeley
UCLA	Department of Special Collections, Charles E. Young Research Library, University of California, Los Angeles
UMM	K. Ross Toole Archives, Mansfield Library, University of Montana, Missoula
UOE	Special Collections, Knight Library, University of Oregon, Eugene
USHS	Utah State Historical Society, Salt Lake City
UU	Manuscripts Division, J. Willard Marriott Library, University of Utah, Salt Lake City
UVA	Special Collections, Alderman Library, University of Virginia, Charlottesville
VDI	Verein Deutscher Ingenieure (Association of German Engineers)
VGK	Vereinigung der Großkesselbesitzer (Association of High-Performance Steam Boiler Owners)
WaBoLu	Preußische Landesanstalt für Wasser-, Boden- und Lufthygiene (Prussian State Agency for the Sanitation of Water, Ground, and Air)
WCCC	Woman's City Club of Chicago Records, Special Collections, University Library, University of Illinois at Chicago
WRHS	Western Reserve Historical Society, Cleveland

NOTES

Chapter 1

For an extended version of the notes that appear in this volume, please see the German edition of this book: Frank Uekötter, *Von der Rauchplage zur ökologischen Revolution: Eine Geschichte der Luftverschmutzung in Deutschland und den USA 1880–1970* (Essen, 2003).

1. F. G. Cottrell, "Recent Progress in Electrical Smoke Precipitation," *Engineering and Mining Journal* 101 (1916): 388.

2. AIS, 83:7, ser. III, FF 1, p. 103. However, the results of the competition were disappointing, ranging from uninspired proposals ("The Non-Smoky City"), to euphoric ones ("Crystal City"), to those without any apparent logic ("The City with a Thousand Eyes"). "I truly think they are all hopeless," one of the jurors declared (Dermitt to O'Connor, Nov. 2, 1916, AIS, 83:7, ser. III, FF 5). Lacking a prize-worthy entry, the league followed an entrant's proposal and donated the sum to the Western Pennsylvania Institution for the Blind ("Contest Held in Conjunction with Smoke Abatement Week–October 23–28, 1916," p. 1, AIS, 83:7, ser. III, FF 12).

3. "Annual Report of the Smoke Abatement Committee," Apr. 7, 1923, p. 3, WRHS, mss. 3535, box 1, folder 6.

4. Angela Gugliotta, "Class, Gender, and Coal Smoke: Gender Ideology and Environmental Injustice in Pittsburgh, 1868–1914," *Environmental History* 5 (2000): 165–93; Angela Gugliotta, "How, When, and for Whom Was Smoke a Problem in Pittsburgh?" in Joel A. Tarr, ed., *Devastation and Renewal: An Environmental History of Pittsburgh and Its Region* (Pittsburgh, 2003), 110–25.

5. Mancur Olson, *The Logic of Collective Action: Public Goods and the Theory of Groups* (Cambridge, 1965).

6. Adam W. Rome, "Coming to Terms with Pollution: The Language of Environmental Reform, 1865–1915," *Environmental History* 1 (1996): 16.

7. *Baltimore Sun*, Apr. 29, 1928.

8. This transformation of health concepts, and the ensuing "invention" of smoke pollution, has been traced extensively for Great Britain in Peter Thorsheim, *Inventing Pollution: Coal, Smoke, and Culture in Britain since 1800* (Athens, Ohio, 2006).

9. World Bank, *World Development Report 1992: Development and the Environment* (Washington, 1992), 52.

10. Environmental Protection Agency, "Review of the National Ambient Air Quality Standards for Particulate Matter: Policy Assessment of Scientific and Technical Information," Office of Air Quality Planning and Standards, staff paper (Research Triangle Park,

NC, June 2005); H. E. Wichmann, J. Heinrich, and A. Peters, *Gesundheitliche Wirkungen von Feinstaub* (Landsberg, 2002).

11. Charles S. Maier, "Consigning the Twentieth Century to History: Alternative Narratives for the Modern Era," *American Historical Review* 105 (2000): 807–31.

12. My interest in institution building was inspired by Louis Galambos, "The Emerging Organizational Synthesis in Modern American History," *Business History Review* 44 (1970): 279–90, and "Technology, Political Economy, and Professionalization: Central Themes of the Organizational Synthesis," *Business History Review* 57 (1983): 471–93.

13. James Bryce, *The American Commonwealth*, 3rd ed. (New York and London, 1893), 1: 637.

14. Daniel J. Fiorino, *Making Environmental Policy* (Berkeley, 1995), 25.

15. H. H. Meredith Jr., "Industrial Planning for Air Pollution Control," p. 231, NA, RG 90 A 1, entry 11, box 57, "Proceedings of the First New England Conference on Urban Planning for Environmental Health" folder.

16. Frederick Law Olmsted Jr., *Nuisances: Prospectus of the Department,* American Civic Association, Department of Nuisances, leaflet no. 2 (n.l., n.d.), 3.

17. Noga Morag-Levine, *Chasing the Wind: Regulating Air Pollution in the Common Law State* (Princeton and Oxford, 2003), 183.

18. Indur M. Goklany, *Clearing the Air: The Real Story of the War on Air Pollution* (Washington, 1999), 155.

19. Goklany's political intentions are even more transparent in his more recent *The Precautionary Principle: A Critical Appraisal of Environmental Risk Assessment* (Washington, 2001).

20. Henry N. Doyle, "Polluted Air, a Growing Community Problem," *Public Health Reports* 68 (1953): 858; G. Edward Pendray, "Management Aspects of Air Pollution," *Mechanical Engineering* 77 (1955): 581; Helen B. Shaffer, "Poisoned Air," *Editorial Research Reports* (1955): 247; *Steel* 128 (Mar. 5, 1951): 63; Division of Air Pollution Control, "A Review of the New Jersey Air Pollution Control Program," Aug. 1966, p. 7, NA, RG 90 A 1, entry 11, box 34, "LL-2-1–Bills and Acts Part 2" folder.

21. Goklany, *Clearing the Air,* 102. See also p. 88, where Goklany uses the word *automatically* to describe the environmental transition.

22. William D. Ruckelshaus, "Stopping the Pendulum," *Environmental Forum* 12, no. 6 (1995): 25–29.

23. Samuel P. Hays, with Barbara D. Hays, *Beauty, Health, and Permanence: Environmental Politics in the United States, 1955–1985* (Cambridge, 1989), xi. For an example for this kind of bias, see Lennart J. Lundqvist, *The Hare and the Tortoise: Clean Air Policies in the United States and Sweden* (Ann Arbor, 1980).

24. David Vogel, *National Styles of Regulation: Environmental Policy in Great Britain and the United States* (Ithaca and London, 1986), 195, 220.

25. Raymond R. Tucker, "Smoke Prevention in St. Louis," *Industrial and Engineering Chemistry* 33 (1941): 839.

26. My interest in the inherent momentum of the German bureaucracy was one of my reasons for excluding the German Democratic Republic from my discussion of post–World War II Germany. I argue that air pollution–control policies emerge from an interplay among officials, experts, businesspeople, and the general public—an approach that obviously runs into trouble in a totalitarian socialist state.

27. Ernest Renan, "What Is a Nation?" in Geoff Eley and Ronald Grigor Suny, eds., *Becoming National: A Reader* (New York and Oxord, 1996), 53.

Chapter 2

1. *Baltimore Sun,* Aug. 20, 1904.

2. Suellen Hoy, *Chasing Dirt: The American Pursuit of Cleanliness* (New York and Oxford,

1995); Manuel Frey, *Der reinliche Bürger: Entstehung und Verbreitung bürgerlicher Tugenden in Deutschland, 1760–1860* (Göttingen, 1997).

3. "A Year's Record of Usefulness: Annual Report of the President to the Chamber of Commerce of Pittsburgh," May 1907, 8.

4. *Chicago Tribune*, Mar. 9, 1917; *Baltimore Sun*, May 31, 1908.

5. Z. A. Willard, *Report on the Smoke Nuisance* (Boston, 1908), 22; Lockwood to O'Connor, May 16, 1914, AIS, 83:7, ser. III, FF 6; *Baltimore Sun*, Feb. 19, 1902.

6. *Chicago Record-Herald*, Sept. 16, 1909; Civic League of St. Louis, *The Smoke Nuisance: Report of the Smoke Abatement Committee of the Civic League*, Nov. 1906 (n.l., n.d.), 5.

7. *Baltimore Sun*, Dec. 23, 1905.

8. City of St. Louis, *Report of the Special Committee on Prevention of Smoke*, Mar. 8, 1892, 36; *Proceedings of the Engineers' Society of Western Pennsylvania* 8 (1892): 55.

9. Ted Steinberg, "Honest (and Honest to God) Dirt," *Journal of Urban History* 31 (2004): 101. See also David E. Nye, "Smoke Gets in Your Eyes: Pollution, Aesthetics, and Social Class," *Reviews in American History* 28 (2000): 427. For the broader context of Steinberg's argument, see Mary Douglas, *Purity and Danger: An Analysis of Concepts of Pollution and Taboo* (London and New York, 2005); Mary Douglas and Aaron Wildavsky, *Risk and Culture: An Essay on the Selection of Technical and Environmental Dangers* (Berkeley, 1982).

10. It is important to stress this point, as attempts have been made to depict the civic antismoke movement as some kind of inevitable outcome of history. Harold Platt, e.g., argues that "driven by the rapid deterioration of its air quality, Chicago had little choice"—a statement that is all the more surprising since Matthew Crenson noted, as early as 1971, that a society's decision on an environmental problem may be a nondecision (Harold L. Platt, "Invisible Gases. Smoke, Gender, and the Redefinition of Environmental Policy in Chicago, 1900–1920," *Planning Perspectives* 10 [1995]: 80; Matthew A. Crenson, *The Un-Politics of Air Pollution: A Study of Non-Decisionmaking in the Cities* [Baltimore and London, 1971]).

11. See Robert Dale Grinder, "The Battle for Clean Air: The Smoke Problem in Post–Civil War America," in Martin V. Melosi, ed., *Pollution and Reform in American Cities 1870–1930* (Austin and London, 1980), 88. The social structure of the smoke abatement movement was examined using the list of members of the Anti-Smoke League of Baltimore ("Fifth Letter to Members," May 1, 1906, BCA, RG 29 S 1, "Smoke Control 1905–1946") and the list of 161 active women belonging to the Women's Civic League of Baltimore (*History of the Women's Civic League of Baltimore, 1911–1936* [Baltimore, 1937], 75–81). The results confirm the classic interpretation of Samuel P. Hays: see "The Politics of Reform in Municipal Government in the Progressive Era," *Pacific Northwest Quarterly* 55 (1964): 159.

12. Civic League of St. Louis, *A Year of Civic Progress: Annual Address of Henry T. Kent, President of the Civic League of St. Louis* (St. Louis, 1906), 19.

13. Smoke Abatement League of Cincinnati, *Annual Report for 1912* (n.l., n.d.), 3.

14. Anti-Smoke League of Baltimore, "Second Letter to Members," 1906, 3; Syracuse Chamber of Commerce, *Report upon Smoke Abatement: An Impartial Report of the Ways and Means of Abating Smoke* (Syracuse, 1907), 10; Charles H. Benjamin, "Smoke and Its Abatement," *Transactions of the American Society of Mechanical Engineers* 26 (1905): 715n; *Baltimore Sun*, Feb. 17, 1902.

15. Chicago Department of Health, *Report for 1894*, 196; Department of Smoke Inspection, City of Chicago, *Bulletin Number 1*, Feb. 1908, 1; *Chicago Record-Herald*, Apr. 3, Apr. 9, 1902; *Baltimore Sun*, Feb. 25, 1915.

16. Baltimore City General Property Tax Books, 1905, BCA, RG 4, ser. 1.

17. Grinder, "Battle," 88, 96; Platt, "Invisible Gases," 81–86, 91.

18. Sarah B. Tunnicliff, "Smoke Elimination in Chicago," *Educational Bi-Monthly* 10 (1915–16): 392.

19. Woman's Club of Minneapolis, *1916–1917* (n.l., 1916), 31; "Black Smoke Means Waste

of Fuel and Loss of Boiler Efficiency: History of the Work of the Smoke Committee of the Women's Civic League, Mrs. James Swan Frick, Chairman," n.d., AIS, 83:7, ser. IV, FF 9.

20. Benjamin, "Smoke and Its Abatement," 716; J. M. Searle, "Record of the Year in Smoke Abatement, in City of Pittsburgh," *Industrial World* 48 (1914): 131.

21. *Domestic Engineering* 16, no. 7 (1926): 22.

22. Women's Civic League of Baltimore, *History*, 34; *Baltimore Sun*, Jan. 21, 1914.

23. *Chicago Record-Herald*, Apr. 17, 1909.

24. Mrs. David F. Simpson, Chairman, Committee on Abatement of Smoke, Woman's Club of Minneapolis, "Outline of their Work," AIS, 83:7, ser. 1, FF 21; *Bulletin* 3, no. 7 (Dec. 1914): 5, WCCC, folder 6; Woman's Club of Minneapolis, *Year Book 1911* (n.l., n.d.), 17, *Year Book 1916–1917* (n.l., n.d.), 31, and *Year Book, 1917–1918* (n.l., 1917), 30.

25. *Proceedings of the Engineers' Society of Western Pennsylvania* 8 (1892): 22, 302, 311; John O'Connor Jr., *Some Engineering Phases of Pittsburgh's Smoke Problem*, Mellon Institute of Industrial Research and School of Specific Industries, Smoke Investigation Bulletin no. 8 (Pittsburgh, 1914), 14; "Club Meetings," p. 88, CWC, box 22B, 1908–9; Mrs. Ernest R. Kroeger, "Smoke Abatement in St. Louis," *American City* 6 (1912): 907.

26. *Proceedings of the Engineers' Society of Western Pennsylvania* 8 (1892): 302; O'Connor, *Some Engineering Phases*, 13, 15; Civic Club of Allegheny County, *Fifty Years of Civic History 1895–1945* (n.l., n.d. [1945?]), 4; Boston Chamber of Commerce, *Smoke Abatement: Report by the Committee on Fuel Supply of the Boston Chamber of Commerce*, Apr. 1910; William H. Gerrish, "Spirit of Co-Operation Takes Hold in Boston," *Industrial World* 48 (1914): 139; Rochester Chamber of Commerce, Smoke Abatement Committee, *The Smoke Shroud: How to Banish It* (n.l., n.d. [1915?]); Cleveland Chamber of Commerce, *Report of the Municipal Committee on the Smoke Nuisance* (n.l., 1907).

27. Anti-Smoke League of Baltimore, "First Letter to Members"; *St. Louis Post-Dispatch*, Jan. 22, 1893; James Cox, *Old and New St. Louis: A Concise History of the Metropolis of the West and Southwest, with a Review of Its Present Greatness and Immediate Prospects* (St. Louis, 1894), 87, 116; Citizens' Association of Chicago, *Annual Reports*: 1880, 8, 1881, 19, 1882, 6n, 1883, 7, 1884, 5, 1885, 15, 1886, 22, 1887, 13, 1888, 25, 1889, 8, 1890, 13, 1891, 10; Donald L. Miller, *City of the Century: The Epic of Chicago and the Making of America* (New York, 1996), 233 (quotation); Kenneth Finegold, *Experts and Politicians: Reform Challenges to Machine Politics in New York, Cleveland, and Chicago* (Princeton, 1995), 125.

28. Maureen Flanagan argues that "Chicago women . . . urged the city government to enact strict smoke abatement ordinances and to employ smoke inspectors to enforce them" ("The City Profitable, the City Livable: Environmental Policy, Gender, and Power in Chicago in the 1910s," *Journal of Urban History* 22 [1996]: 175). However, she ignores the fact that the Woman's City Club, founded in 1910, supported the smoke ordinance of 1907 (Tunnicliff, "Smoke Elimination in Chicago"). Moreover, Flanagan criticizes the dearth of activity within the City Club after 1910 but fails to mention that its pre-1910 work was instrumental in the reform of 1907 (Flanagan, "City Profitable," 174–76).

29. A. Jacobi, "Smoke in Relation to Health," *Journal of the American Medical Association* 49 (1907): 813; *Journal of the American Medical Association* 44 (1905): 1620.

30. C. W. A. Veditz to Mrs. Franklin P. Iams, Mar. 15, 1912, AIS, 83:7, ser. I, FF 1; *Proceedings of the Eighth Annual Convention of the International Association for the Prevention of Smoke Held at Pittsburgh, Pennsylvania, September 9–12, 1913* (Pittsburgh, n.d.), 5.

31. "Smoke Abatement League News Clippings—1906 to 1915," p. 16, CHS.

32. *Journal of the Franklin Institute* 144 (1897): 318n.

33. Charles H. Benjamin, *Smoke Abatement* (Purdue University, Publications of the Engineering Department, 1915), 11.

34. Chicago Department of Health, *Report for 1881–82*, 34; Z. A. Willard, *Final Report on the Smoke Nuisance* (Boston, 1912), 2; Anti-Smoke League of Baltimore, "Sixth Letter to Members," 1906, 4; O'Connor, *Some Engineering Phases*, 15; *Annual Reports of the*

Departments of Government of the City of Cleveland for the Year Ending December 31, 1899 (n.l., 1900), 777; John W. Krause, "Smoke Prevention," *Proceedings of the Engineers' Society of Western Pennsylvania* 24 (1908): 92; *Proceedings of the Common Council of the City of Milwaukee for the Year Ending April 21st, 1896. Published by Authority of the Common Council* (Milwaukee, 1896), 694; Rochester Department of Public Safety, *Smoke Ordinances of the Common Council, City of Rochester,* ca. 1906, 4; *Report of the Fire-Escape and Smoke Inspector of the City of Cincinnati for the Year Ending December 31st, 1886* (Cincinnati, 1887), 15.

35. *St. Louis Globe-Democrat,* July 12, 1911.

36. *Baltimore Sun,* Nov. 20, 1906.

37. *Baltimore Sun,* Dec. 15, 1905, Feb. 5, Feb. 15, Apr. 12, 1907; D. T. Randall, *The Burning of Coal without Smoke in Boiler Plants: A Preliminary Report,* U.S. Geological Survey Bulletin 334 (Washington, 1908), 8.

38. Anti-Smoke League of Baltimore, "Letters to Members 1–6," 1905–6, and "Seventh Letter, Explaining Present Exasperating Conditions," Jan. 1, 1910, p. 4 (quotation).

39. *Baltimore Sun,* Mar. 15, 1907, Apr. 29, 1928. For similar events, see *Annual Report of the City Departments of the City of Cincinnati, for the Fiscal Year Ending December 31, 1881,* 11n; *Annual Report of the City Departments of the City of Cincinnati for the Fiscal Year Ending December 31, 1882* (Cincinnati, 1883), 9, 736n; "Annual Report of the Smoke Inspector, City of Milwaukee, for the Year 1909," p. 5, HML, acc. 915, box 2, "Misc. Reports, Photos, Pamphlets" folder.

40. Chicago Department of Health, *Report for 1885,* 102, *Report for 1886,* 81, *Report for 1887,* 99.

41. Citizens' Association of Chicago, *Annual Report 1887,* 13; Chicago Department of Health, *Report for 1889,* 119n.

42. Chicago Department of Health, *Report for 1891,* 60.

43. Citizens' Association of Chicago, *Annual Report 1890,* 13.

44. Chicago Department of Health, *Report for 1888,* 21; Citizens' Association of Chicago, *Annual Report 1884,* 5, and *Manual of the Citizens' Association of Chicago* (n.l., 1890), 36; *Chicago Record-Herald,* Oct. 4, 1902; *Baltimore Sun,* June 23, 1906.

45. Christine Meisner Rosen, "Businessmen against Pollution in Late Nineteenth Century Chicago," *Business History Review* 69 (1995): 376; Paul P. Bird, *Report of the Department of Smoke Inspection, City of Chicago* (Chicago, Feb. 1911), 13; Kroeger, "Smoke Abatement," 907; *Public Affairs: A Monthly Record of Civic and Social Progress in St. Louis, Published by the Civic League of St. Louis* 1, no. 5 (1912): 4, 2 no. 3 (1913): 8; *St. Louis Globe-Democrat,* May 13 (quotation), July 12, 1911.

46. *St. Louis Post-Dispatch,* May 15, 1911.

47. *Baltimore Sun,* Jan. 18, 1907; *Chicago Record-Herald,* June 29, 1906.

48. *Industrial World* 47 (1913): 141; Air Pollution Control League of Greater Cincinnati, "Circular to Members from Matthew Nelson," June 28, 1907, Public Library of Cincinnati and Hamilton County, Rare Book Room, Pamphlet File.

49. Pittsburgh Chamber of Commerce, *Year Book and Directory of the Chamber of Commerce of Pittsburgh, Pa.* (n.l., 1900), 63.

50. Stewart to Timanus, Jan. 10, 1906, p. 2, BCA, RG 9 S 13, box 77, Timanus file 99.

51. Charles A. L. Reed, *The Smoke Campaign in Cincinnati: Remarks before the National Association of Stationary Engineers, Cincinnati, July 10, 1906,* 9.

52. Reed, *Smoke Campaign,* 1 (quotation); Matthew Nelson, "Smoke Abatement in Cincinnati," *American City* 2 (1910): 8, 10.

53. For fines and jail terms, see *Chicago Record-Herald,* July 13, 1906. For publication of the accused, see Civic League of St. Louis, *1911 Year Book,* 12, 34; Civic League of St. Louis, *Supplement on Smoke Abatement,* Feb. 8, 1911; Anti-Smoke League of Baltimore, "Second Letter," 2n, and "Sixth Letter," 4; *Baltimore Sun,* Jan. 4, Jan. 7, 1902, June 15, Dec. 14, Dec.

22, 1905, Jan. 11, Jan. 18, 1907; *Chicago Record-Herald,* Nov. 1, Nov. 27, 1901, Mar. 20, Apr. 3, Apr. 17, June 19, Aug. 5, Oct. 4, Dec. 18, 1902, Apr. 25, 1904. For the quotation, see *Chicago Record-Herald,* Feb. 8, 1902.

54. L. L. Carson, "Convention for the Prevention of Smoke Held at Milwaukee," *Industrial World* 41 (1907): 850.

55. *Proceedings of the Engineers' Society of Western Pennsylvania* 8 (1892): 47; *Proceedings of the City Council of the City of Minneapolis from January 1, 1895, to January 1, 1896: Published by Authority of the City Council* (n.l., 1895), 557; *Journal of the Western Society of Engineers* 11 (1906): 731n; *Baltimore Sun,* Nov. 24, 1904, Jan. 24, Dec. 17, Dec. 23, 1905, June 7, 1908, May 13, 1909, Nov. 17, Nov. 27, Dec. 5, 1911; *Chicago Record-Herald,* Nov. 20, 1901.

56. *Chicago Record-Herald,* Nov. 20, 1901; Rochester Chamber of Commerce, Committee on Smoke Abatement, *The Abatement of Smoke* (n.l., 1911), 1.

57. *Chicago Tribune,* Jan. 15, 1897; Smoke Abatement Department of the City of St. Louis, *A Reply to the Civic League Report on the Smoke Nuisance,* Dec. 1906, 2; *Baltimore Sun,* Feb. 25, 1902, Nov. 22, 1904.

58. Chicago Department of Health, *Report for 1891,* 59.

59. Reed, *Smoke Campaign,* 11n.

60. Rochester Chamber of Commerce, *Abatement,* 4.

61. Chicago Department of Health, *Report for 1887,* 100.

62. Inspection of Boilers and Elevators and Smoke Abatement of the City of St. Louis, *Annual Report for the Fiscal Year 1910–1,* 1911, 4.

63. Chicago Department of Health, *Report for 1887,* 101.

64. Civic Improvement League of St. Louis, *Second Annual Report of the Civic Improvement League, March 1904* (St. Louis, 1904), 27.

65. Civic Improvement League of St. Louis, *Second Annual Report,* 27.

66. Reed, *Smoke Campaign,* 5.

67. Civic League of St. Louis, *Smoke Nuisance,* 26.

68. *Chicago Record-Herald,* Nov. 1, Nov. 8, Nov. 9, Nov. 27, 1901; Citizens' Association of Chicago, *Annual Report 1887,* 13.

69. *Chicago Record-Herald,* July 13, 1906.

70. Citizens' Association of Chicago, *Annual Report 1905,* 4; William Nicholson, *Smoke Abatement: A Manual for the Use of Manufacturers, Inspectors, Medical Officers of Health, Engineers, and Others* (London, 1905), 92; Bird, *Report,* 14n; *Chicago Record-Herald,* Nov. 1, 1901, Apr. 17, 1902; Inspection of Boilers, St. Louis, *Annual Report 1910–11,* 2; *St. Louis Post-Dispatch,* Dec. 12, 1910; Joseph M. Lonergan, "Smoke Abatement Activities in American Cities: New York," *Heating and Ventilating Magazine* 14, no. 10 (1917): 28.

71. Smoke Abatement Department, St. Louis, *Reply,* 2.

72. Chicago Department of Health, *Report for 1888,* 21.

73. City of St. Louis, *Report of the Special Committee,* 37–41.

74. CCC, box 9, folder 5, pp. 162–67; *Chicago Record-Herald,* Jan. 15, Feb. 25 (quotation), 1907. David Stradling is incorrect when he describes the sitting official as "a powerful and active ally" of the antismoke movement (*Smokestacks and Progressives: Environmentalists, Engineers, and Air Quality in America, 1881–1951* [Baltimore and London, 1999], 102).

75. CCC, box 9, folder 5, p. 160; City Club of Chicago, "The Smokeless Chicago Meeting," *City Club Bulletin* 1 (1907): 5.

76. City Club of Chicago, "Smokeless Chicago Meeting," 10, 11.

77. CCC, box 9, folder 5, p. 166.

78. Civic League of St. Louis, *Smoke Nuisance,* 15, 18.

79. Bird, *Report,* 10.

80. Bird, *Report,* 50–52, 116n; Department of Smoke Inspection, Chicago, "Notes on Smoke Abatement," p. 3, and "Methods of Approaching the Smoke Problem," p. 5, both HU, Eng 2650.2; Thomas E. Donnelly, "Smoke Abatement in Chicago," in Coal Smoke Abatement

Society, ed., *Papers Read at the Smoke Abatement Conferences March 26, 27 & 28, 1912* (London, n.d.), 52n; Osborn Monnett, "New Methods of Approaching the Smoke Problem," *National Engineer* 16 (1912): 721; Krause, "Smoke Prevention," 94; Raymond C. Benner, "Methods and Means of Smoke Abatement," *American City* 9 (1913): 231; Tunnicliff, "Smoke Elimination," 401; *Chicago Record-Herald*, Sept. 16, 1909.

81. Bird, *Report*, 46–52, 113–26.

82. Department of Smoke Inspection, Chicago, *Bulletin Number 1*, 3n; Bird, *Report*, 19, 91–94, 106–12; Paul P. Bird, "City Supervision of New Boiler Plants," *Power and the Engineer* 29 (1908): 206; Robert H. Kuss, "Construction of Furnaces to Meet the Requirements of the Chicago Smoke Department," *Power and the Engineer* 29 (1908): 207. However, this part of the overall strategy was not unprecedented (Civic League of St. Louis, *Smoke Nuisance*, 18).

83. *Chicago Record-Herald*, Oct. 27, 1909. Robert Dale Grinder is incorrect in depicting the *Record-Herald* as an opponent of smoke inspection ("The Anti Smoke Crusades: Early Attempts to Reform the Urban Environment, 1893–1918" [Ph.D. diss., University of Missouri, 1973], 91).

84. Jos. W. Hays, *Combustion and Smokeless Furnaces* (New York, 1906), 48n; Jos. W. Hays, *Combustion and Smokeless Furnaces*, 2nd ed. (Chicago, 1915), 49 (quotation).

85. *Chicago Record-Herald*, Apr. 17, 1909.

86. CCC, box 9, folder 6, p. 315.

87. George H. Cushing, "A Practical Campaign for Smoke Prevention," *American Review of Reviews* 38 (1908): 62; *Power* 38 (1913): 305.

88. O'Connor to Monnett, June 3, 1915, AIS, 83:7, ser. I, FF 15; Samuel B. Flagg, *Smoke Abatement and City Smoke Ordinances*, Bureau of Mines Bulletin 49 (Washington, 1912), 14, 29–33; *Baltimore Sun*, June 7, July 6, July 27, 1912.

89. Reed, *Smoke Campaign*, 5.

90. "Smoke Abatement League News Clippings—1906 to 1915," p. 94, CHS.

91. Smoke Abatement League of Cincinnati, *Annual Report for 1911* (n.l., n.d.), 10, and *Annual Report for 1912*, 5n.

92. AIS, 83:7, ser. III, FF 1, p. 60; Civic League of St. Louis, *Eleventh Year Book 1911–1912* (St. Louis, 1913), 6.

93. Boston Chamber of Commerce, *Smoke Abatement*, 8

94. Stradling, *Smokestacks*, 103, 106.

95. Inspector of Boilers, Elevators and Smoke Abatement of the City of St. Louis, *Report for Fiscal Year Ending April 7th, 1913* (n.l., 1913), 3.

96. Raymond C. Benner, "The Smoke Investigation of the Industrial Research Department of the University of Pittsburgh," *Industrial World* 46 (1912): 1273.

97. *Chicago Record-Herald*, Mar. 13, 1908.

98. Bird, *Report*, 57.

99. Donnelly, "Smoke Abatement," 58; *Chicago Record-Herald*, Apr. 24, May 5, July 31, Sept. 18, Dec. 18, 1908.

100. Lester D. Breckenridge, "Fuel Loss Caused by Smoking Chimneys," *Metal Worker, Plumber and Steam Fitter* 85 (1916): 130; William M. Barr, *A Catechism on the Combustion of Coal and the Prevention of Smoke: A Practical Treatise* (New York, 1901), 119; F. M. Logan, "Economy in Use of Fuel Results in Elimination of Dense Smoke," *Engineering Record* 72 (1915): 582; F. R. Wadleigh, "Firing Stationary Boilers Economically," *Power and the Engineer* 29 (1908): 959; J. A. Switzer, "The Economy of Smoke Prevention," *Engineering Magazine* 40 (1910–11): 406; Charles H. Benjamin, "Smoke Abatement in Large Cities," *Outlook* 70 (1902): 482.

101. Civic Improvement League of St. Louis, *Second Annual Report*, 28; Cushing, "Practical Campaign," 62.

102. AIS, 83:7, ser. III, FF 1, p. 75.

103. J. W. Henderson, "Up-to-Date Smoke Regulation," in *Proceedings, Eleventh Annual Convention, Smoke Prevention Association, St. Louis, Mo., Sept. 27–29, 1916*, 90.

104. George Ethelbert Walsh, "Smokeless Cities of To-Day," *Harper's Weekly* 51 (1907): 1139.

105. See the businessmen's remarks in Rochester Chamber of Commerce, *Abatement*, 4, 8, 11, 13; J. W. Henderson, "Reducing Smoke in Pittsburgh," *Power* 42 (1915): 152.

106. Boston Chamber of Commerce, *Smoke Abatement*, 11.

107. CCC, box 12, folder 5, p. 321.

108. Searle, "Record of the Year in Smoke Abatement," 131.

109. "Annual Report Smoke Inspection Department Lowell, Massachusetts, for the Year Ending December 31, 1912," p. 11, AIS, 83:7, ser. IV, FF 7.

110. Citizens' Association of Chicago, *Report of the Smoke Committee of the Citizens' Association of Chicago*, 1889, 11.

111. Tunnicliff, "Smoke Elimination," 395n.

112. "Black Smoke Means Waste of Fuel," p. 2, AIS 83:7, ser. IV, FF 9.

113. CCC, box 9, folder 6, p. 314; Bird, *Report*, 18.

114. CCC, box 10, folder 2, p. 240n, box 11, folder 1, pp. 281, 285n, 289n, and box 12, folder 5, pp. 316, 322.

115. "Club Meetings 1909–1910," p. 200, CWC, box 22.

116. Civic League of St. Louis, *Fourteenth Year Book 1915–16* (St. Louis, 1916), 7; Benner to Garland, Feb. 15, 1913, AIS, 83:7, ser. III, FF 17; AIS, 83:7, ser. III, FF 1, p. 55.

117. *Public Affairs: A Monthly Record of Civic and Social Progress in St. Louis*, published by the Civic League of St. Louis 2, no. 2 (1913): 6, and 2, no. 3 (1913): 8.

118. AIS, 83:7, ser. III, FF 1, pp. 67, 108n, 122n; O'Connor to Benner, Apr. 9, 1915, AIS, 83:7, ser. I, FF 2.

119. Henderson to O'Connor, July 7, 1915, AIS, 83:7, ser. III, FF 18.

120. Bird, "City Supervision," 206.

121. *Industrial World* 47 (1913): 134; "Report of the Committee on Smoke Prevention of the Cleveland Chamber of Commerce, May 15th, 1912," p. 5, AIS, 83:7, ser. IV, FF 9; CCC, box 9, folder 5, p. 160; Bird, *Report*, 5.

122. *Proceedings of the Eighth Annual Convention*, 59.

123. Civic League of St. Louis, *Smoke Nuisance*, 25; Donnelly, "Smoke Abatement," 58.

124. Pittsburgh Bureau of Smoke Regulation, *Hand Book 1916*, 3.

125. Rochester Chamber of Commerce, *Abatement*, 1.

126. "Club Meetings 1909–1910," p. 200; *Power* 46 (1917): 718.

127. Walter M. Squires, "Smoke Abatement Activities in American Cities. Cincinnati," *Heating and Ventilating Magazine* 14, no. 10 (1917): 33.

128. Henderson, "Smoke Abatement Activities," 32; Pittsburgh Bureau of Smoke Regulation, *Hand Book 1916*, 5; Lawrence W. Bass, "The Investigation of Air Pollution at the Mellon Institute," *Science* 70 (1929): 186; H. B. Meller, "Whither Smoke and Dust Abatement," Nov. 1, 1933, p. 5n, AIS, 83:7, ser. I, FF 25.

129. *Bulletin* 8 no. 8 (Jan. 1924): 227, WCCC, folder 16.

130. *Baltimore Sun*, Apr. 29, 1928; *Annual Report of the Department of Public Works to the Mayor and City Council of Baltimore for the Year Ending December 31, 1932* (n.l., n.d.), 346; Flagg, *Smoke Abatement*, 11 (quotation); *Baltimore Sun*, Apr. 8, 1911, Apr. 29, 1928.

131. Before a change of the constitution in 1913, only municipal employees could be full voting members; all others had to join as "associate members" without voting rights (Carson, "Convention," 848; *Proceedings of the Eighth Annual Convention*, 6–8).

132. *Industrial World* 47 (1913): 1093. The Pittsburgh Chamber of Commerce recorded some three hundred participants (Pittsburgh Chamber of Commerce, *Annual Report for 1913–1914* [Pittsburgh, 1914], 28).

133. *Proceedings of the Eighth Annual Convention of the International Association for the Prevention of Smoke, Pittsburgh, September 9–12, 1913* (n.d., n.l.), 57.

134. *Proceedings of the Tenth Annual Convention, Smoke Prevention Association, Cincinnati, Ohio, September 8–10, 1915* (n.d., n.l.), 12.

135. Bird, *Report*, 10, 24, 30, 127n.

136. Bird, *Report*, 40; W. F. M. Goss, *Smoke Abatement and Electrification of Railway Terminals in Chicago: Report of the Chicago Association of Commerce Committee of Investigation on Smoke Abatement and Electrification of Railway Terminals* (Chicago, 1915), 178.

137. O'Connor, *Some Engineering Phases*, 71n, 75; Bird, *Report*, 77; Inspection of Boilers, St. Louis, *Report for 1912–13*, 4n, and *Annual Report 1914–15*, 6; Boston Chamber of Commerce, *Smoke Abatement*, 6; Osborn Monnett, *Smoke Abatement*, Bureau of Mines Technical Paper 273 (Washington, 1923), 23.

138. Donnelly, "Smoke Abatement," 53; Clinton Rogers Woodruff, "Railroads and the Smoke Nuisance," *Popular Science Monthly* 74 (1909): 155; Clinton Rogers Woodruff, "The Business Man and the Smoke Nuisance," *World To-Day* 19 (1910): 1124; Krause, "Smoke Prevention," 111; Pittsburgh Chamber of Commerce, *A Year's Record of Great Achievements: Annual Report of the President to the Chamber of Commerce of Pittsburgh*, 1908, 24; J. M. Searle, "The Year's Labor for Smoke Prevention in the City of Pittsburgh," *Industrial World* 47 (1913): 129. See also David Stradling, "Dirty Work and Clean Air: Locomotive Firemen, Environmental Activists, and Stories of Conflict," *Journal of Urban History* 28 (2001): 48.

139. Bird, *Report*, 68; Civic League of St. Louis, *Report of the Committee on Railroad Electrification*, 1911, 11; Inspection of Boilers, St. Louis, *Report for 1912–13*, 5; E. P. Roberts, "Cleveland Reports Thousands Saved in Smoke Abatement, 1913 over 1912," *Industrial World* 48 (1914): 138; Arthur G. Hall, "Results of Eighteen Months of Co-operation by Cincinnati Smoke Department," *Industrial World* 48 (1914): 137; John M. Lukens, "Lively Campaign in Courts Fought in Philadelphia," *Industrial World* 48 (1914): 133; Flagg, *Smoke Abatement*, 21; E. A. Thompson, "Baltimore's New Campaign Actively Under Way," *Industrial World* 48 (1914): 135.

140. *The International Association for the Prevention of Smoke, Ninth Annual Convention, September 9, 10 and 11, 1914, Grand Rapids, Mich.*, 40, 46.

141. Donnelly, "Smoke Abatement," 54; Civic League of St. Louis, *Report of the Committee*, 6; Z. A. Willard, *Smoke Nuisance in Europe and America: Its Menace and Its Cure* (n.l. [Boston?], n.d. [1909?]), 42; John Llewellyn Cochrane, "Government Solves the Smoke Problem," *American Review of Reviews* 39 (1909): 193; Civic League of St. Louis, *Smoke Nuisance*, 23; Carl W. Condit, *The Port of New York: A History of the Rail and Terminal System from the Grand Central Electrification to the Present* (Chicago and London, 1981), 6–10; Michael Bezilla, *Electric Traction on the Pennsylvania Railroad, 1895–1968* (University Park and London, 1980), 28

142. *Chicago Record-Herald*, Sept. 18, 1908.

143. D. F. Crawford, "The Abatement of Locomotive Smoke," *Industrial World* 47 (1913): 1100.

144. Civic League of St. Louis, *Report of the Committee*, 5n; Crawford, "Abatement," 1100; A. W. Gibbs, "The Railway Smoke Problem," in Frederick Law Olmsted, Harlan Page Kelsey, et al., eds., *The Smoke Nuisance* (Harrisburg, 1911), 48; "The Financial Practicability of the Electrification of Steam Railway Terminals," Oct. 24, 1912, p. 5, HCAP, box 21, "Chicago Smoke Abatement Case" folder, no. 2; "Midday Luncheon May 22, 1912, Remarks by Horace G. Burt," p. 4, HCAP, box 2, "Chicago Smoke Abatement Case" folder, no. 12.

145. Civic League of St. Louis, *Report of the Committee*, 11.

146. Cf. Eugene M. Tobin, "'Engines of Salvation' or 'Smoking Black Devils': Jersey City Reformers and the Railroads, 1902–1908," in Eugene M. Tobin and Michael H. Ebner, eds., *The Age of Urban Reform: New Perspectives on the Progressive Era* (Port Washington and

London, 1977), 144, 155; Mark Aldrich, *Safety First: Technology, Labor, and Business in the Building of American Work Safety 1870–1939* (Baltimore and London, 1997), 282; Albro Martin, *Enterprise Denied: Origins of the Decline of American Railroads, 1897–1917* (New York and London, 1971), 231.

147. Lucius H. Cannon, *Smoke Abatement: A Study of the Police Power as Embodied in Laws, Ordinances and Court Decisions* (St. Louis, 1924), 258; Flagg, *Smoke Abatement*, 19, 24n.

148. See Lonergan, "Smoke Abatement Activities," 27; Flagg, *Smoke Abatement*, 22; Krause, "Smoke Prevention," 92; Henry Obermeyer, *Stop That Smoke!* (New York and London, 1933), 212; *Industrial World* 47 (1913): 141; *Power and the Engineer* 31 (1909): 117; *Power* 38 (1913): 174; *American City* 40, no. 1 (1929): 138; *New York Times*, Aug. 15, Aug. 21, 1913, June 7, 1924.

149. Flagg, *Smoke Abatement*, 22; Cannon, *Smoke Abatement*, 228n, 272; "Minutes of the Board of Estimates," 1914, p. 244, and 1915, pp. 27, 140, both BCA, RG 36 S 1; *Baltimore Sun*, July 8, July 9, 1914, Feb. 27, Mar. 13, Mar. 14, Apr. 14, Apr. 15, Apr. 16, May 8, May 13, May 23, 1915; Carl V. Harris, *Political Power in Birmingham, 1871–1921* (Knoxville, 1977), 230.

150. *Rauch und Staub* 1 (1910): 1.

151. *Mitteilungen der Österreichischen Gesellschaft zur Bekämpfung der Rauch- und Staubplage*, published since 1906.

152. StADr, 3.1, R 24, vol. 1, doc. 37, p. 374.

153. StAB, 4,14/1 IX.c.2.f, case Besselstr. 44, letter of June 1, 1907.

154. StAS, Depot B, C XVIII 4, vol. 1, no. 1, doc. 65a.

155. *Münchener Gemeinde-Zeitung* 35 (1906): 1177.

156. StAL, Hauptverwaltungsamt, Kap. 19, no. 11, vol. 1, pp. 65–67; StADr, 3.1, R 24, vol. 1, pp. 122–25; NieHStA Hann., 122a, no. 3113, pp. 104–6.

157. StAF, Magistratsakte, R 1528, vol. 1, pp. 30b, 30c.

158. Frank Uekötter, "Luftverschmutzung im Stuttgart der Jahrhundertwende: Von der Verwaltung eines Problems," *Zeitschrift für Württembergische Landesgeschichte* 60 (2001): 244n.

159. StAC, Kap. V, sect. VII, no. 583, vol. 1, pp. 55–60, 68–76R, 98–103.

160. Frank Uekötter, "Konsens ohne Strategie: Der Kampf gegen die großstädtische Kohlenrauchplage in Braunschweig und Hannover," in Carl-Hans Hauptmeyer, ed., *Mensch—Natur—Technik: Aspekte der Umweltgeschichte in Niedersachsen und angrenzenden Gebieten* (Bielefeld, 2000), 119–27; *Bayerisches Industrie- und Gewerbeblatt* 23 (1891): 3, 126, 132, 24 (1892): 3, 147n, 25 (1893): 140, 47 (1915): 23, 48 (1916): 97, and 49 (1917): 57.

161. Martin Göpfert to Magistrat der Stadt Frankfurt, Nov. 19, 1928, StAF, R 1528, vol. 1.

162. StAS, Depot B, C XVIII 4, vol. 1, no. 1, doc. ad 1b; StADr, 3.1, R 24, vol. 1, p. 124.

163. *Sitzungsberichte des Vereins zur Beförderung des Gewerbfleißes 1899*, 134.

164. *Neues Tagblatt* (Stuttgart), no. 245, Oct. 20, 1891, 9.

165. Cf. *Bremer Bürger-Zeitung* no. 206 (Sept. 3, 1913), and no. 170 (July 23, 1910); HAK, Best. 424, no. 547, p. 162R.

166. StAB, 4,14/1 V.A.4.bi, case 3, petition of May 20, 1882.

167. *Münchener Gemeinde-Zeitung* 35 (1906): 1057, 1177.

168. StADr, 3.1, R 24, vol. 1, doc. 35, p. 349.

169. John O'Connor Jr., "Civic Phases of the Smoke Problem," *National Municipal Review* 5 (1916): 302.

170. Konrad W. Jurisch, *Zwei Denkschriften über Luftrecht, dem Ausschuß des Bundes der Industriellen in Berlin für das Studium der Errichtung einer gewerblich-technischen Reichsbehörde mit Benutzung der Ergebnisse der vom Ausschuß veranstalteten Umfrage*, Sammlung von Abhandlungen über Abgase und Rauchschäden, ed. H. Wislicenus, vol. 4, reprinted in *Waldsterben im 19. Jahrhundert* (Düsseldorf, 1985), 20.

171. StAS, Depot B, C XVIII 4, vol. 3, no. 12, doc. 226.

172. Gewerberat to Polizeidirektion Braunschweig, Sept. 22, 1890, StABr, E 32,6, no. 5.
173. G. de Grahl, "Ueber die technischen Maßnahmen zur Verhütung der Ruß- und Rauchplage in Großstädten," *Rauch und Staub* 1 (1910–11): 393.
174. GStA, Rep. 120, BB II a 2, no. 28, Adh. 1, vol. 5, p. 5.
175. *Mittheilungen aus der Praxis des Dampfkessel- und Dampfmaschinen-Betriebes* 21 (1898): 495.
176. v. Pasinski, "Der gegenwärtige Stand der Rauchbekämpfungsfrage," *Rauch und Staub* 1 (1910–11): 72.
177. Gewerbeinspektion Bonn to Regierungspräsident, Feb. 1, 1904, HStAD, Regierung Köln no. 2213; Gewerbekammer Bremen to Gewerbekommission des Senats, Sept. 13, 1890, StAB, 3-G.4.a no. 49; Mitteilung des Senats, Mar. 8, 1889, p. 117, 6,40 E 1 no. 17; case 8, note of Feb. 15, 1905, StAB, 4,15 II. W; note of May 18, 1907, StABi, GS 12,482.
178. R. Weinlig, "Die Rauchplage in den Städten und die Mittel der Abhilfe," *Zeitschrift des Vereines deutscher Ingenieure* 28 (1884): 918 (emphasis added).
179. Bäckerinnung zu Braunschweig to Herzogliche Polizei-Direction Braunschweig, Dec. 20, 1883, StABr, E 32,6, no. 5; Gewerbekammer Bremen to Gewerbekommission des Senats, Sept. 13, 1890, 3-G.4.a, no. 49.
180. GStA, Rep. 120, BB II a 2, no. 28, Adh. 1, vol. 6, p. 3.
181. *Sitzungsberichte des Vereins zur Beförderung des Gewerbfleißes 1899*, 133.
182. Michael Karl, *Fabrikinspektoren in Preußen: Das Personal der Gewerbeaufsicht 1854–1945: Professionalisierung, Bürokratisierung und Gruppenprofil* (Opladen, 1993), 98; note of Jan. 12, 1884, StABr, E 32,6, no. 5; BrLHA, Rep. 30, Berlin C, no. 1936, p. 212.
183. *Jahresbericht des Vereins Berliner Kaufleute und Industrieller*, 1901, 240; *Mittheilungen aus der Praxis des Dampfkessel- und Dampfmaschinen-Betriebes* 24 (1901): 578; *Jahresbericht über die Fortschritte und Leistungen auf dem Gebiete der Hygiene* 19 (1901): 623n; *Correspondenz der Aeltesten der Kaufmannschaft von Berlin*, 1901, 49.
184. Edwin T. Layton Jr., *The Revolt of the Engineers: Social Responsibility and the American Engineering Profession* (Baltimore and London, 1986).
185. Benjamin, "Smoke and Its Abatement," 726.
186. Peter Lundgreen, "Die Vertretung technischer Expertise 'im Interesse der gesamten Industrie Deutschlands' durch den VDI 1856 bis 1890," in Karl-Heinz Ludwig, ed., *Technik, Ingenieure und Gesellschaft: Geschichte des Vereins Deutscher Ingenieure 1856–1981* (Düsseldorf, 1981), 75, 85, 115; Tibor Süle, *Preußische Bürokratietraditionen: Zur Entwicklung von Verwaltung und Beamtenschaft in Deutschland, 1871–1918* (Göttingen, 1988), 198–202; Wolfgang König, "Ingenieure in der staatlichen Verwaltung, 1870–1945," in Peter Lundgreen and André Grelon, eds., *Ingenieure in Deutschland, 1770–1990* (Frankfurt and New York, 1994), 146–48.
187. Michael Stolberg offers a different reading of the experts' role that stresses their proximity to industrial interests (*Ein Recht auf saubere Luft? Umweltkonflikte am Beginn des Industriezeitalters* [Erlangen, 1994]). However, this interpretation ignores the fact that almost every expert in the smoke abatement debate took a stance *against* pollution, often chastising industrialists for their scant attention.
188. Hood to O'Connor, Jan. 4, 1917, AIS, 83:7, ser. III, FF 5.
189. M. Gerbel, "Die Ökonomie der Feuerung und die Rauchbelästigung," *Zeitschrift für Gewerbehygiene, Unfallverhütung und Arbeiterwohlfahrtseinrichtungen* 12 (1905): 646.
190. Grabau, "Rauchverbrennung," *Zeitschrift des Vereines deutscher Ingenieure* 40 (1896): 642.
191. Ferdinand Haier, "Die Rauchfrage, die Beziehungen zwischen der Rauchentwicklung und der Ausnutzung der Brennstoffe, und die Mittel und Wege zur Rauchverminderung im Feuerungsbetrieb," *Zeitschrift des Vereines deutscher Ingenieure* 49 (1905): 88.
192. Bird, *Report*, 30 (quotation), 39n.

193. For measurements indicating the significant contribution of domestic polluters, see E. von Esmarch, "Vergleichende Untersuchungen über den Rauch- und Rußgehalt verschiedener Städte," *Rauch und Staub* 1 (1910–11): 248; Hermann Rasch, *Der Schutz der Nachbarschaft gewerblicher Anlagen in Hamburg* (Hamburg, 1911), 52; Ludwig Dietz, *Ventilations- und Heizungsanlagen mit Einschluss der wichtigsten Untersuchungs-Methoden: Ein Lehrbuch für Ingenieure, Architekten, Studierende, Besitzer von Ventilations- und Heizungsanlagen* (Munich and Berlin, 1909), 286.

194. M. Neisser, "Stadtverrußung und Hausfeuerung," *Umschau* 15 (1911): 118 (quotation); v. Pasinski, "Großstadtentwicklung und Rauchbelästigung," in Stadtverwaltung Düsseldorf, ed., *Verhandlungen des Ersten Kongresses für Städtewesen Düsseldorf 1912* (Düsseldorf, 1913), 2: 112.

195. Carl Bach and Theodor Peters, "Zur Frage der Rauchbelästigung," *Zeitschrift des Vereines deutscher Ingenieure* 37 (1893): 1236; *Zeitschrift des Vereines deutscher Ingenieure* 34 (1890): 162, 1100, 1125n, and 40 (1896): 522; GStA, Rep. 120, BB II a 2, no. 28, Adh. 1, vol. 5, pp. 189–90.

196. NieHStA Hann., 122a, no. 3113, pp. 5, 29; StADr, 3.1 R 24, vol. 1, doc. 20, p. 44, doc. 31, p. 23n; Frank Uekötter, "Die Kommunikation zwischen technischen und juristischen Experten als Schlüsselproblem der Umweltgeschichte: Die preußische Regierung und die Berliner Rauchplage," *Technikgeschichte* 66 (1999): 1–31.

197. *Bericht über die V. Sitzung der Kommission zur Prüfung und Untersuchung von Rauchverbrennungs-Vorrichtungen zu Berlin am 24. November 1898* (Stettin, 1899), 65.

198. "Geschäftsbericht des Württembergischen Dampfkessel-Revisions-Vereins über das Vereinsjahr 1883," p. 5, GStA, Rep. 120, BB II a 5, no. 37, vol. 1.

199. Carl Bach, "Ueber den Stand der Frage der Rauchbelästigung durch Dampfkesselfeuerungen," *Zeitschrift des Vereines deutscher Ingenieure* 40 (1896): 494; Reinert, *Dampfkesselanlagen*, 66; von Düsing, "Die Rauchplage und die Vorkehrungen zu ihrer Verringerung," *Zentralblatt für Bauverwaltung* 33 (1913): 239; H. C. Nußbaum, "Die Rußplage in den Städten und die häuslichen Feuerungsstätten," *Zeitschrift für Architektur und Ingenieurwesen* 45 (1899): 533; O. Krell, *Über Rauchverhütung in Nürnberg*, 1903, 3n.

200. Bach and Peters, "Frage," 1237.

201. Cochrane, "Government," 192; J. W. Henderson, "Conservation of Fuel through Smoke Regulation," *Railway Review* 62 (1918): 193; Henderson, "Smoke Abatement Activities," 32; C. S. Sale, "Atmospheric Sanitation," *American City* 20 (1919): 253; J. M. Searle, "What Not to Do in Attempting the Conservation of Fuel," *Industrial World* 47 (1913): 355.

202. "Memorandum on a Proposed National Smoke Abatement Conference, April 1909," p. 2 (quotation), FWTC, folder 10A.

203. "Some of the Signers of Petition for Conference on Smoke Prevention," n.d., MLCP, box 168, "A.S.M.E.—Commercial Control" folder. On the utility's resistance, see Cooke to Bryan, Oct. 11, 1909, pp. 3–5, MLCP, box 168, "A.S.M.E.—Commercial Control" folder; Cooke to Low, Mar. 27, 1917 (quotation), MLCP, box 168, "A.S.M.E.—Boston Meeting" folder; Taylor to Cooke, Dec. 13, 1909, FWTC, folder 114.

204. Executive Committee Meeting Minutes, Aug. 14, 1930, MoHS, Citizens' Smoke Abatement League Minutes of Meetings 1926–1934.

205. Ferdinand Haier, *Dampfkesselfeuerungen zur Erzielung einer möglichst rauchfreien Verbrennung: Zweite Auflage im Auftrage des Vereines deutscher Ingenieure bearbeitet vom Verein für Feuerungsbetrieb und Rauchbekämpfung in Hamburg* (Berlin, 1910).

206. Fabrikeninspektor Wegener to Polizeidirektion, Oct. 22, 1893, StAB, 4,14/1 IX.D.1.bc (emphasis original); Tschorn, "Die Arbeiten der Kommission zur Prüfung von Rauchverbrennungs-Vorrichtungen," *Sitzungsberichte des Vereins zur Beförderung des Gewerbfleißes* (1899): 124.

207. StADr, 3.1, R 24, vol. 1, p. 93.

208. GStA, Rep. 120, BB II a 2, vol. 1, no. 28, p. 32; "Anweisung, betreffend die Geneh-

migung und Untersuchung der Dampfkessel," p. 9, GStA, Rep. 120, BB II a 5, vol. 21, no. 1 (quotation).

209. *Allgemeines Landrecht für die Preußischen Staaten von 1794* (Frankfurt and Berlin, 1970), 620; Rasch, *Schutz*, 38; Tschorn, "Das Rauchen der Schornsteine," *Technisches Gemeindeblatt* 1 (1898): 51; A. Reich, *Leitfaden für die Rauch- und Russfrage* (Munich and Berlin, 1917), 249; F. Hoffmann, *Die Gewerbe-Ordnung mit den gesamten Ausführungsbestimmungen für das deutsche Reich und Preußen*, 7th and 8th eds. (Berlin, 1910), 44; GStA, Rep. 76 VIII B, no. 2082, p. 105.

210. StAF, Magistratsakte, T 730, doc. 72, doc. 126, §§26–28.

211. GStA, Rep. 120, BB II a 2, vol. 4, no. 28, pp. 106–10, 220–20R, 260n; *Sammlung gerichtlicher Entscheidungen auf dem Gebiete der öffentlichen Gesundheitspflege* 4 (1905): 42n, 52n, 60n, and 5 (1908): 74.

212. Gudrun Lies-Benachib, *Immissionsschutz im 19. Jahrhundert* (Berlin, 2002), 441n.

213. Civic League of St. Louis, *Smoke Nuisance*, 8n; *St. Louis Globe-Democrat*, Nov. 17, 1897, Mar. 22, 1905; "Report 1911," p. 3, AIS, 83:7, ser. III, FF 19; *American City* 9 (1913): 201; A. L. H. Street, "Soft-Coal Smoke as Legal Nuisance," *Power* 40 (1914): 376n; A. L. H. Street, "Question of Validity of Anti-Smoke Ordinance," *Power* 42 (1915): 478; John O'Connor Jr., "The History of the Smoke Nuisance and of Smoke Abatement in Pittsburgh," *Industrial World* 47 (1913): 354.

214. *Engineering News* 75 (1916): 194.

215. Christine Rosen, "Differing Perceptions of the Value of Pollution Abatement across Time and Place: Balancing Doctrine in Pollution Nuisance Law, 1840–1906," *Law and History Review* 11 (1993): 303–81. See also Christine Meisner Rosen, "Noisome, Noxious, and Offensive Vapors: Fumes and Stenches in American Towns and Cities, 1840–1865," *Historical Geography* 25 (1997): 49–82; Christine Meisner Rosen, "'Knowing' Industrial Pollution: Nuisance Law and the Power of Tradition in a Time of Rapid Economic Change, 1840–1864," *Environmental History* 8 (2003): 565–97.

216. NieHStA Hann., 122a, no. 3113, p. 42R.

217. "Polizeidirektion Braunschweig to Kreisdirektion Helmstedt," Mar. 5, 1887, StABr, E 32,6, no. 5.

218. StAD, III, 19508, p. 211; Magistrat München to Präsidium der Regierung von Oberbayern, June 17, 1893, StAM, RA 58425; StAS, Depot B, C XVIII 4, vol. 2, no. 11, doc. 3, vol. 3, no. 12, doc. 210; Haupt- und Residenzstadt Stuttgart, *Bericht über die Verwaltung und den Stand der Gemeinde-Angelegenheiten in den Jahren 1896 bis 1898 bezw. 1895/96–1897/98* (Stuttgart, 1900), 185.

219. NieHStA Hann., 122a, no. 3113, p. 7. At times, administrative procedures even ran counter to the letter of the law. For one such case, see Bündel 680, doc. 26, Bündel 682, case Reibedanz & Co., Stadtdirektion Stuttgart to Stadtpolizeiamt, Apr. 13, 1910, StALu, F 201.

220. Direktion der Gas-, Wasser- und Elektrizitätswerke Heidelberg to Stadtrat, Jan. 19, 1909, StAHei, Altaktei Archiv, no. 255, Fasc. 4.

221. "Report of Stadtbauamts," Oct. 1, 1894, StAM, RA 58425.

222. GStA, Rep. 76 VIII B, no. 2082, pp. 90, 135R, 149R, 218, 221R, 232R, 355R, 364, 367R, 371, 393, 396R, 433R.

223. StAD, III, 19508, pp. 26R, 28.

224. StAD, III, 19508, p. 27R.

225. Uekötter, "Kommunikation," 13.

226. *Hygienische Rundschau* 6 (1896): 918 (emphasis original).

227. Karl Hauser, "Versuche mit einer rauchfreien Hausbrandfeuerung," *Rauch und Staub* 3 (1912–13): 159.

228. See Frank Uekoetter, "Solving Air Pollution Problems Once and for All: The Potential and the Limits of Technological Fixes," in Lisa Rosner, ed., *The Technological Fix* (New York and London, 2004), 155–74.

229. StAS, Depot B, C XVIII 4, vol. 1, no. 1, Fasc. 4, doc. 5.
230. GStA, Rep. 120, BB II a 2, no. 28, Adh. 1, vol. 1, p. 133.
231. GStA, Rep. 120, BB II a 2, no. 28, Adh. 1, vol. 1, p. 133R.
232. StAS, Depot B, C XVIII 4, vol. 3, no. 12, doc. 209.
233. GStA, Rep. 76, VIII B, no. 2006, p. 253R.
234. StAD, III, 19508, pp. 163, 168–68R, 171R–72 (quotation on 172).
235. StAD, III, 19508, p. 171R.
236. StADr, 3.1, R 24, vol. 1, p. 93.
237. Polizeidirektion to Staatsministerium in Braunschweig, June 18, 1885, StABr, E 32,6, no. 5.
238. StAD, III, 19508, pp. 195R, 199, 208, 211, 214, 245; StAL, Hauptverwaltungsamt, Kap. 19, vol. 1, no. 11, p. 6; StAHH, 111-1, Cl. VII Lit. F d, vol. 52, no. 1, doc. 77.
239. Uekötter, "Luftverschmutzung," 245; Uekötter, "Konsens," 115; StADr, 3.1, R 24, vol. 1, doc. 1, pp. 68R–69.
240. GStA, Rep. 120, BB II a 2, no. 28, Adh. 1, vol. 5, p. 31n, vol. 6, pp. 150R–51; "Protokolle des Ausschusses zur Prüfung des Antrags von Prahl und Genossen betreffend die Beseitigung von Rauch und Ruß," StAHH, 121-3 I C 611; *Protokolle und Ausschuß-Berichte der Bürgerschaft* 1899, Ausschußbericht no. 39.
241. Uekötter, *Rauchplage*, 525–57.
242. StAC, Kap. V, sect. VII, vol. 2, no. 583, p. 230n.
243. "Gutachten des Kreisarztes," May 3, 1907, StABi, GS 12,482 (emphasis original).
244. HAK, Best. 424, no. 547, p. 36R.
245. Gewerbeinspektion Cöln I to Regierungspräsident, July 23, 1904, HStAD, Regierung Köln, no. 2213.
246. NieHStA Hann., 122a, no. 3113, p. 39R; HAK, Best. 424, no. 547, p. 84R; Karl Hauser, "Die Rauchplage in den Städten," *Deutsche Vierteljahrsschrift für öffentliche Gesundheitspflege* 42 (1910): 136; *Bericht über den Stand der Gemeindeangelegenheiten der kgl. Haupt- und Residenzstadt München für das Jahr 1906: Erster Teil: Verwaltungsbericht* (Munich, n.d.), 108, 1908, 59, 1910, 86, 1912, 45.
247. GStA, Rep. 120, BB II a 2, vol. 3, no. 28, p. 84R; StAC, Kap. V, sect. VII, vol. 1, no. 583, p. 64R.
248. NieHStA Hann., 122a, no. 3113, p. 43; HAK, Best. 424, no. 547, p. 9.
249. StAHH, 111-1, Cl. VII Lit. F d, vol. 52, no. 1, doc. 78.
250. StADr, 3.1, R 24, vol. 1, doc. 31, p. 16.
251. StAC, Kap. V, sect. VII, vol. 1, no. 583, pp. 68R–69.
252. Only two cities—Boston and Providence—pursued smoke abatement on the basis of state laws (Board of Gas and Electric Light Commissioners, *General Law*, 165–70; *Public Laws of the State of Rhode Island and Providence Plantations, Passed at the Session of the General Assembly 1911 and 1912* [Providence, 1912], 380–86).
253. Jürgen Reulecke, *Geschichte der Urbanisierung in Deutschland* (Frankfurt, 1985), 56–62; Wolfgang R. Krabbe, *Die deutsche Stadt im 19. und 20. Jahrhundert: Eine Einführung* (Göttingen, 1989), 99–128.
254. GStA, Rep. 120, BB II a 2, vol. 3, no. 28, pp. 11n, 83n; *Anlage zum Amtsblatt der Stadt Nürnberg*, no. 141 (Dec. 1, 1876): 18; *Sammlung der Statuten der Stadt Braunschweig*, no. 36, "Statut, betreffend den Betrieb von durch Rauch oder Ruß belästigenden Feuerungsanlagen"; StAS, Depot B, C XVIII 4, vol. 1, no. 1, doc. 79; StADr, 3.1, R 24, vol. 1, doc. 20, pp. 40n, 64; *Adreßbuch der Stadt Heidelberg für das Jahr 1915* (Heidelberg, 1915), 571; *Münchener Jahrbuch* 5 (1892): 489; "Ortspolizeiliche Vorschrift, die Verhütung von Belästigung durch Rauch und Ruß betr.," StAFr, C 3, no. 524/2.
255. Citizens' Association of Chicago, *Annual Report 1881*, 19, and *Annual Report 1882*, 7; Chicago Department of Health, *Report for 1881–82*, 34, and *Report for 1883–84*, 137;

Cincinnati City Departments Annual Reports 1881, 668n; John F. Weh, ed., *Ordinances of the City of Cleveland* (Cleveland, 1882), 282–84; O'Connor, *Some Engineering Phases,* 15; Civic League of St. Louis, *Smoke Nuisance,* 8; *St. Louis Post-Dispatch,* Jan. 22, 1893; *Proceedings of the City Council of the City of Minneapolis, Minnesota from January 1, 1894, to January 1, 1895* (n.l., 1894), 75; *Proceedings of the Common Council of the City of Milwaukee,* 720; Willard, *Final Report,* 11n; *Acts and Resolves Passed by the General Assembly of the State of Rhode Island and Providence Plantations, at the January Session, 1902* (Providence, 1902), 58n; Anti-Smoke League of Baltimore, "Sixth Letter," 3n; Rochester Department of Public Safety, *Smoke Ordinances,* 1.

256. StAC, Kap. V, sect. VII, vol. 2, no. 583, pp. 1R–2, 30R–31, 117R–18.

257. *Journal für Gasbeleuchtung und Wasserversorgung* 55 (1912): 144.

258. StAAc, Cap. 15, vol. 1, no. 2, docs. 15, 30–33, 35, 37–42.

259. GStA, Rep. 76, VIII B, no. 2082, p. 400.

260. See Frank Uekötter, "Umweltschutz in den Händen der Industrie—eine Sackgasse? Die Geschichte des Hamburger Vereins für Feuerungsbetrieb und Rauchbekämpfung," *Zeitschrift für Unternehmensgeschichte* 47 (2002): 198–216.

261. *Verwaltungsbericht des Rathes der Königlichen Haupt- und Residenzstadt Dresden für das Jahr 1901* (Dresden, 1902), 139; *Verwaltungsbericht des Rates der Königlichen Haupt- und Residenzstadt Dresden für das Jahr 1902* (Dresden, 1903), 155.

262. StADr, 3.1, R 24, vol. 1, pp. 95R, 103R, 106, 115, 164–64R; Ministerium des Innern no. 11612, "Dienstanweisung des Heizaufsehers für die Feuerungsanlagen in der Stadt," Dresden, May 1, 1902, SHStA; StAD, III, 19508, pp. 106R–7, 195.

263. Karl Hauser, "Aus der Praxis der Rauchbekämpfung," *Rauch und Staub* 2 (1911–12): 113–15; Karl Hauser, "Praktische Erfolge der Rauchbekämpfung in München," *Rauch und Staub* 3 (1912–13): 333–36; Karl Hauser, "Die Rauch- und Rußbekämpfung in München und ihre künftige Ausgestaltung," *Rauch und Staub* 6 (1915–16): 70–77; Hauser, "Versuche"; A. Reich, "Massnahmen zur Verhütung von Rauchschäden in deutschen Städten," *Gesundheit* 43 (1918): 55n.

264. *Bericht über den Stand der Gemeinde-Angelegenheiten der Kgl. Haupt- und Residenzstadt München für das Jahr 1910. Erster Teil: Verwaltungsbericht* (Munich, 1912), 120; *Verwaltungsbericht des Rates der Königl. Haupt- und Residenzstadt Dresden für die Jahre 1909 und 1910* (Dresden, 1912), 401; StAHH, 111-1, Cl. VII, Lit. Q d, vol. 1b, no. 210b, doc. 42b, p. 5.

265. StAD, III, 19508, pp. 195R (quotation), 208.

266. Industriekommission der Handelskammer to Handelskammer in Hamburg, Mar. 26, 1901, StAHH, 111-1, Cl. VII, Lit. F d, vol. 52, no. 1, attachment to doc. 68; "Mitgliederverzeichnis des Vereins für Feuerungsbetrieb und Rauchbekämpfung," StAHH, 111-1, Cl. VII, Lit. Q d, vol. 1b, no. 210b, attachment to doc. 1; *Verwaltungsbericht des Rathes der Königlichen Haupt- und Residenzstadt Dresden für das Jahr 1898* (Dresden, 1900), 165; *Verwaltungsbericht des Rathes der Königlichen Haupt- und Residenzstadt Dresden für das Jahr 1899* (Dresden, 1901), 154; StADr, 3.1, R 24, vol. 1, pp. 103, 105; *Sitzungsberichte der Stadtverordneten in Dresden* 1912, 18; Hauser, "Praxis."

267. E. Nies, "Bericht des Vereins für Feuerungsbetrieb und Rauchbekämpfung in Hamburg über seine Tätigkeit im Jahre 1910," *Rauch und Staub* 1 (1910–11): 275; E. Nies, "The Aims and Works of the Hamburg Abatement Society," in Coal Smoke Abatement Society, ed., *Papers Read at the Smoke Abatement Conferences March 26, 27 & 28, 1912* (London, n.d.), 46–50; *Verhandlungen des Vereins zur Beförderung des Gewerbfleißes* 91 (1912): 376.

268. *Münchener Gemeinde-Zeitung* 35 (1906): 1178; Bach, "Stand," 494.

269. "Direktion des städtischen Elektrizitätswerkes und der Straßenbahn to Stadtrat," May 15, 1908, StAFr, C 3, no. 524/2; Wilhelm Buddëus, *Die Lösung der Rauchfrage in München* (Diessen, 1907), 4; Krell, *Rauchverhütung,* 4n; H. S. Vassar, "Smoke Prevention in

Large Power Stations," *Power* 34 (1911): 173; Obermeyer, *Stop That Smoke!* 140; *Münchener Gemeinde-Zeitung* 35 (1906): 1057; GStA, Rep. 76 VIII B, no. 2082, p. 427R; StADr, 3.1, R 24, vol. 1, p. 93R.

270. *Württemberger Zeitung* no. 78 (Dec. 6, 1907); StADr 3.1, R 24, vol. 1, pp. 92, 102.

271. *Münchener Gemeinde-Zeitung* 35 (1906): 1056.

272. Smoke Abatement League of Cincinnati, *Annual Report for 1911*, 9.

273. Deutsches Konsulat Cincinnati to Reichskanzler, May 20, 1914, BayHStA, MWi 660.

274. "Board of Governors Minutes," Oct. 26, 1914, p. 8, UU, acc. 854, box 16, folder 1. Interestingly, the cliché of the German official diligently implementing the law survives even in some recent studies (e.g., Morag-Levine, *Chasing*, 105).

275. Gustav Lang, *Der Schornsteinbau. Viertes Heft: Sockel, Grundbau, Fuchs und Einstiegöffnungen, Bekämpfung der Rauch- und Russplage* (Hanover, 1911), 543.

Chapter 3

1. H. A. Garfield, *Final Report of the United States Fuel Administrator, 1917–1919* (Washington, 1921), 246–52.

2. Monnett, *Smoke Abatement*, 1; Gordon D. Rowe, "Experiences in Smoke Abatement Work," *Manual of Smoke and Boiler Ordinances and Requirements in the Interest of Smoke Regulation, 1922 Edition* (n.l., 1922), 55; Civic Club of Allegheny County, *Report of Activities for Eighteen Months Ending May 1st, 1919* (Pittsburgh, n.d.), 11; *Power* 47 (1918): 6; *Pittsburgh Post*, July 15, 1917; O'Connor to Hillman, Jan. 19, 1918, AIS, 83:7, ser. III, FF 6.

3. John G. Clark, *Energy and the Federal Government: Fossil Fuel Policy, 1900–1946* (Urbana and Chicago, 1987), 50, 52, 70.

4. David M. Kennedy, *Over Here: The First World War and American Society* (New York, 1982), 124; James P. Johnson, *The Politics of Soft Coal: The Bituminous Industry from World War I through the New Deal* (Urbana, 1979), 65–67; James P. Johnson, "The Wilsonians as War Managers: Coal and the 1917–18 Winter Crisis," *Prologue* 9 (1977): 193–208; James P. Johnson, "The Fuel Crisis, Largely Forgotten, of World War I," *Smithsonian* 7, no. 9 (1976): 64–68, 70n.

5. J. Clark, *Energy*, 72; Johnson, *Politics*, 70–75.

6. Pittsburgh Bureau of Smoke Regulation, *Hand Book for Nineteen-Eighteen*, 3.

7. *Journal of the Proceedings of the City Council of the City of Chicago for the Council Year 1917–18*, 1381n.

8. Lonergan, "Smoke Abatement Activities," 29; *Survey* 40 (1918): 45; *Bulletin* 6, no. 7 (Nov. 1917): 12, WCCC, folder 9; AIS, 83:7, ser. III, FF 1, p. 120; O'Connor to Hamerschlag, Sept. 29, 1917, AIS, 83:7, ser. III, FF 3; Henderson to Earhart, Apr. 29, 1918, AIS, 83:7, ser. III, FF 10; Pittsburgh Bureau of Smoke Regulation, *Hand Book for Nineteen-Seventeen*, 3, 7; *Ohio State Journal*, Nov. 18, 1917; J. W. Henderson, "Smoke Abatement Means Economy," *Power* 46 (1917): 127; Henderson, "Conservation," 193–95; Henderson, "Smoke Abatement Activities," 32.

9. On the Salt Lake City campaign, see Osborn Monnett, G. St. J. Perrott, and H. W. Clark, *Smoke-Abatement Investigation at Salt Lake City, Utah*, Bureau of Mines Bulletin 254 (Washington, 1926), 1; H. W. Clark, "Results of Three Years of a Smoke Abatement Campaign," *American City* 31 (1924): 343n; O. P. Hood, "Practical Suggestion on Smoke Abatement," *City Manager Magazine* 8, no. 6 (1926): 23; *Municipal Record* (Salt Lake City) 8, no. 12 (1919): 3–6, and 9, no. 10 (1920): 3–8. On Chicago, see *Bulletin* 8 no. 2 (Oct. 1919): p. 6, WCCC, folder 11; John Dill Robertson, "Smoke Abatement Activities in Chicago," *American City* 23 (1920): 513.

10. *Power Plant Engineering* 27 (1923): 1173.

11. Smoke Abatement League of Cincinnati, *Annual Report 1912*, 11 (emphasis original).

12. O'Connor to Carmalt, Apr. 28, 1919, AIS, 83:7, ser. III, FF 6.

13. Monnett to O'Connor, Sept. 19, 1912, AIS, 83:7, ser. I, FF 15.

14. *Power* 38 (1913): 316.
15. *Power* 43 (1916): 21, 45, (1917): 6, and 50 (1919): 435.
16. *Bulletin* 8 no. 8 (Jan. 1924): 227, WCCC, folder 16.
17. *Bulletin* 17 no. 7 (Feb. 1928): 508, and 18 no. 1 (May 1928): 15, both WCCC, folder 20; *American City* 40, no. 1 (1929): 138.
18. *Bulletin* 18, no. 6 (Jan. 1929): 213n, WCCC, folder 21; *Journal of the Proceedings of the City Council of the City of Chicago for the Council Year 1928–1929*, 4166–68.
19. *Bulletin* 18, no. 8 (Mar. 1929): 274, WCCC, folder 21; Chicago Public Library, "Comparative Statement of Activities of the Department of Smoke Inspection and Abatement for Years 1929, 1930, 1931, 1932 and 1933" (typewritten manuscript).
20. "Report of Smoke Abatement Committee," Dec. 1, 1923, WRHS, mss. 3535, box 1, folder 6.
21. Ellerton J. Brehaut, "The Future of the Smoke Situation in Boston," *Our Boston*, Apr. 1929, 13.
22. Monnett, "Status," 1284; Smoke Prevention Association, *Manual of Smoke and Boiler Ordinances and Requirements in the Interest of Smoke and Air Pollution Regulation and Fuel Combustion* (n.l., 1936), 28n.
23. William G. Christy, "The Human Side of Smoke Abatement," *Mechanical Engineering* 55 (1933): 349.
24. Harvey N. Davis, "How the Problem of Smoke Abatement has been Attacked," *Stevens Indicator* 49 (1932): 71; William Culbert, "Practical Procedure for Inaugurating and Conducting Smoke Abatement Work in the Smaller Cities," *Proceedings of the Thirtieth Annual Convention of the Smoke Prevention Association of America, Atlanta, Georgia, June 2–5, 1936*, 82; Martin A. Rooney, "Experiences in Smoke Abatement Work," *Manual of Smoke and Boiler Ordinances and Requirements in the Interest of Smoke Regulation, 1922 Edition* (n.l., 1922), 60; Obermeyer, *Stop That Smoke!* 135, 147n; Hood, "Practical Suggestion," 20; Henderson, "Smoke Abatement in Pittsburgh," 46; O'Connor, *Some Engineering Phases*, 17n.
25. Burrows to Gundlach, July 17, 1924, MoHS, Citizens' Smoke Abatement League Papers, box 1, "Papers 1924" folder.
26. Smoke Prevention Association, *Manual* (1936), 26n.
27. Muhlhauser to Gephart, Nov. 15, 1923 (quotation), SHSW, Parkside SC 49; "Minutes of Smoke Abatement Committee Meeting," Nov. 15, 1923, WRHS, mss. 3535, box 1, folder 6; "Executive Committee Meeting Minutes," May 14, 1926, p. 1, MoHS, Citizens' Smoke Abatement League Minutes of Meetings, 1926–1934; "General Plan," p. 2n, MoHS, Citizens' Smoke Abatement League, Miscellaneous Records, Publicity Material, Pamphlets, etc.
28. C. G. Buder, "Smoke Abatement Work in St. Louis," *Journal of the American Society of Heating and Ventilating Engineers* 33 (1927): 108.
29. *Baltimore Sun*, Mar. 20, 1931.
30. "Women's City Club History," p. 9, WRHS, mss. 3535, box 1, folder 4.
31. "Minutes of Smoke Abatement Committee Meeting," Nov. 15, 1923, WRHS, mss. 3535, box 1, folder 6.
32. "Fuels and Steam Power Section," paper no. 21, *Transactions of the American Society of Mechanical Engineers* 50 (1928): 168.
33. Junior Chamber of Commerce of Cincinnati, *Smoke over Cincinnati: A Smoke Abatement Survey with Recommendations* (Cincinnati, 1938), 42n; Charles N. Howison, "The History of a Citizens Organization for Cleaner Air," July 27, 1954, p. 1, CHS, mss. qA 298, box 1, folder 4.
34. *Bulletin* 18, no. 6 (Jan. 1929): 213, WCCC, folder 21.
35. Lund to Horsley, Oct. 22, 1940, p. 2, BYUL, mss. 276, box 1, folder 4; "President's Report to the Officers and Members of the Salt Lake City Women's Chamber of Commerce," Mar. 1, 1937, p. 1n, USHS, mss. B-164.
36. Ormsby to Hocker, Jan. 19, 1926, and Christy to Gundlach, Jan. 26, 1926, both MoHS,

Citizens' Smoke Abatement League Papers, box 1, "Papers 1926" folder, part 1; "Pro Forma Decree of Incorporation," MoHS, Citizens Smoke Abatement League Papers, box 1, "Papers 1927 Mar. 27; n.d." folder; Buder, "Smoke Abatement Work," 105.

37. "General Plan," 2.

38. "Minutes of the Board of Directors Annual Meeting," Feb. 18, 1932, p. 1, MoHS, Citizens' Smoke Abatement League Minutes of Meetings, 1926–1934. Pledges for funding ran even higher, allegedly exceeding $250,000 ("Board of Directors Meeting Minutes," May 6, 1927).

39. O'Connor to Roberts, Aug. 17, 1914, AIS, 83:7, ser. IV, FF 7.

40. "Board of Directors Meeting Minutes," July 7, 1927, Jan. 10, Jan. 27, Feb. 14 (pp. 3–5), Nov. 7 (pp. 1, 3), 1928, Jan. 16 (p. 1), 1929, "Executive Committee Meeting Minutes," June 24, Aug. 12, Dec. 16 (p. 1), 1926, Jan. 13, Jan. 25, 1927, May 22 (p. 1), July 8, 1928, all MoHS, Citizens' Smoke Abatement League Minutes of Meetings, 1926–1934; "Statement by Citizens' Smoke Abatement League of St. Louis on Completion of Second Year of Campaign, May 1st, 1929," esp. p. 3, MoHS, Scrapbook File, Smoke Abatement, 1925–1932.

41. "Executive Committee Meeting Minutes," Mar. 7, Mar. 29, 1929, Mar. 24, May 26, Oct. 1, Oct. 9, 1931, Jan. 5, Mar. 10, Apr. 5, Oct. 6, 1932, "Board of Directors Meeting Minutes," Mar. 15 (p. 2n), 1929, Mar. 3, 1931, Mar. 24, 1932, Apr. 11 (p. 1), 1933, May 31, Oct. 23 (p. 1), 1934, Feb. 20, Apr. 24 (p. 1), 1935, all MoHS, Citizens' Smoke Abatement League Minutes of Meetings, 1926–1934.

42. Frantz, "Problem," 202.

43. Rowe, "Experiences," 57.

44. "Report of Smoke Abatement Committee," Dec. 1, 1923, WRHS, mss. 3535, box 1, folder 6.

45. H. B. Meller, "What Is Ahead in Smoke Abatement," *Chemical and Metallurgical Engineering* 38 (1931): 512.

46. O. P. Hood, "Gains Against the Nuisances," *National Municipal Review* 12 (1923): 112; Harvey N. Davis, "The Air Pollution Problem," *National Electric Light Association Proceedings* 88 (1931): 104; H. B. Meller, "Some Features of Smoke Regulation in Pittsburgh," *Municipal and County Engineering* 59 (1920): 128; Charles S. Rhyne and William G. Van Meter, *City Smoke Control and Air Pollution Programs—Model Ordinance Annotated,* National Institute of Municipal Law Officers Report No. 120 (Washington, 1947), 4; "Proceedings of the Sixth Meeting of the Pittsburgh Smoke Commission," Apr. 7, 1941, p. 6, PCCA.

47. William G. Christy, "Smoke Abatement in Coal Burning Plants," *Proceedings of the Sixteenth Fuel Engineers' Meeting: Sponsored by Fuel Engineering Division of Appalachian Coals, Inc., Queen City Club, Cincinnati, Ohio, October 5, 1936,* 8.

48. Monnett, Perrott, and Clark, *Salt Lake City,* 92. Due to the city's topography, early-morning smoke was a key problem for Salt Lake City.

49. *Power Plant Engineering* 30 (1926): 1128.

50. Morgan B. Smith, "The Smoke Nuisance from the Manufacturer's View-Point," *Factory* 35 (1925): 404.

51. Monnett, Perrott, and Clark, *Salt Lake City,* 95.

52. Christy, "Smoke Abatement," 6; A. W. Akins, "Organization of the Industrial Smoke Abatement Association," *Proceedings of the Thirty First Annual Convention of the Smoke Prevention Association, May 31st–June 5th, 1937, New York City,* 41–44; *Proceedings, Thirty-Second Annual Convention, Smoke Prevention Association, Nashville–Tennessee, May 17–20, 1938,* 165n; *Power Plant Engineering* 40 (1936): 437.

53. Akins, "Organization," 43.

54. H. J. Meyer, "Smoke Abatement," *Minnesota Municipalities* 9 (1924): 175.

55. H. W. Evans, "The Smoke Nuisance and Its Abatement," *Power Plant Engineering* 37 (1933): 401.

56. O. P. Hood, "Keeping the Atmosphere Clean," *Forge: Bulletin of St. Louis Section, the American Society of Mechanical Engineers* 49 (June 1926): 3.

57. Ralph A. Sherman, "Paths to Smoke Abatement," *Mechanical Engineering* 67 (1945): 521.

58. Parr, "Smoke Combustion Problems," 455; Obermeyer, *Stop That Smoke!* 38, 99; P. H. Hardie, "Defining Equitable Limits of Dust Emission from Stacks," *Mechanical Engineering* 61 (1939): 895.

59. *Bulletin* 21 no. 2 (June 1931): 38, WCCC, folder 23; Monnett, "Status," 1284.

60. Flagg, *Smoke Abatement*, 32, 35; Monnett, *Smoke Abatement*, 8; Monnett, Perrott, and Clark, *Salt Lake City*, 71; Obermeyer, *Stop That Smoke!* 100; H. B. Meller, "A Modern Plan for a Community Campaign against Air Pollution," *American Journal of the Medical Sciences* 186 (1933): 160; H. B. Meller, "The Chemist's Interest in Air Pollution," *Industrial and Engineering Chemistry* 27 (1935): 950; *Municipal Record* (Salt Lake City) 8, no. 12 (1919): 5.

61. *Bulletin* 16, no. 5 (1927), p. 185 (quotation), WCCC, folder 19; "Annual Report of the Smoke Abatement Committee," Apr. 7, 1923, p. 2; "Mr. Erle Ormsby, Speech Delivered at Public Affairs Meeting, A.S.M.E.," Jan. 15, 1926, p. 2, MoHS, Citizens' Smoke Abatement League Papers, box 1, "Papers 1926" folder, part 2; Meller to Weidlein, Jan. 6, 1930, AIS, 83:7, ser. I, FF 29; "Report of the Smokeless City Committee of the City Federation 1927–29," p. 1, UU, acc. 1069, box 2, folder 5; Obermeyer, *Stop That Smoke!* 204, 217; Buder, "Smoke Abatement Work," 107.

62. Monnett, "Status," 1284.

63. Osborn Monnett, "Smoke Ordinances," *Industrial and Engineering Chemistry* 33 (1941): 840.

64. H. D. Blackwell, "Untitled Presentation," *Proceedings of the Fuel Engineer's Meeting, sponsored by the Fuel Engineering Division of Appalachian Coals, Inc., 4th Meeting, Jan. 8, 1935* (n.l., n.d.), 3.

65. Monnett, Perrott, and Clark, *Salt Lake City*, 57; Hood, "Progress," 125; William Culbert, "Smoke Abatement," *Proceedings of the Twenty-Ninth Annual Convention of the Smoke Prevention Association, St. Louis, Missouri, June 4–7, 1935*, 50; C. V. Beck, "Observations of a Coal Man on Smoke Abatement," *Proceedings of the Twenty-Ninth Annual Convention of the Smoke Prevention Association, St. Louis, Missouri, June 4–7, 1935*, 65; Marc Bluth, "Cooperative Effort between Stoker Industry and Smoke Abatement Officials," *Proceedings of the Thirtieth Annual Convention of the Smoke Prevention Association of America, Atlanta, Georgia, June 2–5, 1936*, 74.

66. "Board of Directors Meeting Minutes," Jan. 17, 1928, p. 1, MoHS, Citizens' Smoke Abatement League Minutes of Meetings, 1926–1934.

67. "Board of Directors Meeting Minutes," Jan. 27, 1928, p. 2n, MoHS, Citizens' Smoke Abatement League Minutes of Meetings, 1926–1934.

68. "Board of Directors Meeting Minutes," May 28, 1929, p. 2, MoHS, Citizens' Smoke Abatement League Minutes of Meetings, 1926–1934.

69. "General Plan," 4.

70. "Minutes of the Board of Directors Annual Meeting," Feb. 18, 1932, p. 2, MoHS, Citizens' Smoke Abatement League, Minutes of Meetings, 1926–1934.

71. Merrill to Lund, Dec. 5, 1936, p. 2, BYUL, mss. 276, box 1, folder 1.

72. Joel A. Tarr, "The Metabolism of the Industrial City: The Case of Pittsburgh," *Journal of Urban History* 28 (2002): 524.

73. "Executive Committee Minutes," Mar. 2, 1934 (quotation), and "Board of Directors Meeting Minutes," Mar. 29, 1934, p. 1, both MoHS, Citizens' Smoke Abatement League Minutes of Meetings, 1926–1934.

74. Obermeyer, *Stop That Smoke!* 139.

75. Ormsby to Hocker, Jan. 19, 1926.

76. Raymond R. Tucker, "Smoke Abatement: Its Past, Present, and Future in St. Louis," *Mechanical Engineering* 60 (1938): 379.

77. "Address of Raymond R. Tucker to the Engineers Club of St. Louis," Sept. 23, 1937, p. 2, RRTSAC, ser. 4, box 1, "Speeches—1937" folder.

78. Oscar Hugh Allison, "Raymond R. Tucker: The Smoke Elimination Years, 1934–1950" (Ph.D. diss., Saint Louis University, 1978), 7, 9.

79. Tucker to Christy, Apr. 28, 1941, RRTSAC, ser. 2, box 1, "William G. Christy" folder.

80. Tucker, "Smoke Abatement," 379; Raymond R. Tucker, "New Features in Smoke Abatement Legislation," *Proceedings, Thirty-Second Annual Convention, Smoke Prevention Association, Nashville, Tennessee, May 17–20, 1938*, 56n; Raymond R. Tucker, "Smoke Prevention in St. Louis," *Industrial and Engineering Chemistry* 33 (1941): 838; Frank A. Chambers, A. D. Singh, and I. A. Deutch, "Atmospheric Pollution in Chicago," *Proceedings of the Thirty-First Annual Convention*, 167.

81. Raymond R. Tucker, "A Smoke Elimination Program that Works," *Heating Piping and Air Conditioning* 17–18 (1945–46): 464n; Tucker, "New Features," 57; Tucker, "Smoke Abatement," 379; C. G. Stiehl, interview, May 11, 1939, p. 1, RRTSAC, ser. 4, box 1, "Speeches—1939" folder.

82. K. K. Richmond, "Where the Responsibility Rests for Smoke and Air Pollution Problems," *Proceedings, Thirty-Second Annual Convention, Smoke Prevention Association, Nashville, Tennessee, May 17–20, 1938*, 16.

83. J. Clark, *Energy*, 10, 23; Johnson, *Politics*, 20; Sam H. Schurr and Bruce C. Netschert, *Energy in the American Economy, 1850–1975* (Baltimore, 1960), 74.

84. *Heating and Ventilating Magazine* 24, no. 8 (1927): 88.

85. *Kansas City Public Affairs*, June 24, 1937, 3; Leita Thompson, Chairman, Smoke Abatement Committee, attachment to letter, p. 2, AHC, Metropolitan Business and Professional Women's Club—Atlanta Chapter, Subject File; Charles W. Gruber, "Requirements That Should Be Incorporated in a Smoke Ordinance," *Proceedings of the Thirty-Third Annual Convention of the Smoke Prevention Association, Milwaukee, Wisconsin, June 13–16, 1939*, 45.

86. Raymond R. Tucker, address to the Engineers Club of St. Louis, Sept 23, 1937, p. 5, RRTSAC, ser. 4, box 1, "Speeches—1937" folder.

87. Tucker, "Smoke Abatement," 380.

88. "Report of Operations of Division of Smoke Regulation, Department of Public Safety, October 15, 1937 to October 15, 1938," p. 12, MoHS, Citizens' Smoke Abatement League, Miscellaneous Records, Publicity Material, Pamphlets, etc.

89. Tucker to Dickmann, Dec. 2, 1939, RRTSAC, ser. 1, box 1, "Mayor's Office, Correspondence with" folder; James Neal Primm, *Lion of the Valley: St. Louis, Missouri, 1764–1980*, 3rd ed. (St. Louis, 1998), 449; Allison, "Raymond R. Tucker," 43.

90. "Minutes of Meeting of Special Citizens' Committee with City Officials for Consideration of the Smoke Elimination Problem, held Tuesday, December 5, 1939," p. 17, RRTSAC, ser. 2, box 2, "Smoke Elimination Committee" folder.

91. Tucker, "Smoke Prevention," 836; *Report of the St. Louis Committee on Elimination of Smoke. Presented to Mayor Bernard F. Dickmann, February 24, 1940*, 7.

92. Tucker, "Smoke Prevention," 839; Tucker, "Smoke Elimination Program," 521.

93. "Recommendations for Smoke Elimination in St. Louis by the Special Smoke Committee of the Associated Engineering Societies of St. Louis, February 1940," p. 1, RRTSAC, ser. 2, box 1, "'A' Miscellaneous" folder.

94. "Smoke Elimination Committee, Ninth Meeting," Jan. 19, 1940, p. 2, RRTSAC, ser. 2, box 2, "Smoke Elimination Committee" folder.

95. "Statement of Richard F. Wood, representing the Coal Exchange of St. Louis," p. 3, MoHS, Luther Ely Smith Papers, box 15, folder 4.

96. Tucker, "Smoke Elimination Program," 469; Raymond R. Tucker, "You Must Mean Business to Make Smoke Control Work," *American City* 65, no. 11 (1950): 113; Tucker to

Dickmann, Dec. 2, 1939; "Smoke Elimination Committee Meeting," Dec. 20, 1939, pp. 62, 69, ser. 2, box 2, "Smoke Elimination Committee" folder.

97. Tucker to Riley, Apr. 26, 1941, RRTSAC, ser. 1, box 1, "Mayor's Office, Correspondence with" folder.

98. Tucker to Ford, Mar. 20, 1946, RRTSAC, ser. 8, box 1, "James L. Ford, Jr." folder.

99. Tucker, "Smoke Abatement," 378.

100. RRTSAC, ser. 2, box 2, "Smoke Elimination Committee" folder, esp. meeting of Dec. 13, 1939.

101. "Minutes of Meeting of Special Citizens' Committee with City Officials for Consideration of the Smoke Elimination Problem, held Tuesday, December 5, 1939," p. 8, RRTSAC, ser. 2, box 2, "Smoke Elimination Committee" folder.

102. "Board of Directors Meeting Minutes," Jan. 16, 1929, p. 2, MoHS, Citizens' Smoke Abatement League Minutes of Meetings, 1926–1934.

103. "Smoke Elimination Committee Meeting," Dec. 13, 1939, p. 31, RRTSAC, ser. 2, box 2, "Smoke Elimination Committee" folder.

104. Carter to Ducas, May 10, 1946, p. 2, RRTSAC, ser. 2, box 2, "'H' Miscellaneous" folder. Stradling thus gives a misleading impression when he describes Tucker's approach as "shifting the regulatory focus from the boiler room to the coal yard." After all, traditional smoke controls of industrial furnaces remained a routine part of administrative work, with 14,970 inspections being recorded for 1940–41 (Stradling, *Smokestacks*, 167; St. Louis Department of Public Safety, Division of Smoke Regulation, *Annual Report for Fiscal Year 1940–41* [St. Louis, Apr. 1941], 1).

105. Carter to Ducas, May 10, 1946, p. 2.

106. *Business Week*, Apr. 29, 1944, 39.

107. Tucker, "You Must," 113; *Business Week*, Aug. 31, Dec. 14, 1940; J. H. Carter, "Highlights of St. Louis Smoke Elimination Program," July 1944, p. 2, RRTSAC, ser. 2, box 1, "Fly Ash Material" folder.

108. Joel A. Tarr and Carl Zimring, "The Struggle for Smoke Control in St. Louis: Achievement and Emulation," in Andrew Hurley, ed., *Common Fields: An Environmental History of St. Louis* (St. Louis, 1997), 215n; Allison, "Raymond R. Tucker," 115n.

109. Tucker, "Smoke Elimination Program," no. 1, p. 102.

110. "Smoke Meeting," Feb. 20, 1941, p. 7, AIS, 70:2, Add. 1971, box 11, folder 182.

111. Tucker, "Smoke Elimination Program," 468.

112. "Smoke Elimination Committee Meeting," Dec. 13, 1939, p. 28, RRTSAC, ser. 2, box 2, "Smoke Elimination Committee" folder.

113. *St. Louis Commerce*, Mar. 6, 1940, 2.

114. Tucker to Dickmann, Oct. 24, 1940, RRTSAC, ser. 1, box 1, "Mayor's Office, Correspondence with" folder.

115. "Smoke Elimination Committee Meeting," Dec. 13, 1939, p. 33, RRTSAC, ser. 2, box 2, "Smoke Elimination Committee" folder.

116. Tucker to Riley, Apr. 26, 1941.

117. Tucker, address to the Engineers Club of St. Louis.

118. *Life*, Jan. 15, 1940, 8n, 11, Jan. 13, 1941, 6n, 9; *Business Week*, Apr. 6, 1940, 33n, Apr. 20, 1940, 48n, Aug. 31, 1940, 20n, Dec. 14, 1940, 29n; *American City* 55, no. 5 (1940): 105, 55, no. 10 (1940): 115; Tucker to Dickmann, Feb. 28, 1940 (quotation), RRTSAC, ser. 1 box 1, "Mayor's Office, Correspondence with"folder.

119. *Business Week*, Feb. 22, 1941, 30.

120. Joel A. Tarr, *The Search for the Ultimate Sink: Urban Pollution in Historical Perspective* (Akron, 1996), 232–53; Sherie R. Mershon and Joel A. Tarr, "Strategies for Clean Air: The Pittsburgh and Allegheny County Smoke Control Movements, 1940–1960," in Joel A. Tarr, ed., *Devastation and Renewal: An Environmental History of Pittsburgh and Its Region* (Pittsburgh, 2003): 145–73.

121. Pittsburgh Bureau of Smoke Prevention, "Report on Stationary Stacks: Year 1947," pp. 5–8, AIS, 80:7, box 1, FF 5; Pittsburgh Bureau of Smoke Prevention, "1950 Report on Stationary Stacks," p. 12, AIS, 70:2, add. 1971, box 8, folder 151.

122. Pittsburgh Bureau of Smoke Prevention, "Report on Stationary Stacks: Year 1947," 8, 10n; "Suggestions by Dr. Mellor [sic] for Smoke Control in Pittsburgh and Allegheny County," Aug. 22, 1940, p. 2, AIS, 70:2, add. 1971, box 11, folder 182; Tucker, "Smoke Elimination Program," no. 1, p. 102.

123. "Smoke Memo," Jan. 28, 1941, p. 1, AIS, 70:2, add. 1971, box 11, folder 182.

124. Pittsburgh Bureau of Smoke Prevention, "Report on Stationary Stacks: Year 1948," p. 7, AIS, 80:7, box 1, FF 5. With that, it is doubtful that the 55 percent approval rate that Jay Ream, the future chair of the Smoke Abatement Committee of the Allegheny Conference on Community Development, reported in a 1941 hearing was authentic ("Hearing before the Pittsburgh Smoke Commission," Mar. 28, 1941, p. 24n, PCCA, Proceedings).

125. Tarr, *Search*, 255; Stefano Luconi, "The Enforcement of the 1941 Smoke-Control Ordinance and Italian Americans in Pittsburgh," *Pennsylvania History* 66 (1999): 580–94.

126. Tarr and Zimring, "Struggle," 218; Tarr, *Search*, 251–53; John H. Herbert, *Clean, Cheap Heat: The Development of Residential Markets for Natural Gas in the United States* (New York, 1992), 141, 166; Schurr and Netschert, *Energy*, 127.

127. George T. Love, "Coal: The Backbone of Pittsburgh's Economy. Address before Rotary Club of Pittsburgh, June 5, 1946," pp. 8–10, AIS, 69:14, box 2, folder 26; Wurts to Tucker, Dec. 11, 1947, RRTSAC, ser. 2, box 2, "H. B. Lammers" folder.

128. "Public Hearing of the Council of the City of Pittsburgh Re Enforcement Date of Smoke Ordinance Provisions Affecting Dwellings," Apr. 17–18, 1946, p. 31n, PCCA.

129. "Hearing before the Pittsburgh Smoke Commission," Apr. 15, 1941, p. 49, PCCA, Proceedings.

130. Wurts to Tucker, Dec. 11, 1947.

131. Ford to Condie, Oct. 15, 1947, RRTSAC, ser. 8, box 1, "James L. Ford, Jr." folder; Tucker, "Smoke Elimination Program," 105.

132. Clarence A. Mills, "A Practical Approach to the Problem of Urban Air Pollution," *Cincinnati Journal of Medicine* 22 (1941–42): 502; Cooney to Tucker, Nov. 10, 1941, RRTSAC, ser. 2, box 2, "'M' Miscellaneous" folder; Berl to Tucker, Nov. 16, 1944, RRTSAC, ser. 2, box 1, "'B' Miscellaneous" folder.

133. Monnett, "Smoke Ordinances," 840.

134. Pendray, "Management Aspects," 581.

135. Kugel to Tucker, July 28, 1941, RRTSAC, ser. 1, box 1, "K" folder.

136. H. B. Lammers, "'Low Volatile' Smoke Ordinance Does Not Solve Problem," *Heating, Piping and Air Conditioning* 19, no. 11 (1947): 80n; H. B. Lammers, "Employ Engineering to Correct the Individual Smoke Problem," *Heating, Piping and Air Conditioning* 19, no. 12 (1947): 100n; Carroll F. Hardy, "Review of Surveys of Heating and Power Plants in Various Cities with Recommendations for Smoke Elimination," *Proceedings of the Smoke Prevention Association of America, 1943*, 68; *Rotarian* 70, no. 4 (1947): 57; H. B. Lammers, "Special Report to the Coal Producers Committee for Smoke Abatement: Air Pollution in St. Louis," CHS, mss. 562, box 73, folder 2. The author of a *Reader's Digest* article reported that he had been flooded with defamatory letters (Detzer to Tucker, Aug. 15, 1941, and Aug. 20, 1941, RRTSAC, ser. 2, box 2, "'R' Miscellaneous" folder).

137. Charles N. Howison, "The History of a Citizens Organization for Cleaner Air," July 27, 1954, p. 1n, CHS, mss. qA 298, box 1, folder 4; Charles W. Gruber, "Air Pollution Control in Cincinnati 1964," p. 25, CGC, group 1.

138. AIS, 83:7, ser. III, FF 1, pp. 73, 92, 95; *Year Book Issue 1928–29*, p. 72, *Bulletin* 18, no. 5 (1928): 163, 180, WCCC, folder 20; *Bulletin* 18, no. 7 (1929): 246n, WCCC, folder 21.

139. Allen J. Johnson and George H. Auth, eds., *Fuels and Combustion Handbook* (New York, 1951), 471.

140. William G. Christy, "Let's Make Air Pollution Control Effective," *American City* 64, no. 11 (1949): 78.
141. Robert T. Griebling, "Cleaning Our Air," *Smokeless Air* 22 (1952): 164.
142. Ford to Condie, Oct. 15, 1947.
143. Dyktor to Anderson, July 20, 1950, p. 1, WRHS, mss. 4035, box 1, folder 2.
144. Uekötter, "Umweltschutz."
145. StADr, 2.3.15, no. 1197, p. 41R.
146. *Verwaltungsbericht der Landeshauptstadt München 1924–1926 (1. April 1924 bis 31. März 1927)* (Munich, n.d.), 64.
147. Polizeipräsident Berlin to Minister für Volkswohlfahrt, Apr. 7, 1925, BArch, R 154/64. On the earlier attempts to enact a smoke ordinance, see Uekötter, "Kommunikation."
148. StABr, E 32,1, no. 301, pp. 1–7 (quotation on p. 1).
149. *Sitzungsberichte des Preußischen Landtags, 3. Wahlperiode, vol. 5* (Berlin, 1929), 6414; HStAD, BR 1050, no. 8, p. 72.
150. A. Crone, "Ursachen und Bekämpfung der Rauch- und Rußbelästigungen," *Gesundheits-Ingenieur* 51 (1928): 755.
151. Raymond H. Dominick III, *The Environmental Movement in Germany: Prophets and Pioneers, 1871–1971* (Bloomington, 1992), 42.
152. Regierungspräsident Köln to Book, Feb. 28, 1933, HStAD, Regierung Köln, no. 8314.
153. LHAM, Rep. C 28 I f, no. 291, vol. 12, pp. 201 (quotation), 202, 207R, 220.
154. LHAM, Rep. C 28 I f, no. 291, vol. 13, p. 46.
155. Albert Stoll to Gewerbeaufsichtsamt Karlsruhe, June 7, 1940, BadGLA, 455/Zug. 1991-49, no. 1336.
156. Gewerberat Witten to Polizeiverwaltung Linden-Dahlhausen, May 8, 1928, StABo, A L-D 4. When the company balked, officials even threatened to close it, as it was now technically an unlicensed enterprise (Polizeiverwaltung des Amtes Linden-Dahlhausen to Vereinigte Press- und Hammerwerke Dahlhausen-Bielefeld, July 3, 1928, and notes of Mar. 20, Mar. 23, 1929, all StABo, A L-D 4).
157. HStAD, Regierung Düsseldorf, no. 34218, p. 93.
158. Gewerbeaufsicht to Bezirksamt Villingen, July 17, 1928, June 25, 1929, both BadGLA, 455/Zug. 1991-49, no. 1367.
159. Memorandum, Nov. 27, 1929, StABi, MBV 041.
160. StAHe, VII/247 a und b, p. 40R.
161. Certificate, Oct. 25, 1924, p. 2, StADo, Bestand 3, no. 5872.
162. Bürgermeister Würselen to Regierungspräsident Aachen, Nov. 10, 1930, HStAD, Regierung Aachen, no. 13633; Reichsgesundheitsamt to Bezirksamt Neustadt an der Aisch, Mar. 22, 1927, BArch, R 86, no. 2332, vol. 3.
163. Reichsarbeitsminister to Sozialministerien der Länder, Aug. 21, 1924, attachment C, p. 10, BadGLA, Abt. 233/26102.
164. Der Preußische Minister für Volkswohlfahrt to the Regierungspräsidenten, July 27, 1931, StABi, MBV 037.
165. Der Preußische Minister für Volkswohlfahrt to Regierungspräsident Hannover, Apr. 25, 1923, StAH, HR 23, no. 7.
166. StAL, Stadtgesundheitsamt, no. 225, p. 137R.
167. H. Nehls, "Untersuchungsverfahren für die Bestimmung der in Niederschlagswässern enthaltenen Verunreinigungen," *Kleine Mitteilungen für die Mitglieder des Vereins für Wasser-, Boden- und Lufthygiene* 14 (1938): 112.
168. Wilhelm Liesegang, "Die Bedeutung der chemischen Luftuntersuchung für die gewerbepolizeiliche Genehmigung von Industrieanlagen," *Kleine Mitteilungen für die Mitglieder des Vereins für Wasser-, Boden- und Lufthygiene* 12 (1936): 413.
169. Die Polizeiverwaltung, Der Bürgermeister, Eschweiler to Hastenrather Kalkwerke, Oct. 9, 1929, HStAD, Regierung Aachen, no. 13633.

170. Norman Fuchsloch, *Sehen, riechen, schmecken und messen als Bestandteile der gutachterlichen und wissenschaftlichen Tätigkeit der Preußischen Landesanstalt für Wasser-, Boden- und Lufthygiene im Bereich der Luftreinhaltung zwischen 1920 und 1960* (Freiberg, 1999), 12; M. Beninde, *Die Preußische Landesanstalt für Wasser-, Boden- und Lufthygiene zu Berlin-Dahlem im Laufe der Zeiten: Rückblick und Ausblick anläßlich der Vierteljahrhundertfeier ihres Bestehens* (Berlin, 1926), 17; M. Beninde, *Vierteljahrhundertfeier der Preußischen Landesanstalt für Wasser-, Boden- und Lufthygiene zu Berlin-Dahlem* (Berlin, 1927), 21; Der Preußische Minister für Volkswohlfahrt to Regierungspräsident Hannover, Apr. 25, 1923, StAH, HR 23, no. 7. Norman Fuchsloch's study of WaBoLu's air pollution–control work is somewhat contested; for my criticism, see *Technikgeschichte* 68 (2001): 397–98.

171. Hans Lehmann, "Entwicklung, Zweck und Ziel der Lufthygiene im Hinblick auf die menschliche Gesundheit und öffentliche Gesundheitspflege," *Kleine Mitteilungen für die Mitglieder des Vereins für Wasser-, Boden- und Lufthygiene* 8 (1932): 331.

172. Minutes of meeting, Dec. 18, 1930, p. 9, BArch, R 154/12103.

173. Wilhelm Liesegang, "Die Bedeutung der Industrieabgase im Rahmen der deutschen Volkswirtschaft," *Kleine Mitteilungen für die Mitglieder des Vereins für Wasser-, Boden- und Lufthygiene* 13 (1937): 111.

174. Reichsgesundheitsamt to Bezirksamt Neustadt an der Aisch, Mar. 22, 1927, BArch, R 86, no. 2332, vol. 3.

175. Internal note, June 14, 1963, NieHStA, Nds. 300, Acc. 188/81, no. 26.

176. Notes, Jan. 14, Apr. 9, 1957, both BArch, B 136/5343; Der Bundesminister für Arbeit to Arbeitsminister der Länder, Feb. 10, 1955, p. 2, BArch, B 149/10407; HStAD, NW 50, no. 1214, p. 71n; NW, 66, no. 430, p. 67; Note, Aug. 1, 1963, p. 1, HStAS, EA 8/301, Büschel 552.

177. *Sitzungsberichte des Vereins zur Beförderung des Gewerbfleißes 1899*, 129.

178. Ferdinand Fischer, *Taschenbuch für Feuerungstechniker*, 5th ed. (Stuttgart, 1904), iii.

179. StAMs, Regierung Arnsberg, no. 1566, pp. 104R–5.

180. A. Althammer, "Wege zur Brennstoffersparnis im Hausbrand unter Berücksichtigung der bestehenden Verhältnisse," *Rauch und Staub* 11 (1920–21): 77; Hans Schulze-Manitius, "Rauchgasanalyse," *Illustrierte Technik für Jedermann* 5 (1927): 39.

181. Wilhelm Liesegang, "Die Bekämpfung von Rauch, Staub und Abgasen als hygienische Aufgabe," *Zeitschrift für Desinfektions- und Gesundheitswesen* 22 (1930): 332; G. von Meyeren, "Die Rechtslage bei der Bekämpfung von Luftverunreinigungen," *Kleine Mitteilungen für die Mitglieder des Vereins für Wasser-, Boden- und Lufthygiene* 6 (1930): 318; M. Hahn, "Zur großstädtischen Verkehrshygiene," *Gesundheits-Ingenieur* 51 (1928): 232; W. Schweisheimer, "Rauchplage und Rußbekämpfung in der Stadt," *Rauch und Staub* 19 (1929): 83; Lehmann, "Entwicklung," 332; Hermann Ilzhöfer and Hans-Joachim Giese, "Straßenluftuntersuchungen in München," *Archiv für Hygiene und Bakteriologie* 113 (1935): 215; P. Albert Kratzer, *Das Stadtklima* (Braunschweig, 1937), 20n; Arnold Heller, "Die Bedeutung des Kampfes gegen Rauch, Ruß und Flugstaub für die deutsche Gesamtwirtschaft," *Kleine Mitteilungen für die Mitglieder des Vereins für Wasser-, Boden- und Lufthygiene* 13 (1937): 129; Siedlungsverband Ruhrkohlenbezirk, *Rauchbekämpfung im Ruhrkohlenbezirk: Bisherige Tätigkeit des Ausschusses für Rauchbekämpfung beim Siedlungsverband Ruhrkohlenbezirk* (Essen, 1928), 36.

182. Cf. Tarr, *Search*, 7–35.

183. Arnold Heller, "Feuerungsanlagen und Luftbeschaffenheit," *Gesundheits-Ingenieur* 60 (1937): 387; W. Arend, "Flugstaub im neuzeitlichen Feuerungsbetrieb: Technische, wirtschaftliche und rechtliche Auswirkungen," *Rauch und Staub* 20 (1930): 78; P. Noss, "Grundsätzliche Betrachtungen über die Staubentwicklung in Feuerungen," *Die Wärme* 64 (1941): 310; Hans Weise, "Die mechanische Abscheidung von Flugasche und Ruß aus Abgasen," *Die Wärme* 52 (1929): 104.

184. Georg Herberg, *Handbuch der Feuerungstechnik und des Dampfkesselbetriebes*, 4th ed. (Berlin, 1928), 128; Johannes Schubert, *Planungsgrundlagen für Rauchgasentstauber:*

Herausgegeben von der Wirtschaftsgruppe Elektrizitätsversorgung W.E.V. (Berlin, 1940), 14, 17; Heller, "Feuerungsanlagen," 387; W. Arend, "Wirtschaftlichkeit der Flugstaubabscheidung," *Archiv für Wärmewirtschaft und Dampfkesselwesen* 12 (1931): 313n; W. Arend, "Flugstaubbildung und -beseitigung," *Archiv für Wärmewirtschaft und Dampfkesselwesen* 11 (1930): 278; O. Knabner, "Die Entwicklung der Rauchgas-Entstaubungsanlagen," *Feuerungstechnik* 25 (1937): 127; Robert Meldau, *Der Industriestaub: Wesen und Bekämpfung* (Berlin, 1926), 273; R. Tröger, "Die Richtlinien für den Entwurf der Anlage," *Zeitschrift des Vereines deutscher Ingenieure* 71 (1927): 1836.

185. StAMs, Regierung Arnsberg, no. 1568, p. 249; Oberbergamt Dortmund, no. 1372, p. 223.

186. Letter of the von der Bevölkerung gewählten und bevollmächtigten Ausschusses zur Bekämpfung der Staub- und Lärmplage, Feb. 21, 1957, p. 5, LASb, AA 320.

187. R. Durrer, "Elektrische Ausscheidung von festen und flüssigen Teilchen aus Gasen," *Stahl und Eisen* 39 (1919): 1428n; Hubert Thein, "Gasreinigung durch Elektrizität," *Zeitschrift für technische Physik* 2 (1921): 177; J. Körting, "Staubabscheidung aus Gasen durch Elektrizität," *Zeitschrift des Vereines deutscher Ingenieure* 66 (1922): 719; P. E. Landolt, "Progress in the Art of Electrical Precipitation since 1900," *Transactions of the American Electrochemical Society* 51 (1927): 194; H. C. Murphy, "Taking the Impurities Out of Air," *Heating and Ventilating* 35, no. 11 (1938): 41.

188. Meldau, *Industriestaub*, iii; R. Heinrich, "Reinigung von Gasen mittels Elektrofilter," *Zeitschrift des Vereines deutscher Ingenieure* 74 (1930): 193; Robert Meldau, "Einige Ziele und Lösungen der Staubtechnik," *Zeitschrift des Vereines deutscher Ingenieure* 79 (1935): 686; F. Prockat, "Neuzeitliche Entstaubungsanlagen," *Staub* 6 (July 1937): 249; H. B. Rüder, "Ausführung der Elektrofilter für Großkesselanlagen und die Sichtwirkung der Reingase," *Technische Mitteilungen* 34 (1941): 215; Fritz Wellmann, "Bemerkenswerte Rauchgasentstaubungen der Bauart van Tongeren," *Feuerungstechnik* 25 (1937): 109.

189. StAHe, VII/247 a and b, pp. 1–4, and V/5008, p. 41; Kreisarzt des Landkreises Dortmund to Landesanstalt, Nov. 21, 1927, BArch, R 154/64; Landesanstalt to Deutscher Städtetag, Mar. 21, 1930, BArch, R 154/2; Friedrich Münzinger, "Die Kesselanlage des Großkraftwerks Klingenberg," *Zeitschrift des Vereines deutscher Ingenieure* 71 (1927): 1863; Friedrich Münzinger, *Kesselanlagen für Großkraftwerke: Betrachtungen und Richtlinien* (Berlin, 1928), 77; Wilhelm Wietfeldt, "Der Arbeiter- und Nachbarschutz bei Kohlenstaubfeuerungen," *Zentralblatt für Gewerbehygiene* 17 (1930): 146.

190. StAHe, VII/247 a and b, pp. 50, 69; internal note, June 23, 1930, BArch, R 154/12103; "Urteil der 19. Zivilkammer des Landgerichts II in Berlin," July 4, 1930, BArch, R 36/1284; Oberbürgermeister Berlin to Deutscher Städtetag, Nov. 5, 1931, BArch, R 36/1284.

191. Münzinger, *Kesselanlagen*, vii; J. Ehmig, "Zur Geschichte der Kohlenstaubfeuerung," *Die Wärme* 62 (1939): 379n; *Zeitschrift des Vereines deutscher Ingenieure* 71 (1927): 1829–912.

192. Erich Heitmann, *Theorie und Technik der Flugaschenabscheidung mit besonderer Berücksichtigung der Kohlenstaubanlagen im europäischen Auslande* (Berlin, 1929), 5.

193. Volkmar Kohlschütter, *Nebel, Rauch und Staub: Vortrag, gehalten vor der naturf. Gesellschaft in Bern, am 15. Dez. 1917* (Bern, 1918), 35; Walther Deutsch, "Elektrische Gasreinigung," *Zeitschrift für technische Physik* 6 (1925): 436.

194. N. A. Halbertsma, "Die elektrische Niederschlagung von Rauch und Staub," *Technik für Alle* 7 (1916–17): 283.

195. Eugen Feifel, "Über Zyklonentstauber," *Archiv für Wärmewirtschaft und Dampfkesselwesen* 20 (1939): 15.

196. Siemens-Schuckertwerke to Landesanstalt, Apr. 1, 1922, BArch, R 154/2.

197. Hans Schweitzer, "Elektrisch gereinigter Rauch," *Die Koralle* 5 (1929/30): 127; Roland Nagel, "Zusammenhänge zwischen Gesamtentstaubungsgrad, Teilentstaubungsgrad und Fraktionsentstaubungsgrad unter besonderer Berücksichtigung der Fliehkraftabscheider," *Die Wärme* 59 (1936): 735.

198. Bericht über die Sitzung des Fachausschusses für Staubtechnik, Sept. 27, 1929, BArch, R 154/92.

199. StAMs, Regierung Arnsberg, no. 1568, pp. 286, 289n, 396.

200. Arend, "Wirtschaftlichkeit," 315.

201. W. Boie, "Die Aschenschmelzkammer: Ein Beitrag zur Flugaschenfrage," *Die Wärme* 54 (1931): 828.

202. Schubert, *Planungsgrundlagen*, 46.

203. Internal note, June 19, 1941, StAE, Rep. 102, Abt. XIV, no. 93.

204. W. Schöning, "Entaschung von Großkesselanlagen. Bedeutung, Stand und Zukunftsaufgaben," *Archiv für Wärmewirtschaft und Dampfkesselwesen* 23 (1942): 121; W. Herrmann, "Neuzeitliche Entaschungsanlagen," *Braunkohle* 41 (1942): 1.

205. Prockat, "Neuzeitliche Entstaubungsanlagen," 271; C. Hahn, "Entstaubung und Entnebelung von Gasen durch Elektrofilter," *Das Gas- und Wasserfach* 71 (1928): 272.

206. Landesanstalt to Gewerbeaufsichtsamt Celle, Oct. 30, 1931, BArch, R 154/65.

207. Heller, "Bedeutung," 127; Heller, "Feuerungsanlagen," 389.

208. E. Rammler, "Neuere Flugaschenabscheidungsanlagen," *Archiv für Wärmewirtschaft und Dampfkesselwesen* 11 (1930): 276.

209. Fritz Wellmann, "Gewährleistung bei Trocken-Fliehkraft-Entstaubern," *Die Wärme* 59 (1936): 105.

210. Vereinigte Oberschlesische Hüttenwerke AG to Landesanstalt, Feb. 23, 1934, and response of Landesanstalt, Mar. 3, 1934, both BArch, R 154/66.

211. Thein, "Gasreinigung," 209; Voigt, "Ueber den Stand des Elektrofilterbaues in Braunkohlen-Brikettfabriken," *Braunkohle* 25 (1926–27): 462; Verein zur Überwachung der Kraftwirtschaft der Ruhrzechen, "Bericht des Vereins zur Überwachung der Kraftwirtschaft der Ruhrzechen zu Essen über das Geschäftsjahr 1928/29," *Glückauf* 65 (1929): 1171; Heitmann, *Theorie*, 44; Rammler, "Neuere Flugaschenabscheidungsanlagen," 273.

212. Gräflich Schaffgotsch'sche Werke to Gewerbeaufsichtsamt Gleiwitz, Nov. 21, 1941, BArch, R 154/12114.

213. Heller, "Bedeutung," 126.

214. Frank Uekoetter, "A Look into the Black Box: Why Air Pollution Control Was Undisputed in Interwar Germany," in Christoph Bernhardt, Geneviève Massard-Guilbaud, eds., *Le démon moderne: La pollution dans les sociétés urbaines et industrielles d'Europe* (Clermont-Ferrand, 2002), 248n.

215. HStAD, NW 85, no. 163, p. 49R.

216. Noss, "Grundsätzliche Betrachtungen," 311; E. Rammler and K. Breitling, "Erfahrungen und Beobachtungen bei Flugstaub- und Entstaubungsgradmessungen," *Feuerungstechnik* 25 (1937): 97.

217. Manuscript, "Kampf mit dem Staube," p. 4, BArch, R 154/84.

218. Letter of the Gewerbeaufsichtsamt Hannover, Sept. 10, 1925, StAH, HR 23, no. 834.

219. Aktiengesellschaft Sächsische Werke Dresden to Reichsanstalt für Wasser- und Luftgüte, Feb. 26, 1943, and note, June 21, 1943, both BArch, R 154/48; letter of Bergwerksverwaltung Oberschlesien der Reichswerke, "Hermann Göring," Mar. 20, 1943, BArch, R 154/11896; Gutachten über Flugstaub- und Abgasfragen beim Kraftwerk "Wilhelm" auf dem Gelände der Dachsgrube nordwestlich von Jaworzno, Kreis Krenau/OS, p. 3, BArch, R 154/12016; Kurt Guthmann, "Wie bekämpft man heute den Industriestaub?" *Das Industrieblatt* 40, no. 12 (Apr. 25, 1935): 7; Schubert, *Planungsgrundlagen*, 45.

220. Letter of Berliner Bezirksverein deutscher Ingenieure, Dec. 1, 1927, and *Zwanglose Mitteilungen des Fachausschusses für Staubtechnik im Verein deutscher Ingenieure* 1 (June 1928): 1, both BArch, R 154/92.

221. *Zwanglose Mitteilungen des Fachausschusses für Staubtechnik im Verein deutscher Ingenieure* 1 (June 1928): 1, BArch, R 154/92.

222. *Zwanglose Mitteilungen des Fachausschusses für Staubtechnik im Verein deutscher Ingenieure* 1 (June 1928): 1n.

223. Fachausschuß für Staubtechnik, Bericht über die Vollsitzung, Mar. 3, 1937, BArch, R 154/92.

224. Fachausschuß für Staubtechnik im VDI, "Staubtechnische Begriffsbestimmungen," *Archiv für Wärmewirtschaft und Dampfkesselwesen* 11 (1930): 321n; Robert Meldau, "Richtlinien für Leistungsversuche an Entstaubern," *Zeitschrift des Vereines deutscher Ingenieure* 80 (1936): 69.

225. Fachausschuß für Staubtechnik, Bericht über die Vollsitzung, Mar. 3, 1937, BArch, R 154/92.

226. Guthmann, "Wie bekämpft man," 6; Nagel, "Zusammenhänge," 737; Barkow, "Staubtechnische Fragen der Braunkohlenindustrie," *Zeitschrift des Vereines deutscher Ingenieure* 74 (1930): 736; W. Reerink, "Flugstaubabscheidung aus Rauchgasen. Kritische Zusammenfassung eines Berichtes der Electricity Commission," *Zeitschrift des Vereines deutscher Ingenieure* 77 (1933): 770.

227. Letter of Berliner Bezirksverein deutscher Ingenieure, Dec. 1, 1927, BArch, R 154/92.

228. Fachausschuß für Staubtechnik, Bericht über die Vollsitzung, Nov. 5, 1931, BArch, R 154/92.

229. Fachausschuß für Staubtechnik, Bericht über die Vollsitzung, Nov. 17, 1930, BArch, R 154/92.

230. Fachausschuß für Staubtechnik, Bericht über die Vollsitzung, Nov. 5, 1931, BArch, R 154/92.

231. Fachausschuß für Staubtechnik, Bericht über die Vollsitzung, Mar. 19, 1930, BArch, R 154/92.

232. Frank Uekötter, "Der unvermeidliche Korporatismus: Zum Verhältnis von Staat und Industrie in der Dampfkesselüberwachung," in Jürgen Büschenfeld, Heike Franz, and Frank-Michael Kuhlemann, eds., *Wissenschaftsgeschichte heute: Festschrift für Peter Lundgreen* (Bielefeld and Gütersloh, 2001), 178–91.

233. StAC, Kap. V, sect. VII, no. 583, vol. 1, p. 67; Minister für Handel und Gewerbe, Anweisung, betreffend die Genehmigung und Untersuchung der Dampfkessel, Dec. 16, 1909, p. 16, GStA, Rep. 120, BB II a 5, no. 1, vol. 21.

234. Uekötter, "Korporatismus," 187.

235. StAAc, Cap. 15, no. 2, vol. 1, doc. 77.

236. "Württembergischer Revisions-Verein, Geschäfts-Bericht über das Vereinsjahr 1920 zur 46. ordentlichen Haupt-Versammlung am 30. Mai 1921," p. 40, GStA, Rep. 120, BB II a 5, no. 37, vol. 4.

237. Sächsisch-Thüringischer Dampfkessel-Revisions-Verein Halle to Landesanstalt, Feb. 3, 1936, and response of Landesanstalt, Feb. 14, 1936, both BArch, R 154/17.

238. Arend, "Flugstaub im neuzeitlichen Feuerungsbetrieb," 85.

239. Memorandum, June 23, 1930, BArch, R 154/12103.

240. I discussed these efforts extensively in *The Green and the Brown: A History of Conservation in Nazi Germany* (New York, 2006). For a summary of the ongoing debate, see also Frank Uekötter, "Green Nazis? Reassessing the Environmental History of Nazi Germany," *German Studies Review* 30 (2007): 267–87.

241. Frank Uekoetter, "Polycentrism in Full Swing: Air Pollution Control in Nazi Germany," in Franz-Josef Brüggemeier, Mark Cioc, and Thomas Zeller, eds., *How Green Were the Nazis? Nature, Environment, and Nation in the Third Reich* (Athens, 2005): 101–28. I would like to thank Ohio University Press for allowing the reuse of material from this article.

242. Uekoetter, *Green and the Brown*, 202–6.

243. Walther Hofer, *Der Nationalsozialismus: Dokumente 1933–1945* (Frankfurt, 1957), 31.

244. Uekoetter, *Green and the Brown*, 65–67, 142–45.

245. Lies-Benachib, *Immissionsschutz*.

246. Dennis LeRoy Anderson, *The Academy for German Law, 1933–1944* (New York, 1987), 250 (emphasis original).

247. Ernst Eiser, "Die Behandlung industrieller Einwirkungen in der neuen Rechtsprechung des Reichsgerichts," *Zeitschrift der Akademie für Deutsches Recht* 5 (1938): 112.

248. Heinz Schiffer, "Immissionen: Ein Beitrag zur Neugestaltung des Nachbarrechts," *Zeitschrift der Akademie für Deutsches Recht* 3 (1936): 1079.

249. Schiffer, "Immissionen," 1083.

250. Friedrich Klausing, "Immissionsrecht und Industrialisierung," *Juristische Wochenschrift* 36 (1937): 72.

251. Klausing, "Immissionsrecht," 68.

252. Eiser, "Behandlung," 116; Heinz Schiffer, "Zum Verfahren in Immissionssachen," *Zeitschrift der Akademie für Deutsches Recht* 4 (1937): 276n.

253. BArch, R 61/144, pp. 59, 37, 43.

254. BArch, R 61/144, p. 46n.

255. Grund, "Das Gesetz über die Beschränkung der Nachbarrechte vom 13. Dez. 1933," *Juristische Wochenschrift* 63 (1934): 204; Hans Frank, ed., *Nationalsozialistisches Handbuch für Recht und Gesetzgebung* (Munich, 1935), 1003; Liesegang, "Bedeutung der chemischen Lufthntersuchung," 404; *Deutsche Justiz* 95 (1933): 862n; *Reichsgesetzblatt* 1933, part 1, 1058n.

256. *Reichsgesetzblatt* 1935, part 1, 1247n.

257. Grund, "Gesetz," 203.

258. Der Reichs- und Preußische Wirtschaftsminister to the Preußische Regierungspräsidenten, Nov. 14, 1935, HStAD, Regierung Aachen, no. 13633.

259. Liesegang, "Bedeutung der chemischen Luftuntersuchung," 404.

260. Klausing, "Immissionsrecht," 68.

261. Uekötter, *Rauchplage*, 234–39.

262. Attachment to a letter of the Wirtschaftsgruppe Metallindustrie, Mar. 11, 1940, p. 8, BArch, R 154/31.

263. Der Reichsbauernführer, Verwaltungsamt, to the Gutehoffnungshütte Oberhausen, May 11, 1937, p. 1, HStAD, NW 354, no. 42.

264. Der Reichswirtschaftsminister to Reichsanstalt für Wasser- und Luftgüte, May 25, 1944. BArch, R 154/11969.

265. Wirtschaftsgruppe Chemische Industrie to Reichswirtschaftsministerium, Nov. 12, 1941, BArch, R 154/12026.

266. "Reichsnährstand, Kreisbauernschaft Kirchheimbolanden, Rundschreiben," no. 11/35, May 21, 1935, StALh, Bestand Oppau, no. 1084.

267. John E. Farquharson, *The Plough and the Swastika: The NSDAP and Agriculture in Germany 1928–1945* (London and Beverly Hills, 1976), 120; Friedrich Grundmann, *Agrarpolitik im "Dritten Reich": Anspruch und Wirklichkeit des Reichserbhofgesetzes* (Hamburg, 1979), 151–54; Gustavo Corni, *Hitler and the Peasants: Agrarian Policy of the Third Reich, 1930–1939* (New York, 1990), 152.

268. Farquharson, *Plough*, 106.

269. Joachim Lehmann, "Agrarpolitik und Landwirtschaft in Deutschland 1939 bis 1945," in Bernd Martin and Alan S. Milward, eds., *Agriculture and Food Supply in the Second World War* (Ostfildern, 1985), 29–49.

270. Der Reichsbauernführer, Verwaltungsamt, to the Gutehoffnungshütte Oberhausen, May 11, 1937, p. 2, HStAD, NW 354, no. 42.

271. Ernst Eiser, "Die industriellen Einwirkungen auf die Umgebung und ihre Behandlung in der bisherigen Rechtsprechung zu § 906 BGB," *Zeitschrift der Akademie für deutsches Recht* 5 (1938): 86.

272. Michael Kloepfer, *Zur Geschichte des deutschen Umweltrechts* (Berlin, 1994), 42n. For the juridical status quo around 1900, see Kurt von Rohrscheidt, *Die Gewerbeordnung*

für das Deutsche Reich mit sämmtlichen Ausführungsbestimmungen für das Reich und für Preußen (Leipzig, 1901), 53–90, 776–808.

273. *Reichsgesetzblatt 1934*, part 1, 566.

274. Der Reichswirtschaftsminister und Preussische Minister für Wirtschaft und Arbeit to the Regierungspräsidenten, Oct. 30, 1934, HStAD, Regierung Aachen no. 12974.

275. Der Reichswirtschaftsminister und Preussische Minister für Wirtschaft und Arbeit to the Regierungspräsidenten, Oct. 30, 1934, HStAD, Regierung Aachen, no. 12974.

276. Vorschläge für die Änderung des Genehmigungsverfahrens, attachment to Reichsarbeitsminister to the Sozialministerien der Länder, Feb. 1, 1928, p. 5, BadGLA, Abt. 233/26102.

277. HStAD, BR 1050, no. 8, p. 93n. On the Four Year Plan, see Dieter Petzina, *Autarkiepolitik im Dritten Reich: Der nationalsozialistische Vierjahresplan* (Stuttgart, 1968).

278. StAMs, Regierung Arnsberg, no. 1569, pp. 5–8.

279. Niederschrift über die 17. Tagung der Gewerberechtsreferenten der Länder in Bonn, Apr. 24–25, 1958, p. 16, BayHStA, MWi 26077.

280. HStAS, EA 8/301, Büschel 464.

281. StAL, Stadtgesundheitsamt, no. 234, p. 57.

282. HStAD, BR 1015/101, pp. 233n, 253–53R (quotation on 253R).

283. On the impact of Speer's appointment, see Alan S. Milward, *The German Economy at War* (London, 1965), 72n; Ludolf Herbst, *Der Totale Krieg und die Ordnung der Wirtschaft: Die Kriegswirtschaft im Spannungsfeld von Politik, Ideologie und Propaganda 1939–1945* (Stuttgart, 1982), 176n; R. J. Overy, *War and Economy in the Third Reich* (Oxford, 1994), 356n.

284. Nobel-Kaffee KG to Oberbürgermeister der Stadt Essen, June 11, 1943, StAE, Rep. 102, Abt. XIV, no. 94.

285. "Vergleich zwischen der Forstgenossenschaft Lüttgenberg und den Unterharzer Berg- und Hüttenwerken GmbH," Nov. 22, 1943, Niedersächsisches Staatsarchiv Wolfenbüttel, 12 Neu 15, no. 3868.

286. Reisebericht zum Termin, May 19, 1944, StAMs, Regierung Arnsberg 6, no. 217.

287. Zentral-Bauleitung der Waffen-SS und Polizei Auschwitz to Reichsanstalt für Wasser-, Boden- und Luftgüte, Mar. 1, 1943, BArch, R 154/48.

288. Response, Sept. 23, 1943, BArch, R 154/48.

289. Raul Hilberg, *Die Vernichtung der europäischen Juden* (Frankfurt, 1990), 948.

290. Hannah Arendt, *Eichmann in Jerusalem: A Report on the Banality of Evil* (New York, 1994), 231.

291. Runderlaß des Reichswirtschaftsministers, Feb. 18, 1942, BArch, R 154/12026.

292. Landesanstalt to Reichswirtschaftsminister durch die Hand des Reichsministers des Innern, Jan. 11, 1942, BArch, R 154/12026.

293. Der Reichswirtschaftsminister to the Reichsstatthalter in den Reichsgauen Danzig-Westpreussen und Wartheland, die Landesregierungen, die Preußische Regierungspräsidenten and the Polizeipräsident in Berlin, June 25, 1941, BayHStA, MWi 655.

294. Arnold Heller, "Soll man in Deutschland das Abgasproblem, insbesondere bei Feuerstätten, gesetzlich regeln?" *Gesundheits-Ingenieur* 75 (1954): 390; Helmut Köhler, "Die durch die Technische Anleitung zur Reinhaltung der Luft zu erwartende rechtliche Situation," *IWL-Forum* 2 (1964): 308.

295. Memorandum of Liesegang, June 20, 1942, BArch, R 154/12026.

296. Reichsanstalt für Wasser- und Luftgüte to Rüder, Jan. 14, 1944, BArch, R 154/39.

297. Memorandum on "Arbeitsgebiet Trinkwasserhygiene, Abwasserbeseitigung, Luftreinhaltung," ca. Aug. 1943, p. 2, BArch R 18/3754.

Chapter 4

1. John Opie, *Nature's Nation: An Environmental History of the United States* (Fort Worth, 1998), 455; Robert Gottlieb, *Forcing the Spring: The Transformation of the American Environmental Movement* (Washington, 1993), 78, 126–28; Scott Hamilton Dewey, *Don't*

Breathe the Air: Air Pollution and U.S. Environmental Politics, 1945–1970 (College Station, 2000), 113, 163; Walter A. Rosenbaum, *Environmental Politics and Policy*, 4th ed. (Washington, 1998), 58; Richard N. L. Andrews, *Managing the Environment, Managing Ourselves: A History of American Environmental Policy* (New Haven and London, 1999), 232n; Kirkpatrick Sale, *The Green Revolution: The American Environmental Movement, 1962–1992* (New York, 1993), 25n; Hal K. Rothman, *The Greening of a Nation? Environmentalism in the United States since 1945* (Fort Worth, 1998), 115; Gary C. Bryner, *Blue Skies, Green Politics: The Clean Air Act of 1990 and Its Implementation*, 2nd ed. (Washington, 1995), 98n; Gregg Easterbrook, *A Moment on the Earth: The Coming Age of Environmental Optimism* (New York, 1996), 186; Richard B. Stewart, "Pyramids of Sacrifice? Problems of Federalism in Mandating State Implementation of National Environmental Policy," *Yale Law Journal* 86 (1977): 1197.

2. Ronald Inglehart, *The Silent Revolution: Changing Values and Political Styles among Western Publics* (Princeton, 1977), esp. 28, 31, 82.

3. W. C. L. Hemeon, "Air-Pollution Control," *Chemical Engineering Progress* 44, no. 11 (1948): 18.

4. Griebling, "Cleaning Our Air," 165; *Iron Age* 164, no. 22 (1949): 115.

5. *Chemical and Engineering News* 32 (1954): 1110; *Fortune* 51, no. 4 (1955): 143.

6. Melvin Nord, "Pollution Control—an Industry Must," *Chemical Engineering* 58, no. 5 (1951): 117.

7. Perk to Du Pont de Nemours, Sept. 24, 1954, WRHS, mss. 4456, box 7, folder 109; UOE League of Women Voters of Eugene, *The Problem of Local Air Pollution and Its Proposed Control*.

8. Laura Fermi, "The Cleaner Committee," 1960, p. 1, CACR, folder 1; "Statement of the Cleaner Air Committee of Hyde Park–Kenwood before the Special Subcommittee on Air and Water Pollution of the Senate Committee on Public Works," Jan. 30, 1964, p. 1, CACR, folder 4; "Rebuttal Testimony of Mrs. Chauncy D. Harris," ca. 1964, p. 1n, CACR, folder 5.

9. Meeting, Sept. 24, 1953, WRHS, mss. 4074, folder 6.

10. Inglehart, *Silent Revolution*, 43.

11. Syracuse Chamber of Commerce, *Report*, 5.

12. The Donora disaster is discussed extensively in Lynne Page Snyder, "'The Death-Dealing Smog over Donora, Pennsylvania': Industrial Air Pollution, Public Health, and Federal Policy, 1915–1963" (Ph. D. diss., University of Pennsylvania, 1994). See also Lynn Page Snyder, "Revisiting Donora, Pennsylvania's 1948 Air Pollution Disaster," in Tarr, *Devastation and Renewal*, 126–44; and Devra Davis, *When Smoke Ran like Water: Tales of Environmental Deception and the Battle against Pollution* (New York, 2002), 5–30, which notes that an extra fifty people died within a month of the smog lifting.

13. In fact, even the political repercussions of Donora look modest compared to those of the 1952 London disaster. See Erich Ashby and Mary Anderson, *The Politics of Clean Air* (Oxford, 1981), 104; Peter Brimblecombe, *The Big Smoke: A History of Air Pollution in London since Medieval Times* (London and New York, 1988), 165–71; Thorsheim, *Inventing Pollution*, 161–84; Peter Thorsheim, "Interpreting the London Fog Disaster of 1952," in E. Melanie DuPuis, ed., *Smoke and Mirrors: The Politics and Culture of Air Pollution* (New York and London, 2004), 154–69.

14. Wilhelm Liesegang, *Die Reinhaltung der Luft* (Leipzig, 1935), 9.

15. "Report of the Mayor's Commission for the Elimination of Smoke," pp. 23–30, AIS, 69:14, box 2, folder 26A.

16. Pittsburgh Bureau of Smoke Control, "Report on Stationary Stacks, Year 1948," p. 7, AIS, 80:7, box 1, FF 5.

17. W. C. McCrone and Guenther Baumgart, "Industry's Attack on the Air Pollution Problem in Chicago," Apr. 7, 1950, p. 8, LMDP, box 87, folder 3.

18. "Released to the Bulletin-Index for use immediately," June 26, 1942, AIS, 70:2, add. 1971, box 14, folder 218.

19. *Dragnet News*, no. 4 (Mar. 1956): 4, WRHS, mss. 4074, folder 6.
20. *Cleveland Press*, June 19, 1956.
21. "Citizens Air Purification Committee Meeting at the Statler Hotel," Oct. 10, 1951, WRHS, mss. 3535, cont. 6, folder 7; "First Meeting of the Allegheny County Smoke Control Advisory Committee," June 1, 1950, AIS, 80:7, box 1, FF 2; Thomas C. Wurts, "The Pittsburgh Plan and Its Implementation," p. 3, HSWPA, mss. 285, "Smoke Control—ASME Meeting April 20–21, 1959" file.
22. W. A. Raleigh Jr., "Coal's Stake in Reducing Smog," *Coal Age* 61, no. 10 (1956): 59; Hemeon, "Air-Pollution Control," 18.
23. *Dun's Review and Modern Industry* 68, no. 4 (1956): 127; *Factory Management and Maintenance* 113, no. 8 (1955): 97.
24. Philip Sadtler, "Smoke, Dust, Fumes, and Their Fellow Travelers," *Chemical and Engineering News* 25 (1947): 2827; Nord, "Pollution Control."
25. *Business Week*, Nov. 20, 1948, 21.
26. Herbert G. Dyktor, "The Community Problem," *Industrial Medicine and Surgery* 19 (1950): 106; "Pollution in the Air We Breathe," *Consumer Reports* 25 (1960): 406; Henry N. Doyle, "Polluted Air, a Growing Community Problem," *Public Health Reports* 68 (1953): 858; John M. Kane, "Air Pollution Ordinances," *Foundry* 80, no. 10 (1952): 104; Roger A. Renwanz and Schaeffer E. Specht, "Survey of Smoke Control," *Steel* 135, no. 21 (1954): 100; Eugene A. Sloane, "How to Protect Yourself from Air Pollution," *House Beautiful* 103, no. 1 (1961): 112; John E. Dever, "Industries Work with City for Smoke Abatement," *Public Management* 40 (1958): 166; A. Rokicki, "Verhütung der Luftverunreinigung: Bericht über eine Studienreise in den USA," *Technische Mitteilungen* 54 (1961): 161; Jay Etlinger, "Ordinance-less Air Pollution Abatement," *American City* 73, no. 10 (1958): 220; R. Emmet Doherty, "Five Cities Say 'No' to Air Pollution," *American City* 77, no. 5 (1962): 109; Arthur C. Stern and Leonard Greenburg, "Air Pollution—the Status Today," *American Journal of Public Health and the Nation's Health* 41 (1951): 37; George E. Best, "A Rational Approach to Air-Pollution Legislation," *A.M.A. Archives of Industrial Hygiene and Occupational Medicine* 5 (1952): 517; W. C. L. Hemeon and T. F. Hatch, "Atmospheric Pollution," *Industrial and Engineering Chemistry* 39 (1947): 568; Sadtler, "Smoke," 2827; Pendray, "Management Aspects," 582n; John C. Reidel, "Air-Pollution Control in Houston Area," *Oil and Gas Journal* 54, no. 18 (1955): 107; Virgil E. Gex, "Remedial Measures in Air Pollution Which Have Been Taken by an Industry, Including a Discussion of Costs," *American Journal of Public Health and the Nation's Health* 45 (1955): 626; *Steel* 124, no. 9 (1949): 51, 128, no. 10 (1951): 63; *Modern Industry* 18, no. 3 (1949): 50; *Chemical and Engineering News* 32 (1954): 1110, 34 (1956): 1697; *Fortune* 51, no. 4 (1955): 143; *Changing Times* 13, no. 9 (1959): 40; *Oil, Paint and Drug Reporter* 175, no. 27 (1959): 7, 39; Charles C. Fichtner to Marion B. Folson, Aug. 23, 1955, NA, RG 90 A 1, entry 36, box 8, "Technical Assistance—New York" folder; "1959 Annual Report Air Pollution Control, City of Dayton, Ohio," p. 1, NA, RG 90 A 1, entry 36, box 8, "Ohio" folder; "Evaluation of Air Pollution in the State of Washington: Report of Cooperative Survey Made July 1 through November 30, 1956," p. viii, NA, RG 90 A 1, entry 36, box 8, "Washington Air—Evaluation of Air Pollution State of Washington" folder; "Reappraisal of Cleveland Air Purification Program," Sept. 24, 1951, pp. 1, 3, WRHS, mss. 3535, cont. 6, folder 7; Thomas C. Wurts, "What Is the Air Pollution Control Association?" Presented at the APCA East-Central Sectional Meeting in Harrisburg, Pennsylvania, Sept. 24, 1953, p. 1, and "Fundamentals for Setting Up a Successful Air Pollution Control Program," Feb. 25, 1957, pp. 5, 7, both AIS, 80:7, add. 1984, FF 5; A. Weinzirl, "Report on Seminar on Air Pollution Problems," Sept. 20–21, 1955, p. 4, PCCR, box 12, "Air Pollution—General" folder.
27. Roland Marchand, *Creating the Corporate Soul: The Rise of Public Relations and Corporate Imagery in American Big Business* (Berkeley, 1998), 357n, 362.
28. *Mill and Factory* 59, no. 1 (1956): 70.
29. Gex, "Remedial Measures," 622.

30. Pendray, "Management Aspects," 584.

31. Meyeren, "Rechtslage," 310n, 314; Hoffmann, *Gewerbeordnung*, 70n; Christian Meisner and Heinrich Stern, *Das in Preußen geltende Nachbarrecht* (Munich, 1927), 188–99, 524–37.

32. Lawrence R. Bloomenthal, "How the Courts Look upon the Air Pollution Nuisance," *Heating, Piping and Air Conditioning* 12 (1940): 112.

33. William L. Prosser, *Handbook of the Law of Torts* (St. Paul, 1941), 549.

34. Allen D. Brandt, "Problems of Air Pollution," *Iron and Steel Engineer* 27, no. 6 (June 1950): 78.

35. Engineering Department to Executive Committee, Mar. 9, 1949, p. 1, HML, acc. 1801, box 10, "Environmental Quality" folder; Raleigh, "Coal's Stake," 54.

36. *Factory Management and Maintenance* 113, no. 8 (1955): 97; Technical Projects Committee to Board of Directors, Apr, 11, 1955, p. 2, UCLA, coll. 1108, box 2, folder 3.

37. Best, "Rational Approach," 517.

38. Richard F. Hansen, "Air Pollution Regulation," *Chemical Engineering* 57, no. 9 (1950): 241.

39. Pendray, "Management Aspects," 582.

40. Doyle, "Polluted Air," 858; Pendray, "Management Aspects," 581; Shaffer, "Poisoned Air," 247; *Steel* 128, no. 10 (1951): 63.

41. *Modern Industry* 18, no. 3 (1949): 50.

42. Arthur J. Benline to Joseph C. Swidler, Oct. 29, 1964, NA, RG 90 A 1, entry 11, box 46, "OCC—Federal Power Commission Part 1" folder.

43. Division of Air Pollution Control, "Review of the New Jersey Air Pollution Control Program," 7.

44. Florida State Board of Health, "Preliminary Summary of a Report on Air Pollution in the State of Florida," Dec. 14, 1960, p. 14, NA, RG 90 A 1, entry 36, box 5, "Federal &/or State Agencies" folder. Therefore, it is misleading to see Florida's air pollution–control policy from a 1990s viewpoint, as exemplified by Scott H. Dewey ("The Fickle Finger of Phosphate: Central Florida Air Pollution and the Failure of Environmental Policy, 1957–1970," *Journal of Southern History* 65 [1999]: 565–603).

45. Edward L. Stockton, "Air and Water Pollution," Apr. 19, 1967, p. 7, HSWPA, mss. 285, "Smoke Control—General" file; Howard L. Weisz, "Air Pollution Control in Allegheny County," Apr. 14, 1972, p. 6, HSWPA, mss. 285, "Smoke Control—General" file; *Cleveland Press*, Aug. 28, 1965; "Meeting of the Air Pollution Advisory Board," May 6, 1966, p. 2, June 10, 1966, p. 1, and "Meeting of the Air Pollution Advisory Board," Apr. 26, 1968, p. 1, MUA, Milwaukee Series 44, box 2, folders 12 and 13.

46. *Cleveland Press*, Aug. 28, 1965.

47. Division of Air Pollution Control, "Review of the New Jersey Air Pollution Control Program," 7.

48. M. C. Sperry to Ralph J. Perk, Oct. 15, 1954, WRHS, mss. 4456, box 7, folder 109.

49. "Pollution in the Air We Breathe," 407; William H. Megonnell to Leslie H. Horn, Dec. 2, 1959, NA, RG 90 A 1, entry 36, box 7, "A Review of Air Pollution in NY State" folder; Jean J. Schueneman, "Trip Report—Struthers, Ohio," Apr. 5, 1960, p. 3, NA, RG 90 A 1, entry 36, box 8, "Ohio" folder.

50. *Chemical and Engineering News* 30 (1952): 985.

51. Kittelton to Rogers, June 29, 1959, NA, RG 90 A 1, entry 36, box 9, "Manufacturing Chemists Association" folder.

52. American Society of Mechanical Engineers, "Example Sections for a Smoke Regulation Ordinance," May 1949, p. 5n, RRTSAC, ser. 5, box 1, "1948" folder.

53. James H. Carter, "Smoke Abatement's Wider Horizons," *Papers Presented at the Forty-First Annual Meeting, Smoke Prevention Association of America, June 7–11, 1948, New York, N.Y.* (n.l., n.d.), 2.

54. *Iron Age* 167, no. 1 (1951): 344.

55. Pendray, "Management Aspects," 581.

56. Florida State Board of Health, "Preliminary Summary of a Report on Air Pollution in the State of Florida," p. 5.

57. Jean J. Schueneman, "Air Pollution Problems and Control Programs in the United States," p. 8, NA, RG 90 A 1, entry 11, box 38, "OCC—Air Pollution Control Association Part 4" folder.

58. "Report Submitted by the Legislative Research Council Relative to Air Pollution in the Metropolitan Boston Area," Feb. 5, 1960, p. 15, NA, RG 90 A 1, entry 36, box 7, "Massachusetts" folder.

59. Allegheny County Bureau of Smoke Control, "Ninth Annual Report of Activities for the Year Ending May 31, 1958," p. 6, AIS, 80:7, box 1, FF 8.

60. "Statement of Cleaner Air Committee Before the Chicago City Council Budget Hearings on Appropriation for the Department of Air Pollution Control," Nov. 22, 1965, HPKCCP, ser. 2, box 10, folder 5; Bird, *Report*, 30.

61. A 1962 study noted that only seven American cities had "intensive" air pollution–control programs (Harry C. Ballman and Thomas J. Fitzmorris, "Local Air Pollution Control Programs—a Survey and Analysis," *Journal of the Air Pollution Control Association* 13 [1960]: 490).

62. "Meeting of the Air Pollution Advisory Board," Jan. 27, 1967, p. 1, MUA, Milwaukee, ser. 44, box 2, folder 13.

63. Department of Air Pollution Control, City of Chicago, "Annual Report 1964," p. 20, LMDP, box 86, folder 3.

64. Allegheny County Bureau of Smoke Control, "Fourth Annual Report of Activities for the Year Ending May 31, 1953," p. 1, AIS, 83:7, ser. III, FF 19.

65. County of Allegheny, Bureau of Smoke Control, "A Review of Program," June 18, 1951, p. 2, AIS, 80:7, box 1, FF 1.

66. "Progress Report, New York State Air Pollution Control Board," vol. 3, no. 3 (1963): 5, HRCSP, cont. 2, "Air Pollution, Printed Materials" folder.

67. Izaak Walton League of America, "Resolution No. 3," Thirty-seventh Annual Convention, Apr. 25, 1959, NA, RG 90 A 1, entry 36, box 9, "Association A–M" folder; "Minute Book," p. 69, HRCSP, cont. 20, "Financial Matters: Minute Book 1956–1968" folder.

68. Wurts, "Fundamentals," 4; "Air Pollution Survey Report: Greater Elmira, August–December, 1958," p. 80, NA, RG 90 A 1, entry 36, box 7, "New York—Elmira" folder; "Progress Report, New York State Air Pollution Control Board," vol. 3, no. 1 (Mar.–Apr. 1963): 7, HRCSP, cont. 1, "Air Pollution" folder; Dyktor, "Community Problem," 104; F. E. Schuchman, "Pittsburgh—'Smokeless City,'" *National Municipal Review* 39 (1950): 493.

69. Etlinger, "Air Pollution Abatement," 220; Dever, "Industries," 166; Katharine N. Gabell, *Clearing the Air: A Regional Challenge* (n.l., n.d. [1963?]), 20.

70. Doyle, "Polluted Air," 867.

71. Allegheny County Bureau of Smoke Control, "Third Annual Report of Activities for the Year Ending May 31, 1952," p. 10n, AIS, 80:7, box 1, FF 7; A. L. Penniman Jr. and E. F. Wolf, "Accomplishments of the Electric Light and Power Industry in the Reduction of Atmospheric Pollution," *Edison Electric Institute Bulletin* 17 (1949): 343; *Modern Industry* 18, no. 3 (1949): 50; *Steel* 124, no. 9 (1949): 51, 128, no. 10 (1951): 63.

72. Schueneman, "Air Pollution Problems," 6.

73. Oregon State Sanitary Authority, "Air Quality Control in Oregon, December 1967," p. 1, OHS, mss. 2386, box 4, folder 13; "Status Report on Air Pollution Programs, December 1966," p. 81, NA, RG 90 A 1, entry 11, box 72, "Reports General Part 2" folder; Samuel M. Rogers, "Air Pollution Legislation—A Review of Current Developments," *American Journal of Public Health and the Nation's Health* 50 (1960): 647; George T. Abed, *Air Pollution Control Policy in the Willamette Basin*, Background Paper Number 6, Willamette Basin Land Use Study (1965), 18. It is thus a common mistake to take the number of state air pollution laws in

postwar America as an indicator of the vibrancy of air pollution control (David Schoenbrod, *Saving Our Environment from Washington: How Congress Grabs Power, Shirks Responsibility, and Shortchanges the People* [New Haven and London, 2005], 10).

74. Flagg, *Smoke Abatement*, 24; Smoke Prevention Association of America, *Manual of Ordinances and Requirements in the Interest of Air Pollution, Smoke Elimination, Fuel Combustion* (Chicago, 1940), 147, 151, 163, 165.

75. Morrison and Packwood to Portland Air Pollution Committee, Jan. 2, 1953, PCCR, box 12, "City Air Pollution—Minutes of Meetings" folder.

76. *Modern Industry* 18, no. 3 (1949): 47.

77. Hansen, "Air Pollution Regulation," 240.

78. Leverett S. Lyon, "Chicago's Program for Cleaner Air," p. 5, LMDP, box 87, folder 3.

79. Benner, "Smoke Investigation," 1272.

80. R. D. MacLaurin, "Abating Black Smoke Will Not Solve the Problem of Atmospheric Pollution," *American City* 40, no. 5 (1929): 136.

81. Arthur C. Stern, "Report on New York City Air Pollution Survey," *Proceedings of the Thirty-First Annual Convention*, 105; L. B. Sisson, "Summary Report for Anthracite Institute," Dec. 31, 1934, p. 2, AIS, 83:7, ser. I, FF 28; Frank A. Chambers, "Smoke Abatement Administration," Apr. 7, 1950, p. 7n, LMDP, box 87, folder 3; "Board of Directors Meeting Minutes," Jan. 19, May 31, 1934, both MoHS, Citizens' Smoke Abatement League Minutes of Meetings, 1926–1934; "Supplementary Report on Smoke Abatement," p. 6, CHS, mss. 617, box 5, folder 15; Work Projects Administration, *Summary of Smoke Abatement Progress, Cleveland, Ohio: Smoke Abatement Survey, Heating, Power Plant & Incinerator Survey* (n.l., n.d. [July 1939]).

82. Cleveland Division of Air Pollution Control, "Annual Report for 1950," p. 3n, WRHS, mss. 4035, box 1, folder 2; "Reappraisal of Cleveland Air Purification Program," Sept. 24, 1951, p. 7, WRHS, mss. 3535, cont. 6, folder 7; Allegheny County Bureau of Smoke Control, "Report of Activities for the Year Ending June 1, 1950," p. 7, AIS 80:7, box 1, FF 6; Benjamin Linsky, "Air Pollution and Man's Health," *Public Health Reports* 68 (1953): 871.

83. Louis C. McCabe, "Air Pollution Review 1949–1954," *Industrial and Engineering Chemistry* 46 (1954): 1646.

84. *Air Repair* 1, no. 1 (1951); *Journal of the Air Pollution Control Association* 5, no. 2 (1955).

85. *APCA Abstracts* 1, no. 1 (1955).

86. "Air Pollution Technical Information Survey, Final Report," Apr. 26, 1965, p. 15, NA, RG 90 A 1, entry 11, box 13, "Contracts—Science Communications, Inc. PH 86-65 13" folder.

87. Public Health Service, "Air Pollution: A National Problem," p. 49, HRCSP, cont. 2, "Air Pollution, Printed Materials" folder.

88. Louis C. McCabe, ed., *Air Pollution: Proceedings of the United States Technical Conference on Air Pollution. Sponsored by the Interdepartmental Committee on Air Pollution* (New York, 1952).

89. Arthur C. Stern, "Present Status of Atmospheric Pollution in the United States," *American Journal of Public Health and the Nation's Health* 47 (1957): 83.

90. Henry F. Hebley, "Factors Rarely Considered in Smoke Abatement," *Mechanical Engineering* 69 (1947): 287.

91. "Fuel Consumption for 1957 in the Chicago Metropolitan Area: A Report to the Indiana-Illinois Bi-State Air Pollution Survey, June, 1958," p. 7, IMAR, box 106, folder 17; Cleveland Division of Air Pollution Control, "Annual Report for 1950," p. 1 (quotation), WRHS, mss. 4035, box 1, folder 2.

92. Harold J. Paulus, "Air Pollution—A Public Health Problem of the Community," *Minnesota Municipalities* 44 (1959): 268; Shaffer, "Poisoned Air," 242; Jeanne Kuebler, "Air Contamination," *Editorial Research Reports* (1964): 9; William Herring and Melvin Nord, "Aerial Pollution: How It Is Controlled," *Chemical Industries* 67 (1950): 398; Hemeon, "Air-Pollution Control," 19; Hemeon and Hatch, "Atmospheric Pollution," 570; Brandt, "Problems," 81.

93. Roland Czada, "Korporatismus/Neo-Korporatismus," in Dieter Nohlen, ed., *Wörterbuch Staat und Politik* (Munich, 1991), 322–26.

94. Louis C. McCabe, "Atmospheric Pollution," *Industrial and Engineering Chemistry* 43 (1951): 87A.

95. United Smoke Council, "More Sunshine . . . Cleaner Living in Allegheny County," AIS, 70:2, add. 1971, box 14, folder 221; *Cleveland Plain Dealer*, Apr. 29, 1964.

96. "1959 Annual Report, Air Pollution Control, City of Dayton, Ohio," p. 1, AIS, 70:2, add. 1971, box 14, folder 221; Air Pollution Control League of Greater Cincinnati, "Protecting Our Air Resource: Annual Report for 1966," p. 4, CHS, mss. qA 298, box 1, folder 10.

97. Morton Corn, "Clean Air and Allegheny County: Statement, Allegheny Conference Meeting," Sept. 8, 1969, p. 2, HSWPA, mss. 285, "Smoke Control—General" file.

98. Stern and Hallett to Christy, Aug. 1, 1950, RRTSAC, ser. 3, box 2, "New York" folder.

99. F. Müller-Voigt, "Gesundheitsschädigungen und Hygiene der Industrieluft," *Städtehygiene* 3 (1952): 185; Protokollauszug der Stadtverordneten-Versammlung, §965, Dec. 13, 1951, StAF, Magistratsakte R 2228, vol. 1.

100. Sturm Kegel, "Die Luft im Industriegebiet muss sauberer werden!" *Chemische Industrie* 6 (1954): 369.

101. Rainer Tross, "Der Gaskrieg des Alltags," *Die Gegenwart* 12 (1957): 558.

102. *Gesundheits-Ingenieur* 77 (1956): 91.

103. Robert Meldau, "Technische und biologische Probleme der industriellen Abluftreinigung," *Chemische Industrie* 6 (1954): 371; Friedrich Portheine, "Verschmutzung der Luft in Industriestädten," *Deutsche Medizinische Wochenschrift* 82 (1957): 1361.

104. Bundesverband der Deutschen Industrie to Bundesministerium des Innern, Aug. 10, 1956, BArch, B 106/38335.

105. HStAD, NW 66, no. 442, p. 90.

106. Lutz to Gesundheitsamt Augsburg, Apr. 7, 1964, StAA, Gewerbeaufsichtsamt Augsburg, Firmenakten, Stadtkreis Augsburg, Prügel-Brauerei Augsburg.

107. Meyer to Pagel, Oct. 12, 1954, LASH, Abt. 761, no. 92.

108. HStAD, NW 50, no. 1255, p. 68.

109. "Resolution der Arbeitsgemeinschaft Ottobrunner Vereine und Verbände," Oct. 16, 1962, StAM, RA 103181.

110. Gerhart Habenicht, "Die Luftverunreinigung: Gesundheitsgefahren," in Paul Vogler and Erich Kühn, eds., *Medizin und Städtebau: Ein Handbuch für gesundheitlichen Städtebau* (Munich, 1957), 2: 98.

111. Deutscher Ärztetag 1958, "Entschließung zur Reinhaltung der Luft," NieHStA, Nds. 300, Acc. 188/81, no. 28; Deutscher Ärztetag 1963, "Entschließung betr: Technischer Fortschritt und Hygiene," p. 1, LHAK, Bestand 930, no. 10321, 66; HStAD, NW 50, no. 1228, p. 13.

112. DGB Nachrichtendienst, "Bundespressestelle des Deutschen Gewerkschaftsbundes," Oct. 20, 1958, HStAS, EA 8/301, Büschel 531.

113. "Tag der deutschen Heimatpflege in Saarbrücken (Sept. 12–15, 1958), Entschliessung über Staubschäden," LAB, Rep. 142/9, Acc. 5/24–30, vol. 5.

114. "Resolution zur Luftverunreinigung des Zentralverbandes der Deutschen Haus- und Grundbesitzer e.V. und des Verbandes Deutscher Bürgervereine," Sept. 1957, NieHStA, Nds. 600, Acc. 153/92, no. 315.

115. Alfred Müller-Armack, *Studien zur Sozialen Marktwirtschaft* (Cologne, 1960), 35n.

116. Heemeyer to Bundesanstalt für Wasser-, Boden- und Lufthygiene, Dec. 2, 1954, BArch, R 154/68.

117. HStAD, NW 50, no. 1228, p. 59R.

118. HStAD, NW 50, no. 1246, p. 87, and NW 66, no. 441, p. 85; "Notgemeinschaft Kleinblittersdorf und Umgebung to Bundesminister des Auswärtigen," Sept. 13, 1957, p. 2,

LASb, AA 320. The initiator of a Hamburg league received more six thousand letters within a few weeks (*Hamburger Abendblatt*, Aug. 14, 1956).

119. "Aktenvermerk über die Besprechung mit Herrn Oberregierungsrat Öls [sic], Arbeits- und Sozialministerium NRW Düsseldorf," July 3, 1961, p. 2, StADu, 503/689.

120. *Welt der Arbeit*, Oct. 24, 1958.

121. *Christ und Welt*, Sept. 18, 1958, p. 3; *Süddeutsche Zeitung*, June 3, 1959); *Frankfurter Allgemeine*, Sept. 9, 1958.

122. *Westfälische Rundschau*, Feb. 10, 1954; *Kölner Stadt-Anzeiger*, May 17, June 21, 1958; *Neue Ruhr Zeitung*, Jan. 6, 1956.

123. *Der Stern*, Mar. 3, 1956; *Der Spiegel*, May 30, 1956; *Frankfurter Neue Presse*, Apr. 24, 1959.

124. *Rheinischer Merkur*, June 30, 1961.

125. HStAD, NW 66, no. 353, p. 90.

126. *Ruhr-Nachrichten*, Jan. 18, 1955; "Tag der deutschen Heimatpflege in Saarbrücken (Sept. 12–15, 1958), Entschliessung über Staubschäden," LAB, Rep. 142/9, Acc. 5/24–30, vol. 5; "Deutscher Ärztetag 1958, Entschließung zur Reinhaltung der Luft," NieHStA, Nds. 300, Acc. 188/81, no. 28.

127. "Bericht des Instituts für Demoskopie no. 407," Oct. 27, 1959, p. 2, BArch, B 106/38731.

128. "Vorlage no. 71/58 für die Sitzung des Senats," StAB, 3-M.1.d, no. 74; Gewerbeaufsichtsamt Neustadt a. d. Weinstraße to Gesundheitsamt Ludwigshafen, Aug. 1952, BArch, R 154/67; note, Apr. 29, 1963, StAA, Bezirksamt Schwabmünchen, Abgabe 1992, no. 1974.

129. See the agreements on ministerial responsibilities in HStAD, NW 50, no. 1214, p. 71n, NW 66, no. 430, p. 67, and NW 50, no. 94, pp. 6–8.

130. HStAD, NW 50, no. 1220, p. 6.

131. HStAD, NW 50, no. 1214, p. 16, and no. 1212, p. 70.

132. HStAD, NW 85, no. 165, p. 107.

133. *Westdeutsche Allgemeine*, Aug. 27, 1952; *Westfälische Rundschau*, Aug. 28, 1952; HStAD, NW 66, no. 430, pp. 102–6, 120–37.

134. "Sitzung des Staubausschusses der Braunkohlekraftwerke im Kölner Bezirk," Aug. 3, 1954, p. 6, BArch, R 154/86.

135. HStAD, NW 50, no. 1212, pp. 241–41R, and NW 66, no. 436, p. 64; StAMs, Regierung Arnsberg, no. 591, p. 68n.

136. HStAD, NW 50, no. 1246, pp. 79, 72.

137. Arbeitsminister des Landes Nordrhein-Westfalen to Bundesminister für Arbeit, June 5, 1953, BArch, B 149/10407.

138. Rohrscheidt, *Gewerbeordnung*, 57.

139. Note, Nov. 1957, BArch, B 136/5343.

140. Deutscher Bundestag, 2. *Wahlperiode 1953, Drucksache 2598*. For the parliamentary vote, see Verhandlungen des Deutschen Bundestages, 2. *Wahlperiode 1953, Stenographische Berichte vol. 34* (Bonn, 1957), 10168. For the government's response, see Deutscher Bundestag, 2. *Wahlperiode 1953, Drucksache 3757*.

141. "Kurzprotokoll über die Vollversammlung der Interparlamentarischen Arbeitsgemeinschaft," May 3, 1955, p. 17, LAB, Rep. 142/9, Acc. 5/24–30, vol. 1; HStAD, NW 50, no. 1215, p. 29R.

142. HStAD, NW 50, no. 1214, p. 109.

143. "Vorlage für den Bundeskanzler," Oct. 27, 1958, BArch, B 136/5364.

144. Bundesminister für Verkehr to Staatssekretär des Bundeskanzleramtes, Mar. 3, 1959, p. 7, and Bundeskanzler to Bundesminister für Verkehr, Apr. 2, 1959, both BArch, B 136/5364.

145. HStAD, NW 50, no. 1214, p. 108.

146. HStAD, NW 50, no. 1214, p. 113.

147. Kommunale Arbeitsgemeinschaft Rhein-Neckar to Bezirksregierung der Pfalz, Oct. 25, 1957, LHAK, Bestand 930, no. 10332.

148. Gewerbeaufsichtsamt München-Stadt to Landesgewerbeaufsichtsbeamter beim Bayerischen Staatsministerium für Arbeit und soziale Fürsorge, Sept. 25, 1958, p. 1, BayHStA, MArb 2596/I.

149. Arnold Heller, "Die Planung neuer und die Erweiterung bestehender Industrieanlagen im Hinblick auf die Reinhaltung der Luft," *Gesundheits-Ingenieur* 71 (1950): 156; Heller, "Soll man," 392.

150. HStAD, NW 50, no. 1246, p. 79.

151. Josef Weißner, "Probleme der Luftverunreinigung," *Industriekurier Wochenausgabe Technik und Forschung* 9 (1956): 176.

152. Vereinigung der Großkesselbesitzer, "Luftverunreinigung durch Rauchgase aus Dampfkesselanlagen. Bericht über den Stand der Technik," *Mitteilungen der Vereinigung der Großkesselbesitzer* 34–35 (Apr. 1955): 457.

153. Bundesverband der Deutschen Industrie, "Notwendigkeit, Möglichkeiten und Grenzen einer gesetzlichen Regelung zur Reinhaltung der Luft in Industriegebieten," Dec. 10, 1954, p. 1 (quotation), LAB, Rep. 142/9, Acc. 5/24-30, vol. 1; Bundesverband der Deutschen Industrie, "Stellungnahme zum Entwurf eines Gesetzes zur Reinhaltung der Luft in Industriegebieten," Feb. 1955, LAB, Rep. 142/9, Acc. 5/24-30, vol. 1; A. Bachmair, "Aus der Arbeit des VGB für die Klärung der Immissionsfragen aus Kesselanlagen: Vortrag vor dem Staubausschuss der Braunkohlekraftwerke im Kölner Bezirk," Aug. 3, 1954, p. 1 (quotation), BArch, R 154/86; HStAD, NW 50, no. 1215, p. 30.

154. Bundesverband der Deutschen Industrie, "Stellungnahme zum Entwurf eines Gesetzes zur Reinhaltung der Luft in Industriegebieten," Feb. 1955, p. 14, LAB, Rep. 142/9, Acc. 5/24-30, vol. 1.

155. Bundesverband der Deutschen Industrie to Stammberger and Hellwig, June 23, 1958, p. 2, BArch, B 106/38374.

156. Deutscher Industrie- und Handelstag to Bundesministerium für Wirtschaft, May 13, 1958, BArch, B 102/43222.

157. A. Bachmair, "Die Dampfkesselanlagen im Rahmen der Aufgabe der Erhaltung der Luftreinheit," *Mitteilungen der Vereinigung der Großkesselbesitzer* 39 (Dec. 1955): 833; E. Ruhland, "Bilanz einer Reihenuntersuchung der Staubemissionen von Zementwerken," *Zement—Kalk—Gips* 9 (1956): 110; note, Oct. 10, 1958, p. 1, StADu, 101/312.

158. Hermann Bohle, "Fertig werden mit dem Schmutz: Das Gesetz zur Reinhaltung der Luft, seine Hintergründe, Folgen und Kosten," *Industriekurier*, Nov. 7, 1959, 5.

159. German farmers focused rather narrowly on improving their legal position for damage claims while showing only lukewarm support for actual reductions in the pollution load (Deutscher Bauernverband, resolution, Sept. 9, 1958, BArch, B 102/43222).

160. *Mitteilungen des Bundesverbandes der Deutschen Industrie* 4, no. 7 (1956): 11.

161. Erdl to Regierung von Oberbayern, Aug. 5, 1963, StAM, RA 103162.

162. Bundesverband der Deutschen Industrie, "Notwendigkeit, Möglichkeiten und Grenzen einer gesetzlichen Regelung zur Reinhaltung der Luft in Industriegebieten," Dec. 10, 1954, p. 5, LAB, Rep. 142/9, Acc. 5/24-30, vol. 1.

163. Bundesverband der Deutschen Industrie, "Stellungnahme zum Entwurf eines Gesetzes zur Reinhaltung der Luft in Industriegebieten," Feb. 1955, p. 15, LAB, Rep. 142/9, Acc. 5/24-30, vol. 1.

164. Bohle, "Fertig werden," 5; Vereinigung der Großkesselbesitzer, "Luftverunreinigung," 459; Deutscher Industrie- und Handelstag to Bundesministerium für Wirtschaft, May 13, 1958, BArch, B 102/43222; Bundesverband der Deutschen Industrie to Stammberger and Hellwig, June 23, 1958, p. 4, BArch, B 106/38374.

165. Bundesverband der Deutschen Industrie, "Stellungnahme zum Entwurf eines Gesetzes zur Reinhaltung der Luft in Industriegebieten," Feb. 1955, p. 14, LAB, Rep. 142/9, Acc. 5/24-30, vol. 1.

166. Vereinigung der Großkesselbesitzer, "Luftverunreinigung."

167. F. Keil, "Über die Arbeiten des Forschungsinstitutes der Zementindustrie," *Zement—Kalk—Gips* 7 (1954): 343.

168. HStAD, NW 50, no. 1212, p. 209n.

169. HStAD, NW 50, no. 1212, p. 214.

170. Leiter der Abteilung II to Leiter der Abteilungen Z, E, I, III, IV, V, and VI, Dec. 22, 1959, p. 4, BArch, B 102/43222.

171. Gewerbeaufsichtsamt Nürnberg-Fürth to Landesgewerbeaufsichtsbeamter beim Bayerischen Staatsministerium für Arbeit und soziale Fürsorge, Oct. 2, 1958, p. 4, BayHStA, MArb 2596/I.

172. Oberbürgermeister der Stadt Mannheim to Bundesanstalt für Wasser-, Boden- und Lufthygiene, May 7, 1953, BArch, R 154/67.

173. HStAD, NW 66, no. 436, p. 54n.

174. HStAD, NW 50, no. 1212, p. 241.

175. Gewerbeaufsichtsamt München-Stadt to Landesgewerbeaufsichtsbeamter beim Bayerischen Staatsministerium für Arbeit und soziale Fürsorge, Sept. 25, 1958, p. 2, BayHStA, MArb 2596/I.

176. HStAD, NW 66, no. 434, p. 163; Gewerbeaufsichtsamt München-Land to Landesgewerbeaufsichtsbeamter beim Bayerischen Staatsministerium für Arbeit und soziale Fürsorge, Sept. 30, 1958, p. 1, BayHStA, MArb 2596/I; "Tätigkeitsbericht der Gewerbeaufsichtsverwaltung für den Monat Mai 1963," p. 10, LASH, Abt. 761, no. 20425; Gewerbeaufsichtsamt Heilbronn to Arbeitsministerium Baden-Württemberg, Nov. 14, 1966, p. 6, Gewerbeaufsichtsamt Karlsruhe to Arbeitsministerium Baden-Württemberg, Nov. 14, 1966, p. 4, and Gewerbeaufsichtsamt Sigmaringen to Arbeitsministerium Baden-Württemberg, Nov. 14, 1966, p. 1, all HStAS, EA 8/301, Büschel 555.

177. Oberbürgermeister der Stadt Mannheim to Stutz, Dec. 30, 1950, BArch, R 154/48.

178. Note, Aug. 24, 1964, StAA, Gewerbeaufsichtsamt Augsburg, Firmenakten, Stadtkreis Augsburg, Prügel-Brauerei Augsburg.

179. HStAD, NW 50, no. 1215, p. 52 and NW 50, no. 1216, p. 41.

180. McCabe, *Air Pollution*.

181. Bundesgesundheitsamt to Niedersächsischer Minister der Finanzen, Aug. 31, 1954, NieHStA, Nds. 300, Acc. 188/81, no. 25.

182. HStAD, NW 50, no. 1254, p. 14.

183. *Mitteilungen des Bundesverbandes der Deutschen Industrie* 4, no. 7 (1956): 11.

184. HStAD, NW 85, no. 165, p. 153.

185. Albert Kuhlmann, "Die Tätigkeit technischer Sachverständiger auf dem Gebiet des Immissionsschutzes," *Technische Überwachung* 6 (1965): 111; Technischer Überwachungs-Verein Bayern, memorandum, Apr. 4, 1960, BayHStA, MArb 2596/I.

186. Fuchsloch, *Sehen*, 22; E. Naumann, *60 Jahre Institut für Wasser-, Boden- und Lufthygiene* (Stuttgart, 1961), 20.

187. Der Bundesminister für Ernährung, Landwirtschaft und Forsten, "Industrielle Immissionen und ihre Auswirkungen in der Land- und Forstwirtschaft," Feb. 27, 1957, p. 20n, NieHStA, Nds. 600, Acc. 153/92, no. 315.

188. H. Meyer, "Herbsttagung der Fachgruppe Staubtechnik im Verein Deutscher Ingenieure," *Glückauf* 92 (1956): 110.

189. Verein deutscher Ingenieure to Anstalt für Boden- und Luftgüte, Nov. 10, 1947, BArch, R 154/85.

190. Verein deutscher Ingenieure to Robert-Koch-Institut für Hygiene und Infektionskrankheiten, Feb. 3, 1953, BArch, R 154/85; Kurt Guthmann, "Probleme der Staubtechnik," *VDI-Zeitschrift* 92 (1950): 658.

191. A. Löbner, "Bericht über die Tagung des Fachausschusses für Staubtechnik im VDI," Oct. 8, 1951, BArch, R 154/85; Robert Meldau, "25 Jahre Fachausschuß für Staubtechnik im VDI," *VDI: Zeitschrift des Vereines Deutscher Ingenieure* 95 (1953): 282.

192. Verein deutscher Ingenieure to Anstalt für Boden- und Luftgüte, Nov. 10, 1947, BArch, R 154/85; HStAD, NW 50, no. 1215, p. 23R.
193. HStAD, NW 50, no. 1213, p. 20 (emphasis original).
194. HStAD, NW 50, no. 1213, p. 30.

Chapter 5

1. Kenneth T. Jackson, *Crabgrass Frontier: The Suburbanization of the United States* (New York, 1987), 183n.
2. William A. Bassett, "The Smoke Problem of Greater Boston," *Our Boston* (April 1929): 9; *Power Plant Engineering* 44, no. 5 (1940): 120.
3. Obermeyer, *Stop That Smoke!* 102; H. B. Meller, "Clean Air, an Achievable Asset," *Journal of the Franklin Institute* 217 (1934): 726; "For Publication in P.M. Newspapers Bearing Date of February 23, 1933," p. 1 (quotation), AIS, 83:7, ser. I, FF 28.
4. Tucker, "Smoke Elimination Program," 105; *Report of the St. Louis Committee 1940*, 2.
5. P. W. Purdom, "Interjurisdictional Problems in Air Pollution Control," *Public Health Reports* 77 (1962): 682.
6. Barr to Tetzlaff, Jan. 7, 1959, MnHS, Health Department, Environmental Health Division, Director's Subject Files, box 1, "Air Pollution" folder; *Dragnet News*, no. 1 (Dec. 1955), p. 1, WRHS, mss. 4074, folder 6.
7. Charles O. Jones, *Clean Air: The Policies and Politics of Pollution Control* (Pittsburgh, 1975), 46; Schuchman, "Pittsburgh," 492n; Wurts, "Pittsburgh Plan and Its Implementation," 1.
8. Jones, *Clean Air*, 47.
9. U.S. Senate, Committee on Public Works, *A Study of Pollution—Air: A Staff Report*, 88th Cong., 1st sess., Sept. 1963 (Washington, 1963), 33. Among the fifteen county-level institutions, two were working in more than one county.
10. Regional Conference of Elected Officials, "Bylaws, Adopted March 30, 1962," PNJDMPR, box 21, URB 24/IV/62; Regional Conference of Elected Officials, "Executive Committee Meeting," Jan. 27, 1965, p. 2n, PNJDMPR, box 21, URB 24/IV/64; Northeastern Illinois Metropolitan Area Planning Commission, "For Immediate Release," May 18, 1962, NIPCR, box 93, folder 9.
11. Denver Regional Council of Governments, *Air Pollution Control Handbook*, 2nd ed. (Denver, 1968), 3; Thomas R. Heaton, "Inter-Governmental Cooperation," *Governors Conference on Air and Water Pollution* (Denver, 1968), 41.
12. Purdom, "Interjurisdictional Problems," 683. The survey did not seek specific information on these "agreements."
13. U.S. Senate, *Study*, 35.
14. Cincinnati Bureau of Air Pollution Control and Heating Inspection, "Annual Report 1957," p. 14, CGC, group 3.
15. Richard B. Lehmkuhl, "1969 Intercommunity Air Pollution Control Program," p. 1, CHS, mss. qA 298, box 1, folder 1.
16. John A. Maga, "A State Approach to Air Pollution," *American Journal of Public Health and the Nation's Health* 51 (1961): 1662.
17. U.S. Senate, *Study*, 31n, 34 (quotation).
18. Hollis S. Ingraham, "Respiratory Disease and Air Pollution," Oct. 30, 1963, p. 7, HRCSP, cont. 38, "New York State Clean Air Committee" folder.
19. Florida State Board of Health, "Preliminary Summary of a Report on Air Pollution in the State of Florida," p. 14.
20. MacKenzie to Surgeon General, July 27, 1964, NA, RG 90 A 1, entry 11, box 4, "Air Pollution Activities—Clean Air Act PL 88-206" folder.
21. "Statement of Orville V. Bergren before the Joint Air Pollution Study Committee," Chicago, Dec. 16, 1968, p. 2, IMAR, box 109, folder 1; Michigan Department of Public Health,

"Air Pollution in Michigan," Jan. 17, 1967, p. 11, GWRGP, box 326, "Air Quality Act of 1967" folder; U.S. Senate, *Study*, 32–34.

22. Hader to Osborn, June 30, 1960, HRCSP, cont. 1, "Air Pollution" folder.

23. Division of Air Pollution Control, "Review of the New Jersey Air Pollution Control Program," 20, 24 (quotation).

24. Bernard B. Bloomfield, "Air Pollution—A Michigan Problem?" p. 6, GWRGP, box 332, "Air Pollution" folder.

25. Council of State Governments, Washington Office, "Air Pollution Control," Mar. 1967, p. 1, GWRGP, box 326, "Air Quality Act of 1967" folder.

26. Florida State Board of Health, "Preliminary Summary of a Report on Air Pollution in the State of Florida," 15.

27. Schueneman, "Trip Report," 3.

28. "Statement by Arthur C. Stern before the Subcommittee on Science, Research, and Development, U.S. House of Representatives, Committee on Science and Astronautics," July 21, 1966, p. 12, NA, RG 90 A 1, entry 11, box 35, "LL-3—Hearings" folder.

29. U.S. Senate, *Study*, 32; Abraham Ribicoff, "We Are Poisoning the Air," *Look* 27, no. 21 (1963): 132.

30. Schueneman, "Air Pollution Problems," p. 12.

31. Purdom, "Interjurisdictional Problems," 685; Schueneman, "Air Pollution Problems," 17.

32. Megonnel to Surgeon General, Apr. 1, 1959, p. 2, NA, RG 90 A 1, entry 36, box 11, "Committee A–M" folder; Purdom, "Interjurisdictional Problems," 685.

33. Jones, *Clean Air*, 88; James E. Krier and Edmund Ursin, *Pollution and Policy: A Case Essay on California and Federal Experience with Motor Vehicle Air Pollution 1940–1975* (Berkeley, 1977), 107; Sidney Edelman, "Air Pollution Control Legislation," in Arthur C. Stern, ed., *Air Pollution*, vol. 3, *Sources of Air Pollution and Their Control*, 2nd ed. (New York and London, 1968), 569, 592; Council of State Governments, Washington Office, "Air Pollution Control," Mar. 1967, p. 5, GWRGP, box 326, "Air Quality Act of 1967" folder; "News from National Association of Manufacturers," Dec. 10, 1962, HML, acc. 1411, ser. I, box 30, "Cinderella City" folder.

34. Bryce Nelson, "Air Quality Act of 1967: A Step Forward, but Don't Expect Immediate Improvement of Your Air," *Science* 158 (1967): 356; Krier and Ursin, *Pollution*, 183; Robert G. Dyck, "Evolution of Federal Air Pollution Control Policy, 1948–1967" (Ph.D. diss., University of Pittsburgh, 1971), 77; Lundqvist, *Hare*, 72n; Christopher J. Bailey, *Congress and Air Pollution: Environmental Policies in the USA* (Manchester and New York, 1998), 135n; Richard J. Tobin, *The Social Gamble: Determining Acceptable Levels of Air Quality* (Lexington, 1979), 62.

35. "City of Pittsburgh Ordinance no. 344," p. 4, AIS, 69:14, box 2, folder 25.

36. Pittsburgh Bureau of Smoke Prevention, "1953 Report," p. 12, AIS 70:2, add. 1971, box 9, folder 151.

37. Columbus Division of Smoke Regulation and Inspection, "1958 Annual Report," NA, RG 90 A 1, entry 36, box 8, "Ohio" folder.

38. Frank A. Chambers, "Smoke Abatement Administration," Apr. 7, 1950, pp. 2, 7, LMDP, box 87, folder 3; Chicago Department of Air Pollution Control, "Annual Report 1963," p. 2, HPKCCP, ser. 1, box 19, folder 1; M. A. Fisher and Radner, "Chicago Upgrades Air Purity Rules," *American City* 74, no. 7 (1959): 141.

39. Cleveland Division of Air Pollution Control, "Annual Report 1967," pp. 31, 49, WRHS, mss. 4370, box 74, folder 1424.

40. "First Meeting of the Allegheny County Smoke Control Advisory Committee," June 1, 1950, p. 3, AIS, 80:7, box 1, FF 2.

41. Andrew Abbott, *The System of Professions: An Essay on the Division of Expert Labor* (Chicago and London, 1988), 20.

42. *Proceedings of the Thirty-Third Annual Convention of the Smoke Prevention Association, Milwaukee, Wisconsin, June 13–16, 1939* (n.l., n.d.), 7, 190n; *Proceedings of the Thirty-Fourth Annual Convention, Smoke Prevention Association, St. Louis, Missouri, May 21–24, 1940* (n.l., n.d.), 2, 88.

43. William G. Christy, "History of the Air Pollution Control Association," *Journal of the Air Pollution Control Association* 10, no. 2 (1960): 134.

44. Charles W. Gruber, "U.S. Association's Membership Growth," *Smokeless Air* 24 (1953): 74; Stern and Greenburg, "Air Pollution," 29.

45. Air Pollution and Smoke Prevention Association of America, *Minutes, 45th Annual Meeting, June 9–12, 1952, Cleveland, Ohio*, 6, 8.

46. Air Pollution and Smoke Prevention Association of America, *Minutes, 1952*, 9n. Today the association continues under the name "Air and Waste Management Association."

47. *Proceedings of the 39th Annual Meeting, Smoke Prevention Association of America, Conference on Smoke Prevention and Conservation of Fuels 1946* (St. Joseph, 1947), 60.

48. Smoke Abatement League of Cincinnati, *S.A.L. Bulletin*, Nov. 24, 1948, p. 3, RAKA, box 46, grouping 16.

49. Air Pollution and Smoke Prevention Association of America, *Minutes, 1952*, 7; Gruber, "Membership Growth," 75.

50. Gruber, "Membership Growth," 75.

51. Division of Air Pollution Control, "Review of the New Jersey Air Pollution Control Program," 21.

52. Arie Jan Haagen-Smit, "Air, Water and People," *Research Management* 10 (1967): 193.

53. Schueneman, "Air Pollution Problems," 9.

54. Griebling, "Cleaning," 166; Cleveland Division of Air Pollution Control, "Annual Report for 1950," p. 4, WRHS, mss. 4035, box 1, folder 2; Cleveland Division of Air Pollution Control, "Annual Report for 1967," p. 7, WRHS, mss. 4370, box 74, folder 1424; "Report Submitted by the Legislative Research Council Relative to Air Pollution in the Metropolitan Boston Area," Feb. 5, 1960, p. 21, NA, RG 90 A 1, entry 36, box 7, "Massachusetts" folder; Portland Chamber of Commerce, "Minutes of Air and Water Standards Committee," Aug. 2, 1967, p. 1, OHS, mss. 1900, box 17, folder 14.

55. Toledo Division of Air and Water Pollution Control, *1962 Annual Report* (Toledo, 1963), 6.

56. Elliott H. Whitlock, "Smoke-Abatement Methods Used in Cleveland," *Mechanical Engineering* 49 (1927): 1071.

57. Victor J. Azbe, "Smoke Abatement Progress," *Power Plant Engineering* 34 (1930): 764. Azbe had been a leading member of the Citizens' Smoke Abatement League in St. Louis.

58. Hood, "Practical Suggestion," 20; Jack Vogele, "Help for the Smoking Chimney," *American City* 63, no. 6 (1948): 114; Monnett, Perrott, and Clark, *Salt Lake City*, 92; Azbe, "Smokeless Combustion," 764; Sherman, "Paths," 520; H. B. Meller, "For Columbus, Ohio—May 24, 1929," p. 2, AIS, 83:7, ser. I, FF 24.

59. "Minutes of the Smoke Abatement Committee Luncheon," Mar. 14, 1924, WRHS, mss. 3535, box 1, folder 6.

60. Smoke Prevention Association, *Manual* (1924), 7; Smoke Prevention Association, *Manual of Smoke and Boiler Ordinance Requirements: Official Publication of the Smoke Prevention Association 1931–1932* (n.l., n.d.), 4.

61. County of Allegheny, Bureau of Smoke Control, "A Review of Program," June 18, 1951, p. 4, AIS, 80:7, box 1, FF 1.

62. Marvin Brienes, "The Fight against Smog in Los Angeles, 1943–1957" (Ph.D. diss., University of California Davis, 1975), 179; Gebhart to Colodny, Nov. 9, 1955, JAFC, box 25, B III 5 a aa, folder 1.

63. Sidlow to Perk, Mar. 15, 1956, p. 2, WRHS, mss. 4456, box 7, folder 112.

64. "Minutes of the September 27–28 Meeting of the National Advisory Health Council,"

p. 4, NA, RG 90 A 1, entry 11, box 56, "OCC—National Advisory Environmental Health Committee Part 2" folder.

65. "Air Pollution Control in California, 1968 Annual Report—Air Resources Board," p. 29, CSA, F 3935, folder 3.

66. Frank Uekoetter, "The Strange Career of the Ringelmann Smoke Chart," *Environmental Monitoring and Assessment* 106 (2005): 11–26.

67. "Statement of Orville V. Bergren before the Committee on Health, Chicago City Council," Feb. 16, 1968, p. 5, IMAR, box 109, folder 1; Michigan Manufacturers' Association, *Bulletin*, no. 2143 (Oct. 31, 1966), p. 4, GWRGP, box 339, "Air Pollution Rules" folder; "Report of the Subcommittee of the Assembly Interim Committee on Governmental Efficiency and Economy Relative to a Study and Analysis of the Facts Pertaining to Air Pollution Control in Los Angeles County," p. 6, UCLA, coll. 1108, box 39, folder 1; Hebley, "Factors Rarely Considered," 283; Johnson and Auth, *Fuels and Combustion Handbook*, 468; John M. Hodges, "The Ringelmann Chart Method of Determining Smoke Violations," in *Proceedings of the Thirty-Third Annual Convention of the Smoke Prevention Association, Milwaukee, Wisconsin, June 13–16, 1939* (n.l., n.d.), 114.

68. H. G. Dyktor, "The Role of Industrial Hygiene Agencies in Air Pollution Control," *Public Health News* 31 (Apr. 1950): 103; Carey P. McCord, "The Physiologic Aspects of Atmospheric Pollution," *Industrial Medicine and Surgery* 19 (1950): 98; Hansen, "Air Pollution Regulation," 241; G. J. McManus, "Air Control: Effective but Expensive," *Iron Age* 178, no. 17 (1956): 71; Robert A. Kehoe, "The Limitations of Air Quality Standards," *Journal of the Air Pollution Control Association* 14 (1964): 16.

69. "City of Pittsburgh Ordinance no. 344," p. 4, AIS, 69:14, box 2, folder 25.

70. Doyle, "Polluted Air," 866.

71. Morris Katz, "Atmospheric Pollution: A Growing Problem in Public Health," *American Journal of Public Health and the Nation's Health* 45 (1955): 298; Robert A. Kehoe, "Effects of Industrial Atmosphere on Workmen and on Community," *Heating, Piping and Air Conditioning* 24, no. 5 (1952): 124; Malcolm H. Merrill, "Atmospheric Standards in Metropolitan Areas," *California's Health* 17 (1959–60): 34; Nord, "Pollution Control," 114; Brandt, "Problems," 79; McCabe, "Atmospheric Pollution," 88A.

72. Hemeon, "Air-Pollution Control," 18.

73. Schueneman to Chief, Division of Air Pollution, Oct. 11, 1962, NA, RG 90 A 1, entry 11, box 40, "OCC—American Petroleum Institute Part 1" folder.

74. Flanagan to Burke, Apr. 10, 1952, p. 1n, WRHS, mss. 4035, box 1, folder 3.

75. Lee Schreibeis, John J. Grove, and Herbert R. Domke, "Air Pollution Control in Urban Planning," Oct. 21, 1959, p. 13, HSWPA, mss. 285, "Smoke Control—General" file.

76. Wolk to Cohen, Mar. 10, 1955, AIS, 69:14, box 2, folder 24.

77. Manufacturing Chemists' Association, *Air Pollution Abatement Manual* (Washington, 1952).

78. Schreibeis, Grove, and Domke, "Air Pollution Control," 11; Frederick S. Mallette, "Legislation on Air Pollution," *Public Health Reports* 71 (1956): 1070.

79. American Society of Mechanical Engineers, "Example Sections for a Smoke Regulation Ordinance," May 1949, p. 4, RRTSAC, ser. 5, box 1, "1948" folder.

80. American Society of Mechanical Engineers, "Example Sections," 6n. However, an inconclusive result was probably to be expected, as the committee comprised both Raymond Tucker and a coal industry representative (Tucker to Barkley, Oct. 22, 1947, RRTSAC, ser. 5, box 1, "1947" folder).

81. American Society of Mechanical Engineers, "Example Sections," 9.

82. StAMs, *Regierung Arnsberg*, no. 1568 p. 253.

83. Tucker to Barkley, June 29, 1948, RRTSAC, ser. 5, box 1, "1948" folder.

84. Stern to Bodurtha, Mar. 2, 1966, NA, RG 90 A 1, entry 11, box 40, "OCC—ASME" folder.

85. Bodurtha to Members of ASME Air Pollution Standards Committee, Mar. 30, 1966, p. 2, NA, RG 90 A 1, entry 11, box 40, "OCC—ASME" folder.
86. Stern to Bodurtha, Apr. 8, 1966, NA, RG 90 A 1, entry 11, box 40, "OCC—ASME" folder.
87. American Society of Mechanical Engineers, "Example Sections," 3; Stern to Bodurtha, Mar. 2, 1966.
88. Fassett to MaKenzie, Nov. 24, 1964, NA, RG 90 A 1, entry 11, box 39, "OCC—American Industrial Hygiene Association" folder.
89. Hemeon and Hatch, "Atmospheric Pollution," 569.
90. *New York Times*, Dec. 7, 1970.
91. *New York Times*, Dec. 7, 1970.
92. Minutes of Meeting, New York State Action for Clean Air Committee, May 23, 1963, HRCSP, cont. 38, "New York State Clean Air Committee" folder.
93. Smoke Abatement League, "Annual Report for 1949," p. 5, CHS, mss. qA 298, box 1, folder 1; Howison to All Officials in Charge of Air Pollution Control and to Members of the Smoke Prevention Association of America, Sept. 10, 1949, CHS, mss. qA 298, box 1, folder 2; "Resolution of the Smoke Prevention Association of America," May 26, 1949, CHS, mss. qA 298, box 1, folder 5; Howison to Bryson, May 3, 1950, CHS, mss. 562, box 73, folder 3; Stern, "Present Status," 83.
94. "Cleaner Air Week, October 23–29, 1960, Air Pollution Control Association Handbook," pp. 2–8, HSWPA, mss. 285, "Smoke Control—General" file.
95. "Cleaner Air Week, October 23–29, 1966, Help Stop Air Pollution," p. 3, HSWPA, mss. 285, "Smoke Control—General" file.
96. Schneider to the Members of Greater Cincinnati's Cleaner Air Week Committee, Sept. 12, 1969, CHS, mss. 602, box 9, folder 5.
97. "Good Neighbor Award, Cleaner Air Week Program, Pittsburgh, Pennsylvania," Oct. 22, 1969, HSWPA, mss. 285, box 21, "Clean Air Week 1962–1969" folder.
98. Zetzman to LeVander, Dec. 20, 1968, SJGP, box 2, "Minnesota Pollution Control Agency, Gadler's Appointment to" folder; "Cleaner Air Committee Minutes of Meeting," Oct. 31, 1967, MCACR, box 1, "General" folder, no. 1; "1969 Annual Report, Chicago Department of Air Pollution Control," p. 21, LMDP, box 86, folder 3; *National Air Conservation Commission Newsletter*, no. 15 (Dec. 1969): 1n, UMM, mss. 43, box 2, folder 3.
99. "Cleaner Air Week Report: Submitted to the Advisory Committee by Jean Nickeson," Oct. 4, 1968, HSWPA, mss. 285, box 21, "Clean Air Week 1962–1969" folder.
100. "Address by M. P. Venema at 1964 Cleaner Air Week Luncheon," Oct. 19, 1964, p. 14, IMAR, box 107, folder 3.
101. "Cleaner Air Week Report."
102. "1968 Annual Report, Chicago Department of Air Pollution Control," p. 16, LMDP, box 86, folder 3; "Cleaner Air Week . . . Help Stop Air Pollution," 11; "21st Cleaner Air Week October 19–25, 1969," p. 12, CHS, mss. qA 298, box 1, folder 10; Metro Clean Air Committee, "General Meeting Minutes," Oct. 15, 1968, p. 1, MCACR, box 1, "General" folder, no. 1. Adam Rome has recently emphasized the crucial role of middle-class women in 1960s grassroots environmental activism ("'Give Earth a Chance': The Environmental Movement and the Sixties," *Journal of American History* 90 [2003]: 534–41).
103. "1959 Annual Report, Air Pollution Control, City of Dayton, Ohio," p. 9, NA, RG 90 A 1, entry 36, box 8, "Ohio" folder; "Annual Report 1962, Chicago Department of Air Pollution Control," p. 18 (quotation), HPKCCP, ser. 1, box 19, folder 1; Denver Air Pollution Control, *Report to the Mayor and City Council* (Denver, 1967), 18.
104. Griebling, "Cleaning," 165.
105. Allegheny County Bureau of Smoke Control, "Seventh Annual Report of Activities for the Year Ending May 31, 1956," p. 4, AIS, 80:7, box 1, FF 7.
106. Edward L. Stockton, "Air and Water Pollution," Apr. 19, 1967, p. 4, HSWPA, mss. 285,

"Smoke Control—General" file; Herbert J. Dunsmore, "A Local Approach to Air Pollution Control," Nov. 2, 1960, p. 3, AIS, 80:7, box 3, FF 14.

107. Dyktor, "Community Problem," 102.

108. Joseph L. Sax, *Defending the Environment: A Strategy for Citizen Action* (New York, 1971), 54–56, 61, 109n.

109. Vernon G. MacKenzie, "The Evolving National Air Pollution Policy," Jan. 17, 1966, p. 18, NA, RG 90 A 1, entry 11, box 43, "OCC—E" folder.

110. HStAD, NW 66, no. 430, pp. 120–37; draft, "Gesetz zur Reinhaltung der Luft in Industriegebieten," Essen, Sept. 1, 1952, HStAD, NW 354, no. 42.

111. HStAD, NW 66, no. 430, p. 121, 105, 120.

112. HStAD, NW 66, no. 430, pp. 123–25, 130–32.

113. HStAD, NW 66, no. 430, p. 126.

114. HStAD, NW 66, no. 430, pp. 126, 136.

115. Kegel, "Luft im Industriegebiet," 369.

116. HStAD, NW 50, no. 1214, p. 123, and NW 66, no. 430, p. 67.

117. HStAD, NW 50, no. 1215, p. 36n; note, July 15, 1955, p. 1, HStAD, NW 354, no. 42.

118. HStAD, NW 50, no. 1215, pp. 22R, 23R, 20.

119. Bundesverband der Deutschen Industrie, "Stellungnahme zum Entwurf eines Gesetzes zur Reinhaltung der Luft in Industriegebieten," Feb. 1955, pp. 11–13, LAB, Rep. 142/9, Acc. 5/24–30, vol. 1.

120. Note, Dec. 3, 1952, HStAD, NW 354, no. 42; Oberstadtdirektor der Stadt Mülheim to Deutscher Städtetag, Landesverband Nordrhein-Westfalen, July 9, 1953, LAB, Rep. 142/9, Acc. 5/24–30, vol. 1.

121. *Westdeutsche Allgemeine*, Aug. 27, 1952.

122. HStAD, NW 66, no. 430, pp. 47, 125 and NW 50, no. 1215, p. 21R.

123. Karl Schwarz, "Der Staub- und Gasauswurf aus Dampferzeugung und seine Verteilung in der Atmosphäre," *Brennstoff—Wärme—Kraft* 8 (1956): 103; Josef Weißner, "Erfahrungen über Rauchschäden im Ruhrbezirk," *Mitteilungen aus dem Markscheidewesen* 61 (1954): 170n.

124. HStAD, NW 66, no. 430, p. 106.

125. HStAD, NW 85, no. 46, p. 180n.

126. HStAD, NW 66, no. 434, pp. 7–21, 37–61, and NW 85, no. 46, p. 157n.

127. F. Rabeneick, "Die Reinerhaltung der Luft mit Hilfe von Verbänden des öffentlichen Rechts," *Zentralblatt für Arbeitsmedizin und Arbeitsschutz* 11 (1961): 207.

128. Sturm Kegel, "Reinhaltung der Luft," *Kommunalpolitische Blätter* 11 (1959): 376.

129. Institut für Wasser-, Boden- und Lufthygiene, "Vermerk über eine Rücksprache im Bundesgesundheitsamt," Apr. 29, 1955, p. 3, NieHStA, Nds. 100 (05), Acc. 144/81, no. 468.

130. HStAD, NW 85, no. 165, pp. 175–77R.

131. HStAD, NW 50, no. 1214, p. 123.

132. *Verhandlungen des Deutschen Bundestages, 2. Wahlperiode 1953* (Stenographische Berichte vol. 34, Bonn, 1957), 10164.

133. HStAD, NW 85, no. 46, pp. 61–87.

134. HStAD, NW 66, no. 432, pp. 88–92.

135. HStAD, NW 85, no. 46, pp. 88, 111 (quotation).

136. HStAD, NW 85, no. 46, p. 93.

137. HStAD, NW 85, no. 46, p. 96.

138. HStAD, NW 85, no. 46, p. 158.

139. "Niederschrift über die 4. Sitzung des Landesbeirats für Immissionsschutz," Feb. 6, 1964, pp. 2–10, and "Kommunalpolitische Vereinigung der CDU des Landes NW, Fachausschuß für Fragen des Städtebaues, der Regionalplanung und des Mißbrauchs der Technik, Entwurf eines Gesetzes über Luftreinhaltungsverbände, Durchführungsverordnung nebst Erläuterungen," Feb. 21, 1963, both HStAD, NW 354, no. 43.

140. HStAD, NW 85, no. 46, p. 158n.

141. Franz Joseph Dreyhaupt, "Ein neues strategisches Konzept zur Luftreinhaltung in Nordrhein-Westfalen," *Luftverunreinigung* (1973): 3.

142. "Kurzprotokoll über die 2. Sitzung des Länderausschusses für Immissionsschutz in Hanover," Dec. 11, 1964, p. 4, BayHStA, MWi 28361.

143. Note, July 16, 1955, BArch, B 149/10407.

144. Draft, "Technische Anweisung für die Genehmigung gewerblicher Anlagen" (§16 GO.), Feb. 2, 1944, BArch, B 149/10407; note, Aug. 1, 1962, p. 1 (quotation), BArch, B 106/26177.

145. HStAD, NW 50, no. 1215, p. 36, and NW 50, no. 1245, p. 52; Kurzprotokoll über die Vollversammlung der Interparlamentarischen Arbeitsgemeinschaft, May 3, 1955, p. 17, LAB, Rep. 142/9, Acc. 5/24-30. vol. 1.

146. HStAD, NW 50, no. 1215, p. 36; *Die Welt*, May 28, 1955. Mommer entertained some interest in air pollution issues, having visited Pittsburgh's Bureau of Smoke Prevention in 1951 (Pittsburgh Bureau of Smoke Prevention, "1951 Report on Stationary Stacks," back cover, AIS, 70:2, add. 1971, box 9, folder 151).

147. *Die Welt*, May 28, 1955.

148. HStAD, NW 50, no. 1215, p. 36R.

149. HStAD, NW 66, no. 430, p. 29.

150. HStAD, NW 66, no. 430, p. 31.

151. HStAD, NW 66, no. 430, p. 30.

152. HStAD, NW 66, no. 430, pp. 26, 27.

153. Note, Aug. 8, 1955, p. 2 (quotation), LAB, Rep. 142/9, Acc. 5/24-30, vol. 2; HStAD, NW 66, no. 430, p. 24.

154. HStAD, NW 268, no. 433, pp. 11-8, NW 66, no. 331, p. 125, and NW 50, no. 1213, p. 78; Werner Gründer, "Aufgaben und Ziele der VDI-Fachgruppe Staubtechnik," *Glückauf* 94 (1958): 53. Due to a general reorganization of the VDI, the Fachausschuß für Staubtechnik became the "Fachgruppe Staubtechnik" in 1955 (Kurt Mauel, "Die technisch-wissenschaftliche Arbeit des VDI 1946 bis 1981," in Karl-Heinz Ludwig, ed., *Technik, Ingenieure und Gesellschaft. Geschichte des Vereins Deutscher Ingenieure 1856–1981* [Düsseldorf, 1981], 469).

155. HStAD, NW 50, no. 1215, p. 31.

156. HStAD, NW 50, no. 1245, p. 52.

157. HStAD, NW 268, no. 433, p. 109R.

158. Klauth to Wittekind, Feb. 28, 1956, HStAD, NW 354, no. 41.

159. Der Präsident des Bundesgesundheitsamtes to Verein Deutscher Ingenieure, Oct. 28, 1955, BArch, R 154/68.160; HStAD, NW 50, no. 1215, pp. 13R, 14.

161. Note, Feb. 2, 1957, p. 2, HStAD, NW 354, no. 41.

162. HStAD, NW 268, no. 433, pp. 55, 56.

163. HStAD, NW 50, no. 1216, p. 75.

164. "Vorbericht der VDI-Fachgruppe Staubtechnik," Nov. 1956, p. 3, LASb, StK 1491; "Anlage 4 zur Niederschrift über das Ergebnis der 2. Sitzung der Länderarbeitsgemeinschaft 'Reinhaltung der Luft,'" Feb. 7, 1964, p. 6, LHAK, Bestand 930, no. 10322; Verein deutscher Ingenieure to Präsident des Bundesgesundheitsamts, May 22, 1956, BArch, R 154/85.

165. HStAD, NW 268, no. 435, pp. 146-54, and NW 85, no. 165, pp. 150n, 153.

166. "Ergebnisniederschrift über die 7. Sitzung des Länderausschusses für Immissionsschutz in Düsseldorf Apr. 5–7, 1967, zu Punkt 3 a) der Tagesordnung," BayHStA, MWi 28364.

167. U. Franzky, "Was der Besteller einer Entstaubungsanlage beachten sollte," *Wasser, Luft und Betrieb* 7 (1963): 81; H. Petri, "Schädliche Auswirkungen der Luftverunreinigung," *Gesundheits-Ingenieur* 86 (1965): 108.

168. Gerhard Feldhaus and Horst D. Hansel, *Umweltschutz: Luftreinhaltung, Lärmbekämpfung. Rechts- und Verwaltungsvorschriften des Bundes und der Länder mit einer systematischen Einführung* (Cologne, 1971), 45; note, Aug. 26, 1963, p. 3, BArch, B 106/38377; Stellungnahme des Bundesverbandes der Deutschen Industrie, Nov. 19, 1963, p.

9n, and several similar letters, all BArch, B 106/38378; Vereinigung der Großkesselbesitzer to Bundesministerium für Gesundheitswesen, June 18, 1963, p. 4, BArch, B 106/38380; Vereinigung Deutscher Elektrizitätswerke to Mitglieder des Bundesrats, June 24, 1964, p. 4, BayHStA, MInn 91145.

169. HStAD, NW 66, no. 436, pp. 1, 10, 66, 68–77, 85–87.

170. *Jahresbericht der Gewerbeaufsicht des Landes Nordrhein-Westfalen für das Jahr 1962*, 7.

171. HStAD, NW 50, no. 1212, p. 212.

172. Heinrich Lent, "Derzeitiger Stand der Arbeiten der VDI-Kommission Reinhaltung der Luft," *Zentralblatt für biologische Aerosolforschung* 10 (1961–62): 206.

173. HStAD, NW 85, no. 165, p. 260, and NW 50, no. 1215, p. 46.

174. "Kurzniederschrift zur Besprechung im Bundesministerium für Gesundheitswesen," Dec. 7, 1967, p. 2, BArch, B 106/26178.

175. "Kurzniederschrift zur Besprechung im Bundesministerium für Gesundheitswesen," Dec. 7, 1967. See also Manfred Glagow and Axel Murswieck, "Umweltverschmutzung und Umweltschutz in der Bundesrepublik Deutschland," *Aus Politik und Zeitgeschichte* 27 (July 3, 1971): 23.

176. "Niederschrift zur Vorbesprechung über die Gründung eines Unterausschusses 'Warndienst Reinhaltung der Luft' in der Geschäftsstelle der VDI-Kommission Reinhaltung der Luft," Jan. 28, 1963, p. 1, BArch, B 106/38621.

177. Gründer, "Aufgaben," 55n; J. Krämer, "Zur Reinhaltung der Luft," *VDI-Nachrichten* 11, no. 2 (1957): 9; "Interparlamentarische Arbeitsgemeinschaft Drucksache no. 89," HStAD, NW 354, no. 40; attachment to Innenministerium Baden-Württemberg to Bundesministerium für Gesundheitswesen, June 29, 1962, p. 7n, BArch, B 106/35797; Vorbericht der VDI-Fachgruppe Staubtechnik, Nov. 1956, p. 3, LASb, StK 1491.

178. For corresponding public notices, see *Gesundheits-Ingenieur* 86 (1965): 66, 220, 313, and 87 (1966): 175. It is unclear why Susan Rose-Ackerman has nonetheless criticized the VDI Clean Air Commission for its lack of openness. Given the inherent complexity of the matter, the publication of draft standards, along with the invitation for anyone to review, including environmental groups, arguably provided for a degree of transparency rarely found in rule-making processes (Susan Rose-Ackermann, *Controlling Environmental Policy: The Limits of Public Law in Germany and the United States* [New Haven and London, 1995], 64).

179. Deutscher Bundestag, *3. Wahlperiode, Drucksache 301*.

180. HStAD, NW 50, no. 1214, p. 108.

181. "Niederschrift über die Ressortbesprechungen zu dem Entwurf eines Gesetzes zur Änderung der Gewerbeordnung und Ergänzung des Bürgerlichen Gesetzbuchs—BT-Drucks. 301," Apr. 24, 29, 1958, p. 7n, and Bundesminister der Justiz to Bundesminister für Arbeit und Sozialordnung, May 8, 1958, p. 7n, both BArch, B 106/38374.

182. *Bundesgesetzblatt*, part 1 (1959): 781–83; *Gesetz- und Verordnungsblatt für das Land Nordrhein-Westfalen*, edition A 16 (1962): 225–27; HStAD, NW 66, no. 430, p. 1, and NW 66, no. 431, pp. 2–6; "Referenten-Entwurf eines Gesetzes zum Schutze vor Immissionen," Dec. 12, 1957, HStAD, NW 354, no. 40.

183. Note from Referat 9, Dec. 30, 1960, BArch, B 136/5364; Interparlamentarische Arbeitsgemeinschaft to Bundeskanzler Adenauer, Dec. 12, 1960, pp. 1, 2, BArch, B 136/5364.

184. As a result, historians still underestimate the importance of the law of 1959. See, most recently, Christoph Nonn, "Vom Naturschutz zum Umweltschutz: Luftreinhaltung in Nordrhein-Westfalen zwischen fünfziger und frühen siebziger Jahren," *Geschichte im Westen* 19 (2004): 239.

185. HStAD, NW 268, no. 433, p. 109R.

186. Regierung von Oberbayern to Bayerisches Staatsministerium für Wirtschaft und Verkehr, Jan. 7, 1964, StAM, RA 103109.

187. Uekötter, *Rauchplage*, 233–62.

188. Regierungsgewerbedirektor G. Heinze, "Die Aufgaben eines Gewerbeaufsichtsamtes im Rahmen des Umweltschutzes," p. 1, LASH, Abt. 761, no. 20317.

189. HStAS, EA 8/301, Büschel 587–90, 592–98; HStAD, NW 50, no. 1432, NW 66, no. 441, NW 85, no. 162, NW 85, no. 163, NW 85, no. 167, NW 85, no. 168, and Regierung Köln, no. 13647, pp. 140–394.

190. "Tätigkeitsbericht 1961 der VDI-Kommission Reinhaltung der Luft," May 1962, p. 1, BArch, B 106/26177; attachment 4 of VDI-Kommission Reinhaltung der Luft to Arbeitsministerium Baden-Württemberg, Oct. 9, 1967, p. 2, HStAS, EA 8/301, Büschel 485.

191. Frank Uekoetter, "The Merits of the Precautionary Principle: Controlling Automobile Exhausts in Germany and the United States before 194," in DuPuis, *Smoke and Mirrors*, 119–53.

192. Petri, "Schädliche Auswirkungen," 153.

193. "Stellungnahme des Bundesverbandes der Deutschen Industrie," Nov. 19, 1963, p. 13, BArch, B 106/38378.

194. To my knowledge, the first use of the word is from 1964 ("Niederschrift über die Besprechung mit den Ländern," Oct. 15, 1964, p. 2, BayHStA, MArb 2600).

195. Council of Europe, Committee of Ministers, "Resolution (68) 4: Adopted by the Ministers' Deputies on 8th March 1968," p. 2, BArch, B 106/21969.

196. *Rheinischer Merkur*, Sept. 13, 1957.

197. HStAD, NW 66, no. 434, p. 185.

198. HStAD, NW 66, no. 434, pp. 162, 186.

199. HStAD, NW 66, no. 434, pp. 179, 180.

200. HStAD, NW 66, no. 434, p. 163.

201. HStAD, NW 66, no. 434, p. 159.

202. HStAD, NW 66, no. 434, pp. 112–16.

203. HStAD, NW 66, no. 434, p. 78.

204. HStAD, NW 50, no. 1217, p. 30.

205. BayHStA, MWi 28360–70.

206. "Protokoll der 8. Sitzung des Wirtschafts- und Verkehrsausschusses," Nov. 11, 1965, p. 41, LHAK, Bestand 930, no. 10319.

207. "Niederschrift über die 4. Sitzung des 'Beratenden Ausschusses nach § 16 Abs. 3 der Gewerbeordnung,'" Nov. 23, 1962, p. 5, LAB, Rep. 142/9, Acc. 5/24–31.

208. Deutscher Städtetag to Stadtverwaltung Mainz, Oct. 16, 1963, p. 2, LAB, Rep. 142/9, Acc. 5/24–30, vol. 9; LHAK, Bestand 930, no. 10319, "Protokoll der 8. Sitzung des Wirtschafts- und Verkehrsausschusses," Nov. 11, 1965, p. 41, LAB, Rep. 142/9, Acc. 5/24–30, vol. 9; "Vormerkung für Herrn Staatsminister Dr. Pirkl," Oct. 19, 1970, p. 2, BayHStA, MArb 2596/II.

209. *Jahresbericht der Gewerbeaufsicht des Landes Nordrhein-Westfalen für das Jahr 1964*, 29.

210. HStAD, NW 66, no. 436, p. 57.

211. Deutscher Bundestag, 7. *Wahlperiode, Drucksache 2802*, 177.

212. Remington to McCall, May 4, 1967, OSA, ACC 70A-095, box 39, "Pollution—General" folder.

213. Wirtschaftsdienst Studienreisen to Minister des Innern des Landes Rheinland-Pfalz, Mar. 2, 1960, LHAK, Bestand 930, no. 10321.

214. "Report on Du Pont Water and Air Pollution Control," July 19, 1965, pp. 6 (quotation), 12, HML, acc. 1801, box 10, "Environmental Quality" folder.

215. P. von Heimendahl, "Staub und SO_2-Emissionen sowie Verbrennungsrückstände aus Dampfkraftanlagen der Bundesrepublik Deutschland: Ergebnisse einer Erhebung der VGB im Jahre 1962," *Mitteilungen der Vereinigung der Großkesselbesitzer* 87 (Dec. 1963): 411.

216. Helmut Kettner, "Ergebnisse 10jähriger Staubniederschlagsmessungen," *Gesundheits-Ingenieur* 87 (1966): 108.

217. *Gesundheits-Ingenieur* 87 (1966): 122.

218. Manfred Häberle, "Luftreinhaltung in der chemischen Industrie: Eine Übersicht über Abgas/Abluft-Reinigungsverfahren mit Ergebnissen der BASF AG," *Chemiker-Zeitung* 95 (1971): 448.

219. Hans Georg Meyer, "Erfolge bei der Luftüberwachung in einem Chemiewerk," *Chemiker-Zeitung* 95 (1971): 449n, 450 (quotation).

220. H. Brandt and H. Heer, "Besonderheiten bei der Entstaubung in Müllverbrennungsanlagen," *Mitteilungen der Vereinigung der Großkesselbesitzer* 48 (1968): 119; Karl Schwarz, "Maßnahmen zur Reinhaltung der Luft bei Müllverbrennungsanlagen," *Technische Überwachung* 6 (1965): 368, 388; Herbert Pötschke, "Müllverbrennung als Gegenwartsproblem," *Wasser, Luft und Betrieb* 14 (1970): 91n.

221. Heinrich Dratwa, "Die Reinhaltung der Luft von Schwefeloxiden—Problematik und Möglichkeiten," *Technische Überwachung* 8 (1967): 154; W. Strewe, "Wärmeerzeugung mit festen Brennstoffen," *Gesundheits-Ingenieur* 86 (1965): 115; Pieper, "Emissionen," 177n; K. Gasiorowski, "Energieerzeugung aus flüssigen Brennstoffen," *Gesundheits-Ingenieur* 86 (1965): 117; Walther Liese, *Gesundheitstechnisches Taschenbuch*, 2nd ed. (Munich and Vienna, 1969), 72.

222. HStAD, Regierung Köln, no. 13647, pp. 411, 417.

223. Attachment to Arbeits- und Sozialminister des Landes Nordrhein-Westfalen to Members of Länderausschuss für Immissionsschutz, July 17, 1969, BayHStA, MWi 28367.

224. Emil Banik, "Betrachtungen über die Durchführung des neuen Luftreinhaltegesetzes," *Zentralblatt für Arbeitsmedizin und Arbeitsschutz* 11 (1961): 248.

225. Note, Oct. 2, 1958, p. 2, BArch, B 102/43222; "Auszug aus der Niederschrift über die 9: Sitzung des Ausschusses für Gesundheitswesen," Apr. 3, 1964, p. 11, NieHStA, Nds. 300, Acc. 188/81, no. 23.

226. Landtag Nordrhein-Westfalen, *4. Wahlperiode, Stenographische Berichte vol. 3* (Düsseldorf, 1962), 2969.

227. *Gesetzblatt für Baden-Württemberg* (1964): 55–58; *Niedersächsisches Gesetz- und Verordnungsblatt*, edition A 20 (1966): 1–3; *Gesetz- und Verordnungsblatt für das Land Rheinland-Pfalz* (1966): 211–13; *Gesetzblatt der Freien Hansestadt Bremen* (1970): 71–73.

228. Landtag des Saarlandes, *4. Wahlperiode, Drucksache no. 898 and 900*; Landtag des Saarlandes, *5. Wahlperiode, Drucksache no. 90*; Feldhaus and Hansel, *Umweltschutz*, 321–25; "Auszug aus der Niederschrift des Ministerrats," Mar. 3, 1964, p. 12n, BayHStA, StK-GuV 1277; "Jahresbericht der Gewerbeaufsicht des Jahres Schleswig-Holstein 1971," Punkt 6.0 (quotation), LASH, Abt. 761, no. 20556.

229. Landtag Nordrhein-Westfalen, *4. Wahlperiode, Stenographische Berichte vol. 3*, 2969; *Verhandlungen des Landtags von Baden-Württemberg, 3. Wahlperiode 1960–1964, Protokoll*, vol. 6 (Stuttgart, 1964), 7364; *Verhandlungen des Niedersächsischen Landtages, 5. Wahlperiode 1963, Stenographische Berichte vol. 3* (Hanover, 1967), c. 4016; Bremische Bürgerschaft, *7. Wahlperiode, Drucksachenabteilung V (Verhandlungsberichte)*, 2391.

Chapter 6

1. Clemens Dörr, "Ein neues Rauchverbrennungsverfahren für die Industrie," *Gesundheit* 36 (1911): 569.

2. Klocke, "Zur Frage der Rauchverminderung im Industriebezirke," *Stahl und Eisen* 29 (1909): 170.

3. Karl von Wedelstaedt, "Die Rauchplage im Ruhrkohlenbezirk und ihre Bekämpfung," Apr. 11, 1930, p. 24, BArch, R 154/12103.

4. Hermann Bohle, "Fertig werden mit dem Schmutz: Das Gesetz zur Reinhaltung der Luft, seine Hintergründe, Folgen und Kosten," *Industriekurier* 12, no. 173 (1959): 5.

5. Thyssen & Co. to Bürgermeister Styrum, Mar. 26, 1885, StAMü, 1130/50.

Notes to Pages 189-192   321

6. Franz-Josef Brüggemeier and Thomas Rommelspacher, *Blauer Himmel über der Ruhr: Geschichte der Umwelt im Ruhrgebiet 1840–1990* (Essen, 1992), 22.
7. StAMs, Oberbergamt Dortmund, no. 1372, pp. 29, 32n, 36–38, 41–45, 52–54, 56n, 60–63.
8. *Prometheus* 26 (1915): 272.
9. See, e.g., Gesellschaft für Teerverwertung to Oberbürgermeister der Stadt Duisburg, Dec. 2, 1909, StADu, 313/16; StAMs, Regierung Arnsberg, no. 1567, p. 350; Phoenix Aktien-Gesellschaft für Bergbau und Hüttenbetrieb to Oberbürgermeister Duisburg, June 22, 1918, StADu, 306/584.
10. StAHe, VII/247 a and b, p. 6.
11. Niederrheinische Hütte to Oberbürgermeister der Stadt Duisburg, Oct. 25, 1909, p. 2, StADu, 313/44.
12. Gewerbeinspektion Essen to Regierungs- und Gewerberat Düsseldorf, Apr. 18, 1898, p. 4, and Gewerbeinspektion Essen to Regierungs- und Gewerberat Düsseldorf, Mar. 20, 1900, p. 2, both HStAD, Regierung Düsseldorf, no. 25055.
13. Attachment to letter of Wirtschaftsgruppe Metallindustrie, Mar. 11, 1940, p. 11, BArch, R 154/31; HStAD, NW 50, no. 1228, p. 4R.
14. Klaus Tenfelde, "Zur Geschichte der Industriellen Beziehungen im Bergbau," in Klaus Tenfelde and Gerald D. Feldman, eds., *Arbeiter, Unternehmer und Staat im Bergbau: Industrielle Beziehungen im internationalen Vergleich* (Munich, 1989), 8.
15. Walter Hoever to Oberbürgermeister Lehr, Nov. 14, 1913, StADu, 313/17; Walter Hoever to Bezirksausschuß Düsseldorf, Jan. 22, 1914, HStAD, Regierung Düsseldorf, no. 34217.
16. Oberbürgermeister Essen to Regierungspräsident Düsseldorf, Jan. 11, 1907 (quotation), HStAD, Regierung Düsseldorf, no. 34220; Hermann Board to Regierungspräsident Düsseldorf, Jan. 18, 1910, HStAD, Regierung Düsseldorf, no. 34220.
17. Siedlungsverband Ruhrkohlenbezirk, *Rauchbekämpfung*, 22.
18. Attachment to Rheinisch-Westfälischer Schutzverband gegen Rauch- und Bergschäden to Regierungspräsident Düsseldorf, May 28, 1914, p. 2, HStAD, Regierung Düsseldorf, no. 34082.
19. Note by Tiegs, Apr. 16, 1930, p. 1, BArch, R 154/12103.
20. The bureaus of mining were in charge of emissions from coke ovens and power plants affiliated with mines (Der Minister für Handel und Gewerbe to Regierungspräsident Arnsberg, Jan. 14, 1897, StAMs, Regierung Arnsberg 6, no. 457; GStA, Rep. 120, BB II a 2, no. 28, Adh. 1, vol. 8, p. 285).
21. Gewerbeinspektion Essen-Ruhr to Bürgermeister Stoppenberg, Feb. 11, 1908, HStAD, Regierung Düsseldorf, no. 34220; Der Bergrevierbeamte des Bergreviers Ost-Essen to Oberbürgermeisteramt Stoppenberg, Feb. 25, 1908, HStAD, Regierung Düsseldorf, no. 34220.
22. StABo, LA 774, p. 13R.
23. Werner Doenecke, *Der Siedlungsverband Ruhrkohlenbezirk* (Essen, 1926), 16.
24. StAMs, Oberbergamt Dortmund, no. 1372, p. 147n.
25. StAMs, Oberbergamt Dortmund, no. 1372, p. 148; Siedlungsverband Ruhrkohlenbezirk, *Rauchbekämpfung*.
26. StAMs, Oberbergamt Dortmund, no. 1372, pp. 1, 2.
27. Der Verbandsdirektor des Siedlungsverbands Ruhrkohlenbezirk to Regierungspräsident Düsseldorf, Feb. 2, 1925, and response from the Regierungspräsident, Mar. 5, 1925, both HStAD, Regierung Düsseldorf, no. 34221.
28. O. Krawehl, "Die Rauchschadenbekämpfung," *Deutsche Bergwerks-Zeitung* 30, no. 79 (1929): 6. Krawehl was a member of the smoke commission and an engineer for Rheinstahl (StAMs, Regierung Arnsberg, no. 1567, pp. 439, 466).
29. StAMs, Oberbergamt Dortmund, no. 1372, p. 196.

30. Der Verbandsdirektor des Siedlungsverbands Ruhrkohlenbezirk to Preußischer Minister für Volkswohlfahrt, May 30, 1930, p. 6, BArch, R 154/12103.
31. StAMs, Regierung Arnsberg, no. 1567, p. 479.
32. StAMs, Regierung Arnsberg, no. 1567, p. 469. For the memorandum, see Karl von Wedelstaedt, "Die Rauchplage im Ruhrkohlenbezirk und ihre Bekämpfung," Apr. 11, 1930, BArch, R 154/12103.
33. Siedlungsverband Ruhrkohlenbezirk, *Rauchbekämpfung*, 4.
34. Haus- und Grundbesitzerverein Essen to Oberbürgermeister der Stadt Essen, Nov. 30, 1939, p. 2, StAE, Rep. 102, Abt. XIV, no. 93.
35. Lent, "Stand der Arbeiten," 205.
36. Ekkehard Koch, *Der Weg zum blauen Himmel über der Ruhr: Geschichte der Vorläufer-Institute der Landesanstalt für Immissionsschutz* (Essen, 1983), 62; StAMs, Regierung Arnsberg, no. 590, p. 258; HStA, NW 66, no. 430, pp. 85, 87.
37. Arbeits- und Sozialminister des Landes Nordrhein-Westfalen to Bayerisches Staatsministerium für Wirtschaft und Verkehr, July 3, 1964, p. 4, BayHStA, MWi 28360; "Niederschrift über das Ergebnis der 2. Sitzung der Länderarbeitsgemeinschaft 'Reinhaltung der Luft,'" Feb. 7, 1964, p. 2, NieHStA, Nds. 300, Acc. 188/81, no. 26.
38. Regierungspräsident Osnabrück to Niedersächsischer Sozialminister, Mar. 19, 1971, p. 4, NieHStA, Nds. 300, Acc. 188/81, no. 2.
39. *Der Brennpunkt: Informationen der Bürgervereinigung Stadtmitte [Duisburg] für ihre Mitglieder und Freunde* 4 (Nov. 1957): 2n.
40. Central Lane Planning Council, *Crisis: Air* (Eugene, 1968), 10.
41. National Association of Manufacturers, *Cinderella City: How Community Action Transformed Pittsburgh's Smoke-Stained Identity* (New York, 1962).
42. Wolfgang Köllmann, Frank Hoffmann, and Andreas E. Maul, "Bevölkerungsgeschichte," in Wolfgang Köllmann et al., eds., *Das Ruhrgebiet im Industriezeitalter: Geschichte und Entwicklung* (Düsseldorf, 1990), 1: 113.
43. Johann Paul, "Risikodebatten über den Tieftagebau im rheinischen Braunkohlenrevier seit den 1950er Jahren," *Technikgeschichte* 65 (1998): 141–61.
44. Ministerium für Handel und Gewerbe to Regierungspräsident Cöln, May 6, 1906, HStAD, Regierung Köln, no. 2146.
45. Auszug aus dem Protokollbuche des Gemeinderates der Gemeinde Brühl, Feb. 6, 1908, HStAD, Regierung Köln, no. 2147.
46. Zellstoff-Fabrik Düren to Regierungspräsident Aachen, July 1, 1913, and Oberbürgermeister der Stadt Düren to Oberbergamt Bonn, Nov. 13, 1913, esp. p. 6, both HStAD, Regierung Aachen, no. 14244.
47. Oberbürgermeister Köln to Deutscher Städtetag, Feb. 21, 1930, BArch, R 36/1284; Gesundheitsamt der Stadt Köln to Beigeordneter Dr. Vonessen, Aug. 22, 1949, LAB, Rep. 142/9, Acc. 5/24–30, vol. 1.
48. Sülz-Klettenberger Bürgergesellschaft to Ministerpräsident, Apr. 17, 1963, HStAD, NW 310, no. 293.
49. For these negotiations, see BArch, R 154/86; HStAD, NW 50, no. 1245.
50. Bernd Weisbrod, "Arbeitgeberpolitik und Arbeitsbeziehungen im Ruhrbergbau. Vom 'Herr-im-Haus' zur Mitbestimmung," in Gerald D. Feldman and Klaus Tenfelde, eds., *Arbeiter, Unternehmer und Staat im Bergbau: Industrielle Beziehungen im internationalen Vergleich* (Munich, 1989), 154–61.
51. Lent to Deutscher Städtetag, Sept. 29, 1956, LAB, Rep. 142/9, Acc. 5/24–32.
52. Oberstadtdirektor Gelsenkirchen to Deutscher Städtetag, Nov. 9, 1954, LAB, Rep. 142/9, Acc. 5/24–30, vol. 1; HStAD, NW 50, no. 1220, pp. 1–8; "Niederschrift über die Mitgliederversammlung des Vereins zur Bekämpfung der Volkskrankheiten im Ruhrkohlengebiet e.V. in Gelsenkirchen," July 9, 1964, p. 3, StADu, 503/689.
53. GStA, Rep. 76 VIII B, no. 2007, p. 443R.

54. Gesundheitsamt Dortmund to Deutscher Städtetag, Mar. 24, 1930, BArch, R 36/1284; Stadtverwaltung Duisburg-Hamborn to Deutscher Städtetag, Mar. 31, 1930, BArch, R 36/1284; *Kölnische Volkszeitung*, Oct. 5, 1928; *Herner Anzeiger*, Oct. 4, 1928.

55. *Herner Volkszeitung*, Aug. 26, 1929; *Duisburger General-Anzeiger*, Aug. 9, 1928.

56. "Niederschrift über die Besprechung in Essen," Dec. 16–17, 1929, BArch, R 154/12103.

57. Köllmann, Hoffmann, and Maul, "Bevölkerungsgeschichte," 114.

58. *Westdeutsches Tageblatt*, Oct. 27, 1959; *Verhandlungen des Deutschen Bundestages, 2. Wahlperiode 1953*, Stenographische Berichte, vol. 34 (Bonn, 1957), 10165.

59. *Ruhr-Nachrichten* (Essen), Sept. 12–13, 1959; Drucksache 4493, Sept. 16, 1959, StADu, 101/568; "Auszug aus der Tonbandaufnahme über die Sitzung des Rats der Stadt," Sept. 21, 1959, p. 2, StADu, 101/568.

60. HStAD, NW 85, no. 165, p. 169.

61. *Westfälische Rundschau*, Feb. 10, 1954; *Neue Ruhr Zeitung*, Dec. 10, 1956; *Der Brennpunkt: Informationen der Bürgervereinigung Stadtmitte [Duisburg] für ihre Mitglieder und Freunde* 2 (1957): 1; *Neue Ruhr Zeitung* (Duisburg), Jan. 19, 1957.

62. HStAD, NW 66, no. 335, p. 29, and NW 50, no. 1230, p. 92.

63. HStAD, NW 85, no. 46, p. 198.

64. John E. Baur, *The Health Seekers of Southern California, 1870–1900* (San Marino, 1959), 1.

65. James C. Williams, *Energy and the Making of Modern California* (Akron, 1997), 364.

66. Cannon, *Smoke Abatement*, 258.

67. McDonald to Tucker, Jan. 24, 1940, RRTSAC, ser. 1, box 1, "Mc" folder.

68. The city's collective memory has 1943 as the year when smog first appeared (Andrew Rolle, *Los Angeles: From Pueblo to City of the Future*, 2nd ed. [San Francisco, 1995], 158n; Wyn Grant, *Autos, Smog and Pollution Control: The Politics of Air Quality Management in California* [Aldershot and Brookfield, 1995], 29n; South Coast Air Quality Management District, *Fifty Years of Progress toward Clean Air: The Southland's War on Smog* [Diamond Bar, n.d. {ca. 1997}], 3n). While there had been several smog incidents since the early 1940s, none received the kind of public attention that the 1943 episode did (Brienes, "Fight," 33n; Marvin Brienes, "Smog Comes to Los Angeles," *Southern California Quarterly* 58 [1976]: 516; Krier and Ursin, *Pollution*, 52n; Charles L. Senn, "Los Angeles 'Smog,'" *American Journal of Public Health and the Nation's Health* 38 [1948]: 962; Air Pollution Foundation, *Final Report 1961* [San Marino, 1961], 6).

69. "Report to the Los Angeles County Board of Supervisors by the L. A. County Smoke and Fumes Commission," Mar. 13, 1944, p. 1, JAFC, box 25 B III 5 a aa, folder 3.

70. "Mayor's Conference on Control of Smoke and Fumes. Reporter's Transcript of Proceedings," May 8, 1946, p. 6, JAFC, box 25 B III 5 a aa.

71. Richard P. Feynman, *"Sure You're Joking, Mr. Feynman!" Adventures of a Curious Character* (New York and London, 1985), 234.

72. Simon to Ford, Apr. 19, 1950, JAFC, box 26 B III 5 a ff, folder 1.

73. "Testimony of S. Smith Griswold, Dr. Leslie A. Chambers and Hoyt R. Crabaugh before the Select Committee on Small Businesses of the House of Representatives at Los Angeles, Calif.," May 18–19, 1956, p. 8, UCLA, coll. 1108, box 37, folder 4 (emphasis original).

74. Pottenger to Kehoe, Dec. 15, 1950, RAKA, box 46, grouping 13.

75. George A. Gonzalez has stressed this point, criticizing the exceeding emphasis on health arguments pure and simple ("Urban Growth and the Politics of Air Pollution: The Establishment of California's Automobile Emission Standards," *Polity* 35 [2002]: 217). Gonzalez has recently expanded his argument into a general theory of environmental policy being driven by local growth coalitions, but this may overstretch his point (*The Politics of Air Pollution: Urban Growth, Ecological Modernization, and Symbolic Inclusion* [Albany, 2005]).

76. Brienes, "Fight," 117.

77. Silbert to Board of Supervisors, Dec. 7, 1949, JAFC, box 25 B III 5 a aa, folder 8.

78. "Report to the Los Angeles County Board of Supervisors by the L. A. County Smoke and Fumes Commission," Mar. 13, 1944, p. 1, JAFC, box 25, B III 5 a aa, folder 3.

79. Griswold to Brown, June 1, 1959, p. 2, UCLA, coll. 1108, box 11, folder 18.

80. Ayres to Hahn, Oct. 9, 1953, KHC, box 266, folder 1a.

81. "Remarks by Wm. H. T. Holden, CASAC Technical Committee Chairman before the Board of Directors, City of Pasadena," Sept. 6, 1955, p. 1, UCLA, coll. 1108, box 38, folder 1.

82. "Coalition for Clean Air: Representing Citizen Volunteer Groups throughout California," pamphlet, n.d. (ca. 1970), DBC, box 1, folder 14. On the importance of Stamp Out Smog, see Prindle to Chief, Bureau of State Services, June 28, 1960, NA, RG 90 A 1, entry 36, box 9, "Association N–Z" folder; Edward Edelson, *The Battle for Clean Air* (n.l., 1967), 24; Sloane, "How to Protect," 112.

83. Brienes, "Fight," 37–48.

84. Los Angeles County Smoke and Fumes Commission, "First Annual Report to the Los Angeles County Board of Supervisors," Oct. 24, 1944, JAFC, box 25 B III 5 a aa, folder 2; "Report to the Los Angeles County Board of Supervisors by the L. A. County Smoke and Fumes Commission," Mar. 13, 1944, JAFC, box 25 B III 5 a aa, folder 3; "Report to the Los Angeles County Board of Supervisors from the Los Angeles County Smoke and Fumes Commission," Jan. 23, 1945, JAFC, box 25 B III 5 a aa, folder 4.

85. Brienes, "Fight," 92.

86. Senn, "Los Angeles 'Smog,'" 963.

87. Brienes, "Fight," 114–41; Krier and Ursin, *Pollution*, 61–64; Los Angeles County Air Pollution Control District, *Annual Report 1947–48* (n.l., 1949), 1.

88. "Mayor's Conference on Control of Smoke and Fumes: Reporter's Transcript of Proceedings," May 8, 1946, p. 46, JAFC, box 25 B III 5 a aa.

89. Hahn to Larson, Oct. 2, 1953, KHC, box 266, folder 1a.

90. Hahn to Larson, Nov. 19, 1953, KHC, box 266, folder 1a.

91. Brienes, "Fight," 250, 263.

92. U.S. Senate, *Study*, 33.

93. Senn, "Los Angeles 'Smog,'" 965; Brandt, "Problems," 77; "Report of Los Angeles County 1948 Grand Jury Special Committee on Smog," Dec. 14, 1948, p. 1, JAFC, box 25 B III 5 a bb.

94. "Report of the Subcommittee of the Assembly Interim Committee on Governmental Efficiency and Economy Relative to a Study and Analysis of the Facts Pertaining to Air Pollution Control in Los Angeles County," p. 11, UCLA, coll. 1108, box 39, folder 1; Miner L. Hartmann, "Legal Regulation of Air Pollution," p. 16, JRP, cont. 271, "Smog, Folder 2" folder; Brienes, "Fight," 169–71; Senn, "Los Angeles 'Smog,'" 964.

95. Krier and Ursin, *Pollution*, 79n.

96. Hitchcock to Legg, Nov. 8, 1954, JAFC, box 27 B III 5 a ii.

97. Los Angeles County Air Pollution Control District, *Annual Report 1952–53*, 7.

98. "Report of the Subcommittee of the Assembly Interim Committee on Governmental Efficiency and Economy Relative to a Study and Analysis of the Facts Pertaining to Air Pollution Control in Los Angeles County," p. 12, UCLA, coll. 1108, box 39, folder 1.

99. "Testimony of S. Smith Griswold . . . ," 15.

100. Daugherty to Runyon, Apr. 21, 1955, p. 2, KHC, box 266, folder 3a.

101. Harold W. Kennedy, *The History, Legal and Administrative Aspects of Air Pollution Control in the County of Los Angeles: Report Submitted to the Board of Supervisors of the County of Los Angeles* (Los Angeles, 1954), 63.

102. Brienes, "Fight," 118n, 129n; Air Pollution Foundation, *Final Report*, 7.

103. Stewart to McCabe, Sept. 27, 1948, p. 2, JAFC, box 25 B III 5 a bb, folder 4.

104. Michael Pratch, "Regulatory and Legal Aspects of Air Pollution," Oct. 30, 1957, p. 2, UCLA, coll. 1675, box 82, "Regulatory and Legal Aspects of Air Pollution" folder.

105. W. L. Stewart Jr., "Laws, Morals, and Manners," *American Petroleum Institute Proceedings* 31 M (1951), sections 3, 11.

106. Pratch, "Regulatory and Legal Aspects," 4.

107. "Where We Stand on Smog Problem, What's Been Done, What's Ahead," reprinted from the *Los Angeles Times*, Jan. 8, 1961, p. 4, JRP, cont. 513, "H.R. 3577" folder.

108. *Journal of the Assembly, Legislature of the State of California, 1970 Regular Session*, 1: 311.

109. *Journal of the Assembly*, 1970, 1: 308.

110. Edelson, *Battle*, 20.

111. Stern to Hatch, Dec. 16, 1959, NA, RG 90 A 1, entry 36, box 29, "Public Relations I—Employment, Non-Government" folder.

112. Wolfgang Langewiesche, "How Polluted Is the Air Around Us?" *Reader's Digest* 83 (Sept. 1963): 121n; Division of Air Pollution Control, "Review of the New Jersey Air Pollution Control Program," 7.

113. "Testimony of S. Smith Griswold . . . ," 2; "Where We Stand," 5.

114. Caye to Young, Oct. 9, 1952, JAFC, box 26 B III 5 a ff, folder 3.

115. "Report of Special Committee on Air Pollution," Dec. 5, 1953, p. 2, JAFC, box 25 B III 5 a cc, folder 4; "Smog Data Summarized. by Harold H. Story at the Request of Pure Air Committee, Inc.," Nov. 1954, p. 2, JAFC, box 25 B III 5 a cc, folder 5; "Clean Air for California," 43; Randal F. Dickey, "The Dismal Future of Smog Control," Apr. 12, 1955, p. 3, UCLA, coll. 1108, box 38, folder 1; Martin A. Brower, "Needed—a 1¾ Billion-HP Fan to Blow Away Los Angeles Smog," *Electrical West* 112, no. 4 (1954): 78–81.

116. Larson to Braun, Dec. 5, 1952, JAFC, box 26 B III 5 a ff, folder 3.

117. Edwin Hubble, "Smog: Talk before Sunset Club," Nov. 29, 1950, p. 3, Huntington Library, HUB 64; Tucker to Ainsworth, Apr. 3, 1947, RRTSAC, ser. 2, box 1, "Ainsworth, E., Los Angeles Times" folder; Los Angeles County Smoke and Fumes Commission, "First Annual Report to the Los Angeles County Board of Supervisors," Oct. 24, 1944, p. 3, JAFC, box 25 B III 5 a aa, folder 2; Arnow to Hahn, Oct. 8, 1953, KHC, box 266, folder 1a; "News from Kenneth Hahn," Jan. 11, 1954, p. 1, KHC, box 266, folder 1b.

118. Daugherty to Hahn, Jan. 18, 1954, KHC, box 266, folder 1a; White to Ford, Oct. 15, 1953, JAFC, box 25 B III 5 a cc, folder 4; "Report of Special Committee on Air Pollution," Dec. 5, 1953, pp. 5, 8, JAFC, box 25 B III 5 a cc, folder 4; Turner to Ford, Jan. 21, 1954, JAFC, box 25 B III 5 a dd; "Report of the Subcommittee of the Assembly Interim Committee on Governmental Efficiency and Economy Relative to a Study and Analysis of the Facts Pertaining to Air Pollution Control in Los Angeles County," pp. 5, 16n, UCLA, coll. 1108, box 39, folder 1; Greenburg to Jessup, Dec. 13, 1954, JAFC, box 26, B III 5 a gg; Brienes, "Fight," 225.

119. Will to Fuller, Dec. 30, 1954, JAFC, box 26 B III 5 a gg.

120. Air Pollution Control District, "Annual Report 1954–55," p. 20, KHC, box 266, folder 1b.

121. Will to Fuller, Dec. 30, 1954.

122. Harold Lahn Sims, "The Emergence of Air Pollution as a Political Issue in Southern California: 1940–1970" (Ph.D. diss., University of California Riverside, 1973), 315n.

123. "Where We Stand," 4.

124. Brown to Dorn, Nov. 25, 1958, UCLA, coll. 1108, box 37, folder 3.

125. "S.O.S. before the Air Resource Board, San Diego, for Immediate Release," June 18, 1968, CSA, F 3935, folder 154; Prindle to Chief, Bureau of State Services, June 28, 1960, NA, RG 90 A 1, entry 36, box 9, "Association N–Z" folder; Bridgeman to Richards, Mar. 12, 1959, UCLA, coll. 1108, box 39, folder 1; Casady to Richards, Feb. 29, 1960, UCLA, coll. 1108, box 39, folder 5.

126. Citizens' Anti-Smog Action Committee to Board of Supervisors, Feb. 12, 1958, UCLA, coll. 1108, box 38, folder 2; Packard to Richards, Mar. 1, 1957, UCLA, coll. 1108, box

38, folder 3; Packard to Richards, Feb. 20, 1959, UCLA, coll. 1108, box 39, folder 1; "Report of Special Committee on Air Pollution," Dec. 5, 1953, p. 6, JAFC, box 25 B III 5 a cc, folder 4.

127. Air Pollution Foundation, *Final Report,* 48.

128. Howard J. Carswell, "A New Look for the Chimney," *Public Utilities Fortnightly* 44 (1949): 209.

129. "Clean Air for California," 5; California Department of Health, "Report III: A Progress Report of California's Fight against Air Pollution," Feb. 1957, p. 10, UCLA, coll. 1108, box 8, folder 49; Frank M. Stead, "Air Pollution—A Threat to Our Air Resources," *Looking Ahead* 5, no. 4 (1957): 2; Contra Costa County Health Department, "Air Pollution in the Bay Area," June 1, 1956, ERWP, box 54, folder 75.

130. "Testimony of S. Smith Griswold . . . ," 4; *Engineering News-Record* 155, no. 16 (1955): 35.

131. Michigan Department of Public Health, "Air Pollution," 1.

132. Edward L. Stockton, "How Pittsburgh Coped with Its Air Pollution Problem," Mar. 4, 1967, p. 9, HSWPA, mss. 285, "Smoke Control—General" file.

133. "Fuel Consumption for 1957 in the Chicago Metropolitan Area," 13, IMAR, box 106, folder 17.

134. Air Pollution Control District, "Annual Report 1956–57," pp. 4, 16, UCLA, coll. 1108, box 37, folder 4; "News from Kenneth Hahn," 1; Helen B. Shaffer, "Cleaner Air," *Editorial Research Reports* (1959): 28.

Chapter 7

1. "Report to the Los Angeles County Board of Supervisors by the L. A. County Smoke and Fumes Commission," Mar. 13, 1944, p. 3, JAFC, box 25 B III 5 a aa, folder 3.

2. Brienes, "Fight," 164, 192; Krier and Ursin, *Pollution,* 73.

3. "Report of Special Committee on Air Pollution," Dec. 5, 1953, p. 13, JAFC, box 25 B III 5 a cc, folder 4.

4. "News from Kenneth Hahn," 1.

5. Linville to Larson, Nov. 18, 1954, p. 2, KHC, box 266, folder 1a; Lee to Gaudaen, Mar. 7, 1955, KHC, box 266, folder 3a.

6. Griswold to Board of Supervisors, Sept. 19, 1956, KHC, box 266, folder 3b.

7. "Statement by S. Smith Griswold before the National Advisory Committee on Community Air Pollution," Washington, D.C., Feb. 12, 1958, p. 4, KHC, box 266, folder 4b.

8. "Interim Report of the Smog Committee of the 1958 Grand Jury," July 24, 1958, p. 3, KHC, box 22, folder 1.

9. Hart to Hollis, Feb. 24, 1959, p. 1, NA, RG 90 A 1, entry 36, box 11, "Committee A–M" folder.

10. "Correspondence between Kenneth Hahn, Los Angeles County Board of Supervisors and Automobile Manufacturers since February 1953 Concerning Automotive Exhaust Air Pollution Control," Aug. 1962, p. 1, KHC, box 266, folder 5b.

11. Campbell to Board of Supervisors, Feb. 2, 1954, KHC, box 266, folder 1b.

12. Campbell to Roosevelt, Oct. 23, 1956, p. 2, KHC, box 266, folder 3b.

13. "Report by Automobile Manufacturers Association on Air Pollution Control Devices for Motor Vehicles before the California State Legislature: Transcript of Proceedings," Feb. 16, 1959, pp. 19–21, 58n, UCLA, coll. 1108, box 39, folder 2.

14. Ludwig to MacKenzie, Nov. 27, 1963, NA, RG 90 A 1, entry 11, box 41, "OCC—Automobile Manufacturing Association" folder.

15. Hahn, "Zur großstädtischen Verkehrshygiene," 233; Georg Wolff, "Die Bekämpfung der Automobilauspuffplage," *Die Städtereinigung* 5 (1913): 123.

16. Arnold Marsh, *Smoke: The Problem of Coal and the Atmosphere* (London, 1947), 277.

17. "Clean Air for California," 42.

18. Daugherty to Hahn, Jan. 21, 1954, p. 3, KHC, box 266, folder 1a; "Draft History Bureau of Mines," p. 1530n, NA, RG 70 A 1, entry 10, box 1; Hazel Holly, *What's in the Air?* (New York, 1958), 15.

19. Krier and Ursin, *Pollution*, 88.

20. "Resolution of Los Angeles County Board of Supervisors," Jan. 26, 1965, p. 2, KHC, box 306, folder 15; "APCD News," Jan. 19, 1965, UCLA, coll. 1675, box 60, "News Releases 1965" folder. On the interlocking agreement, see Krier and Ursin, *Pollution*, 87; John C. Esposito and Larry J. Silverman, *Vanishing Air: The Ralph Nader Study Group Report on Air Pollution* (New York, 1970), 41.

21. "Statement by S. Smith Griswold," 7.

22. "Resolution of Los Angeles County Board of Supervisors," 3.

23. For an exhaustive discussion of the lawsuit, see Esposito and Silverman, *Vanishing Air*, 42–47; Scott H. Dewey, "'The Antitrust Case of the Century.' Kenneth F. Hahn and the Fight against Smog," *Southern California Quarterly* 81 (1999): 353–68.

24. Opie, *Nature's Nation*, 457; Jack Doyle, *Taken for a Ride: Detroit's Big Three and the Politics of Pollution* (New York and London, 2000), chap. 3; Dewey, *Don't Breathe*, 78–80.

25. Kenneth Hahn, *A Factual Record of Correspondence between Kenneth Hahn, Los Angeles County Supervisor and the Presidents of General Motors, Ford and Chrysler Regarding the Automobile Industry's Obligation to Meet Its Rightful Responsibility in Controlling Air Pollution from Automobiles* (n.l., 1966), 1.

26. Hahn to Townsend, Aug. 5, 1963 (quotation), KHC, box 266, folder 5b; Nesvig to Townsend, Nov. 22, 1963, KHC, box 266, folder 5b.

27. Hahn to Mitchell, Sept. 4, 1969, p. 2, KHC, box 306, folder 17.

28. Hahn to Mitchell, Feb. 16, 1970, p. 1, KHC, box 304, folder 1b.

29. "Report by Automobile Manufacturers Association," 8.

30. "Report by Automobile Manufacturers Association," 9.

31. "Report by Automobile Manufacturers Association," 26.

32. "Report by Automobile Manufacturers Association," 62.

33. "Report by Automobile Manufacturers Association," 46; Hart to Hollis, Feb. 24, 1959, p. 1, NA, RG 90 A 1, entry 36, box 11, "Committee A–M" folder.

34. Hart to Hollis, Feb. 24, 1959, p. 1; "Report by Automobile Manufacturers Association," 19, 37n.

35. Prindle to Deputy Chief, Bureau of State Services, July 15, 1960, p. 2, NA, RG 90 A 1, entry 36, box 9, "Association—Automobile Manufacturers Assoc." folder.

36. Prindle to Deputy Chief, Bureau of State Services, July 15, 1960, 1n.

37. Prindle to Deputy Chief, Bureau of State Services, July 15, 1960, 2.

38. Matthew Zafonte and Paul Sabatier, "Short-Term Versus Long-Term Coalitions in the Policy Process: Automotive Pollution Control, 1963–1989," *Policy Studies Journal* 32 (2004): 89. On the Advocacy Coalition Framework, see Paul A. Sabatier, "An Advocacy Coalition Framework of Policy Change and the Role of Policy-Oriented Learning Therein," *Policy Sciences* 21 (1988): 129–68.

39. Williams to Hahn, May 17, 1961, p. 3, KHC, box 266, folder 5a.

40. Uekoetter, "Merits."

41. At Griswold's presentation at the June 1964 conference of the Air Pollution Control Association, he chastised the automobile manufacturers' "arrogance and apathy" (S. Smith Griswold, "Reflections and Projections on Controlling the Motor Vehicle," p. 5, CSA, F 3935, folder 150).

42. Hahn, *Factual Record*.

43. Linville to Larson, Nov. 18, 1954, p. 2, KHC, box 266, folder 1a.

44. Krier and Ursin, *Pollution*, 138.

45. "Statement by S. Smith Griswold," 6n.

46. Revealingly, the Los Angeles County Board of Supervisors had previously considered initiating a lawsuit against the automobile industry to mandate sufficient cleaning equipment (Nesvig to Air Pollution Control Officer, Oct. 8, 1964, KHC, box 266, folder 6).

47. "Hearing of Jan. 26, 1965 before the County Board of Supervisors in the Matter of Resolution in re Motor Vehicle Air Pollution," p. 10, KHC, box 266, folder 7a.

48. Blankenship to Hunter, Feb. 25, 1959, NA, RG 90 A 1, entry 36, box 11, "Committee A–M" folder.

49. Krier and Ursin, *Pollution*, 141.

50. *Nation's Business* 56, no. 9 (1968): 58.

51. Attachment to Taylor to Berry, Aug. 11, 1969, UCB, BANC MSS 71/103 c, cont. 122, folder 2.

52. League of Women Voters of Los Angeles, "Study Kit Air Pollution—Environmental Quality Study," Jan. 1971, p. 17, LWVLAC, box II 12, folder 9.

53. Andrews, *Managing*, 234; "Statement by Wilbur J. Cohen before the Subcommittee on Science, Research, and Development, U.S. House of Representatives, Committee on Science and Astronautics," July 21, 1966, p. 14, NA, RG 90 A 1, entry 11, box 35, "LL-3—Hearings" folder.

54. Donald E. Carr, *The Breath of Life* (New York, 1965), 143; Esposito and Silverman, *Vanishing Air*, 26.

55. Howard R. Lewis, *With Every Breath You Take: The Poisons of Air Pollution, How They Are Injuring Our Health, and What We Must Do about Them* (New York, 1965), 252. See also Alan P. Loeb, "Paradigms Lost: A Case Study Analysis of Models of Corporate Responsibility for the Environment," *Business and Economic History* 28, no. 2 (1999): 104.

56. J. A. Holmes, Edward C. Franklin, and Ralph A. Gould, *Report of the Selby Smelter Commission*, U.S. Bureau of Mines Bulletin 98 (Washington, 1915); Uekoetter, "Merits," 127.

57. Snyder, "Death-Dealing Smog," 6; Dyck, "Evolution," 22; Doyle, "Polluted Air," 863; George D. Clayton, "Objectives of the Detroit-Windsor Air Pollution Study," *Public Health Reports* 67 (1952): 658; McCabe, *Air Pollution*, v–vi. Dewey's argument for continuous federal air pollution activity since Donora is incorrect (*Don't Breathe*, 236).

58. Randall B. Ripley, "Congress and Clean Air: The Issue of Enforcement, 1963," in Frederic N. Cleaveland, ed., *Congress and Urban Problems* (Washington, 1969), 229–31; Bailey, *Congress*, 84.

59. Jones, *Clean Air*, 36, 110; Ripley, "Congress," 238, 240; Dyck, "Evolution," 192, 232. In 1966, the PHS Division of Air Pollution became the National Center for Air Pollution Control, which became the National Air Pollution Control Administration within the Department of Health, Education, and Welfare in 1968. Since then, the National Air Pollution Control Administration has no longer formally been a part of the PHS (J. Clarence Davies III, *The Politics of Pollution* [New York, 1970], 104). In order to avoid confusion over terminology, the narrative speaks consistently of PHS officials.

60. Maga, "State Approach," 1662; Mallette, "Legislation," 1073; Dyck, "Evolution," 33; Snyder, "Death-Dealing Smog," 279.

61. Ripley, "Congress," 231; Bailey, *Congress*, 96; Davies, *Politics*, 51.

62. Shaffer, "Cleaner Air," 37.

63. Price to Surgeon General, Feb. 6, 1959, p. 2, NA, RG 90 A 1, entry 36, box 24, "Legal 2—S 441" folder; Ripley, "Congress," 237; Jones, *Clean Air*, 32.

64. Krier and Ursin, *Pollution*, 110, 169.

65. Snyder, "Death-Dealing Smog," 197, 276, 334; Ripley, "Congress," 232n, 246f; Krier and Ursin, *Pollution*, 111; Jones, *Clean Air*, 35; Bailey, *Congress*, 100.

66. Program Officer to Chief, Division of Air Pollution, Oct. 10, 1960, p. 1, NA, RG 90 A 1, entry 36, box 3, "Pollution" folder.

67. MacKenzie to "The file," Apr. 3, 1959, NA, RG 90 A 1, entry 36, box 24, "Legal 2—S 441" folder; Ripley, "Congress," 234; Dyck, "Evolution," 218n.

68. Ripley, "Congress," 248.
69. Krier and Ursin, *Pollution*, 173; Dyck, "Evolution," 51, 146, 194; Jones, *Clean Air*, 120n; Bailey, *Congress*, 108.
70. "Statement by Arthur C. Stern," 11; Ralph C. Graber, "Trends in Air Pollution Control—The National Picture," Dec. 1, 1965, p. 5, NA, RG 90 A 1, entry 11, box 37, "OCC-1 Acceptance Part 4" folder; L. J. Carter, "Air Pollution: Federal Standards Likely unless States and Localities Take Early Action," *Science* 150 (1965): 468.
71. Thomas O. Harris, "A Preliminary Appraisal of the Air Resources of the State of Maine," Oct. 1965, pp. 22–25, NA, RG 90 A 1, entry 11, box 7, "CS+S General Part 5" folder; Stern to Udall, Mar. 31, 1965, NA, RG 90 A 1, entry 11, box 28, "IFF-11-1 Congressional Inquiries Part 3" folder; "Statement by Arthur C. Stern," 59; Vernon G. MacKenzie, "Emission Standards and Air Quality Standards in Relation to Urban Planning," pp. 224–26, NA, RG 90 A 1, entry 11, box 57, "Proceedings of the First New England Conference on Urban Planning for Environmental Health" folder; CGC Group 2, "Air Resource Management Program for Southwestern Ohio—Northern Kentucky, Cincinnati 1967," pp. 3–5.
72. "Statement by Arthur C. Stern," 64.
73. MacKenzie, "Emission Standards," 226.
74. Jones, *Clean Air*, 120n; Krier and Ursin, *Pollution*, 173. Federal officials called only a total of eight "enforcement conferences" before 1970 (Davies, *Politics*, 190).
75. Megonnell to Mathews, Apr. 24, 1964, NA, RG 90 A 1, entry 11, box 35, "Legislation 2-1, New York State Part 3" folder.
76. "Air Pollution Technical Information Survey, Final Report," 13.
77. "Program Officer to Chief, Division of Air Pollution," Oct. 10, 1960, p. 1.
78. Division of Air Pollution Control, "Review of the New Jersey Air Pollution Control Program," 11.
79. Johnson to Metcalf, Dec. 30, 1966, p. 6, NA, RG 90 A 1, entry 11, box 34, "LL-2-1—Bills and Acts Part 1" folder.
80. Kaiser to Stern, Sept. 9, 1964, NA, RG 90 A 1, entry 11, box 26, "IF-9—Guide to Good Practice" folder.
81. Megonnell to "Files," May 7, 1965, NA, RG 90 A 1, entry 11, box 26, "IF-9—Guide to Good Practice" folder.
82. Vernon G. MacKenzie, "Air Pollution Control—Industry's Responsibility," Dec. 8, 1966, p. 8, NA, RG 90 A 1, entry 11, box 40, "OCC—American Petroleum Institute Part 3" folder; "Statement by Arthur C. Stern," 13.
83. Vernon G. MacKenzie, "Conservation of the Air Resource: A National Challenge," Jan. 13, 1965, p. 3, NA, RG 90 A 1, entry 11, box 37, "OCC-1 Acceptance Part 4" folder.
84. "Air Pollution Technical Information Survey, Final Report," 8.
85. John Hornbeak, "Memorandum on Pennsylvania Proposed Air Quality Standards," Aug. 15, 1969, HSWPA, mss. 285, "Smoke Control—General" file; Morris to "Breathing Citizens of Greater Washington D.C.," Sept. 30, 1970, p. 1, UVA, call no. 10030, box 1, "Air Quality—Virginia (folder # 3)"; Garrison to "News Media," Mar. 17, 1970, MCACR, box 1, "General" folder, no. 1; "CF Letter: A Report on Environmental Issues from the Conservation Foundation," Jan. 31, 1968, p. 1, OSA, ACC 70A-095, box 38, "Pollution—Air" folder; Jaskulski and Jacobs to Voigt, Nov. 24, 1970, SHSW, ser. 2550, box 131, folder 1; Graham to Carnow, May 20, 1969, BML, FC 12, Helen T. Graham Papers, ser. 7, 12/3/1/2.
86. Environmental Protection Agency, *The Clean Air Act* (Washington, 1970), 10; Richard H. K. Vietor, *Environmental Politics and the Coal Coalition* (College Station and London, 1980), 161–67.
87. Bauer to Surgeon General, May 11, 1960, p. 2, NA, RG 90 A 1, entry 36, box 24, "Legal 2—S 441" folder; NAM Area Industrial Problems Committee, Facilities Development Subcommittee, "Air Pollution Control, Adopted," Apr. 3, 1963, HML, acc. 1411, ser. I, box 102, "Positions—Air Pollution" folder; "News from National Association of Manufacturers,"

Dec. 10, 1962, HML, acc. 1411, ser. I, box 30, "Cinderella City" folder; "Minutes of the Meeting of the Informal Clean Air Committee," Feb. 20, 1964, p. 9, IMAR, box 107, folder 3; Illinois Manufacturers' Association, "Policy Regarding Air and Water Pollution Control," June 1967, p. 2, IMAR, box 108, folder 3; Kuebler, "Air Contamination," 18; Vietor, *Environmental Politics*, 44, 134; Richard H. K. Vietor, "The Evolution of Public Environmental Policy: The Case of 'No-Significant Deterioration,'" in Char Miller and Hal Rothman, eds., *Out of the Woods: Essays in Environmental History* (Pittsburgh, 1997), 129; Ripley, "Congress," 255, 263; Dyck, "Evolution," 57n.

88. Krier and Ursin, *Pollution*, 174n; Dyck, *Evolution*, 249; Jones, *Clean Air*, 67; Carter, "Air Pollution," 468; Davies, *Politics*, 54.

89. Vernon G. MacKenzie, "Management of Our Air Resources," Oct. 7, 1963, p. 7, IMAR, box 107, folder 4; "Clean Air for Your Community," Public Health Service Publication no. 1544, rev. 1967, rpt. 1969, p. 7, DBC, box 1, folder 14.

90. Holmes to Griswold, May 4, 1966, p. 3, NA, RG 90 A 1, entry 11, box 34, "LL-2-1—Bills and Acts Part 2" folder.

91. Williams to Williams, Aug. 31, 1966, and Williams to Anderson, Jan. 6, 1966, both NA, RG 90 A 1, entry 11, box 54, "OCC–'N'" folder; Stern to Chief, Bureau of State Services, Jan. 10, 1964, p. 2n, NA, RG 90 A 1, entry 11, box 54, "OCC–National Advisory Committee on Air Pollution 1963 Part 1" folder; Levine to Beaty, May 5, 1966, NA, RG 90 A 1, entry 11, box 53, "OCC–'M'" folder.

92. Esposito and Silverman, *Vanishing Air*, 22.

93. Vernon G. MacKenzie, "The Citizens' Role in Air Pollution Control," May 12, 1965, p. 11, NA, RG 90 A 1, entry 11, box 37, "OCC-1 Acceptance Part 5" folder.

94. "Public Awareness and Concern with Air Pollution in the St. Louis Metropolitan Area: Summary Report," Mar. 1964, pp. 12–16 (quotations on 16, 13), NA, RG 90 A 1, entry 11, box 14, "Contracts—Southern Illinois Univ. PH 86-63-131" folder.

95. G.A.S.P., Inc., Washington, D.C., "For Release: July 13, 1969," UVA, call no. 10030, box 22, "D. C. Air Quality" folder; UVA, call no. 10030, box 26, "Greater Washington Alliance to Stop Pollution (GASP)" folder; "Gals Against Smog and Pollution, Missoula, Montana," UMM, mss. 43; "Group Against Smog and Pollution, Pittsburgh," HSWPA, mss. 43; League of Women Voters of Los Angeles, "Study Kit Air Pollution—Environmental Quality Study," Jan. 1971, p. 30, LWVLAC, box II 12, folder 9.

96. Ohio Pure Air Association, "News Release," Dec. 19, 1960, WRHS, mss. 4074, folder 4.

97. "Rebuttal Testimony of Mrs. Chauncy D. Harris," 6.

98. Charles Schaeffer and Art Cosing, "How Much Poison Are You Breathing?" *Harper's Magazine* 219 (October 1959): 64.

99. Wiles to Romney, Sept. 30, 1966 (emphasis original), GWRGP, box 151, "Air Pollution" folder.

100. Shaffer, "Cleaner Air," 23.

101. *Clearing the Air: Monthly Newsletter of the Metro Clean Air Committee* 1, no. 8 (1969), p. 1, MnHS, microfilm 996.

102. Coalition for the Environment, St. Louis Region, "Statement of Purpose," May 13, 1969, BML, FC 12, Helen T. Graham Papers, ser. 7, 12/3/1/7.

103. "News Letter Clear Air Clear Water Unlimited," June–July 1958, p. 1, MnHS, microfilm 1594. On the issue of nuclear fallout, see Allan M. Winkler, *Life under a Cloud: American Anxiety about the Atom* (Urbana and Chicago, 1999), 84–108.

104. Sloane, "How to Protect," 55.

105. "Needed: Clean Air: The Facts about Air Pollution," brochure, p. 2, GPMR, box 12, "Air Pollution" folder.

106. Hyde Park–Kenwood Community Conference, "Report of the Air Pollution Committee," June 13, 1966, p. 1, HPKCCP, ser. 2, box 10, folder 3.

107. Conservation Foundation, "CF Commentary," July 15, 1966, p. 1, MUA, Milwaukee, ser. 44, box 2, folder 12.

108. Todd Gitlin, *The Sixties: Years of Hope, Days of Rage* (New York, 1987), 195.

109. Maurice Isserman and Michael Kazin, *America Divided: The Civil War of the 1960s* (New York and Oxford, 2000). Adam Rome recently noted the dearth of links between the history of environmentalism and narratives of the 1960s ("Give Earth," 525n).

110. Ellery R. Fosdick, "The Pollution of Man's Environment," *National Parks Magazine* 40, no. 228 (1966): 16.

111. Civic Club of Allegheny County, *Fifteen Years of Civic History: October 1895–December 1910* (n.l., n.d.), 33.

112. Civic Club of Allegheny County, *Fifty Years*, 5–7; O'Connor, *Some Engineering Phases*, 17; Robert Dale Grinder, "From Insurgency to Efficiency: The Smoke Abatement Campaign in Pittsburgh before World War I," *Western Pennsylvania Historical Magazine* 61, no. 3 (1978): 194; Iams to Veditz, Mar. 18, 1912, AIS, 83:7, ser. I, FF 1; "Members of the Smoke and Dust Abatement League—1913," AIS, 83:7, ser. III, FF 2; "Hearing before the Pittsburgh Smoke Commission," Mar. 28, 1941, esp. p. 42n, PCCA, Proceedings; Civic Club of Allegheny County, "Directory of War Activities," Oct. 1942, p. 18, HSWPA, MFF 513; David N. Kuhn, Secretary, United Smoke Council, "Smoke Elimination via the United Smoke Council," p. 1, HSWPA, mss. 285, "Smoke Control—General" file; F. E. Schuchman, "How a City Is Losing Its Nickname!" *Smaller Manufacturer* 11, no. 10 (1956): 11.

113. *Monthly Bulletin, Civic Club, Allegheny County* 3 no. 3 (1969): 3, HSWPA, mss. 122, box 1, folder 5.

114. Gordon D. Rowe, *History of the Anti Smoke Movement in Cincinnati, Ohio* (n.l., n.d. [1921?], 1; Charles N. Howison, "The History of a Citizens Organization for Cleaner Air," July 27, 1954, p. 1, CHS, mss. qA 298, box 1, folder 4.

115. Howison to Blu, Feb. 4, 1969, p. 5, CHS, mss. qA 298, box 1, folder 2.

116. Howison to Johnson, Aug. 27, 1965, NA, RG 90 A 1, entry 11, box 67, "PO–General Part 3" folder.

117. Charles N. Howison, "Cleaner Air: The Community Commitment to the Seventies," Oct. 29, 1970, p. 6, CHS, mss. qA 298, box 1, folder 4.

118. "Statement to the Atomic Energy Commission in the Matter of the Environmental Impact of the William H. Zimmer Nuclear Power Station by the Air Pollution Control League of Greater Cincinnati," Sept. 20, 1972, RAKA, box 94, grouping 4. Kehoe, in a remarkable statement for an advocate of tetraethyl lead, argued: "It is exceedingly difficult to develop uniformly safe practices of satisfactory control in the performance of the day's work in many hazardous situations in present-day industry.... Human performance in these matters is notoriously unreliable when considered over prolonged periods of time" (Kehoe to Howison, May 17, 1979, RAKA, box 104, grouping 6).

119. "Minutes of the Board of Trustees of the Air Pollution Control League," Dec. 8, 1981, p. 3, and Hansen and Howison to Waxman, May 21, 1981, both RAKA, box 105, grouping 2.

120. "Citizen Complaints of Air Pollution—Northeastern Illinois," Apr. 1, 1966, p. 3, NIPCR, box 83, folder 3.

121. Madar to "Citizens Concerned with Environmental Issues," Mar. 23, 1970, and "Poison in the Air: National Conference on Air Pollution," Mar. 25–26, 1969, p. 8, both MCACR, box 2, "Unions" folder.

122. Andrew Hurley, *Environmental Inequalities: Class, Race, and Industrial Pollution in Gary, Indiana, 1945–1980* (Chapel Hill and London, 1995), 138.

123. Coalition for the Environment, St. Louis Region, *Alert* 1, no. 9 (1971): 2, BML, FC 12, Helen T. Graham Papers, ser. 7, 12/3/1/7.

124. Stokes to Brown, July 17, 1970, WRHS, mss. 4370, box 21, folder 367.

125. Brockway to Romney, June 19, 1966, GWRGP, box 151, "Air Pollution" folder; Bosselman to Harris, Feb. 25, 1964, CACR, folder 5. Tellingly, the minutes omitted the presentation,

since it would have required filing the handkerchief "as an exhibit," and the lawyers found the stuff simply "too awkward to handle."

126. Shaffer, "Cleaner Air," 21; Harry Heimann et al., "Progress in Medical Research on Air Pollution," *Public Health Reports* 73 (1958): 1055–69; Clarence A. Mills, "Air Pollution and Community Health," *American Journal of the Medical Sciences* 224 (1952): 403, 406; Clarence C. Mills, "Motor Exhaust Gases and Lung Cancer in Cincinnati," *American Journal of the Medical Sciences* 239 (1960): 316; Paul Kotin, "Organization and Operation of a Study of Air Pollution in Relation to Experimental Carcinogenesis," *A.M.A. Archives of Industrial Hygiene and Occupational Medicine* 5 (1952): 553; Paul Kotin, "Air Pollution and Lung Cancer," *California's Health* 11 (1953–54): 134; Lester Breslow and John Goldsmith, "Health Effects of Air Pollution," *American Journal of Public Health and the Nation's Health* 48 (1958): 914n; Vincent Marteka, "Air Pollution—a Growing Menace," *Science News Letter* 80 (1961): 10n; Moyer D. Thomas, "New Understandings from Current Atmospheric Pollution Research," *American Journal of Public Health and the Nation's Health* 49 (1959): 1168; American Medical Association, "Report of the Board of Trustees, AMA Position on Air Pollution," p. 1, NA, RG 90 A 1, entry 11, box 39 "OCC–AMA (American Medical Association)" folder.

127. James H. Cassedy, *Medicine in America: A Short History* (Baltimore and London, 1991), 128n; Elizabeth Fee, "The Origins and Development of Public Health in the United States," in Walter W. Holland, Roger Detels, and George Knox, eds., *Oxford Textbook of Public Health*, vol. 1, *Influences of Public Health*, 2nd ed., (New York, 1991), 13.

128. "Letter from Buhrer Air Pollution Civic Association," Sept. 1961, WRHS, mss. 4074, folder 2; *Buhrer Buzz: Official Newsletter of the Buhrer Air Pollution and Civic Association* (Spring 1962), WRHS, mss. 4456, box 7, folder 114.

129. "Statement of Organization, Ohio Pure Air Association," Dec. 10, 1960, WRHS, mss. 4074, folder 4; "Clear Air Clear Water Unlimited, Who—Where—Why," brochure (n.d. [ca. 1958]), MnHS, microfilm 1594.

130. Churchill to Public Health Service, June 5, 1959, NA, RG 90 A 1, entry 36, box 22, "Information 6—Public Enemy and the Troubled Air" folder; Ferguson to Chief, Division of Air Pollution, Dec. 6, 1960, NA, RG 90 A 1, entry 36, box 9, "Association A–M" folder.

131. *National Tuberculosis Association News*, Nov. 28, 1966, and "Minutes, National Air Conservation Commission," Nov. 28–29, 1966, with attachments, both NA, RG 90 A 1, entry 11, box 54, "OCC–'N'" folder; Delaware Valley Citizens' Council for Clean Air, "Join the Clean Air Corps," brochure (n.d.), GPMR, box 12, "Air Pollution" folder; "Cleaner Air Committee Meeting Minutes," Oct. 31, 1967, and "Metro Clean Air Committee Minutes," Dec. 5, 1967, both MCACR, box 1, "General" folder, no. 1; Denomme to Prindle, Mar. 5, 1965, NA, RG 90 A 1, entry 11, box 59, "OCC–'R'" folder.

132. Hudson River Conservation Society, resolution, June 18, 1959, and Hader to Desmond, June 22, 1959, both HRCSP, cont. 48, "Smoke Abatement" folder. The Sierra Club did not voice an opinion on air pollution legislation until 1968 (Anthrop to Foran, June 21, 1968, UCB, BANC MSS 71/103 c, cont. 117, folder 2).

133. On Rachel Carson, see Linda Lear, *Rachel Carson: Witness for Nature* (New York, 1997). On the broader debate over pesticides, see Edmund Russell, *War and Nature: Fighting Humans and Insects with Chemicals from World War I to Silent Spring* (Cambridge, 2001); Christopher J. Bosso, *Pesticides and Politics: The Life Cycle of a Public Issue* (Pittsburgh, 1987).

134. Bielen to Southeast Community Council, Feb. 18, 1960, WRHS, mss. 4074, folder 6.

135. "News Letter Clear Air Clear Water Unlimited," Jan. 1958, p. 4, MnHS, microfilm 1594; Birch to Prouty, Apr. 20, 1968, UMM, mss. 43, box 1, folder 8.

136. "Minutes of the Metro Clean Air Committee General Meeting," Dec. 17, 1968, p. 3, MCACR, box 1, "General" folder, no. 1; Delaware Valley Citizens' Council for Clean Air, "Join the Clean Air Corps."

137. Ohio Pure Air Association, "For Immediate Release," Nov. 7, 1965, WRHS, mss. 4074, folder 4; Bielen to Perk, May 22, 1961, WRHS, mss. 4456, box 7, folder 113.

138. National Air Conservation Commission, "Air Conservation Program Guide for TB-RD Associations," Apr. 1967, pp. 5 (quotation), 8, MCACR, box 2, "Air Pollution—NACC" folder.

139. Lewis, *Breath*, 266; *Buhrer Buzz* (Spring 1962).

140. Betz to Penjerdel, Nov. 20, 1963, PNJDMPR, box 1, URB 24/I/1.

141. Hader to Dalliba, Aug. 18, 1959, HRCSP, cont. 1, "Air Pollution" folder; "Report of the Joint Air Pollution Study Committee," Oct. 28, 1968, p. 8, IMAR, box 109, folder 1; Laura Fermi, "How Our Community Is Fighting Air Pollution," *New City: Man in Metropolis: A Christian Response* 1, no. 16 (1963): 6.

142. Conservation Foundation, "CF Commentary," July 15, 1966, p. 1, and attachment to Gubser to MacKenzie, Oct. 6, 1966, p. 3, both MUA, Milwaukee, ser. 44, box 2, folder 12.

143. Marteka, "Air Pollution," 10; *Newsletter of the Committee for Clean Air Now*, May 1, 1968, p. 3, UCB, BANC MSS 71/103 c, cont. 117, folder 2; Metro Clean Air Committee, "General Meeting Minutes," Sept. 16, 1969, p. 2, MCACR, box 1, "General" folder, no. 1.

144. "Clear Air Clear Water Unlimited: Who—Where—Why."

145. Despres to Smith, Feb. 16, 1961, LMDP, box 86, folder 6.

146. MnHS, microfilm 1594, *Newsletter Clear Air Clear Water Unlimited* 8, no. 5 (1969): 3.

147. *Dun's Review and Modern Industry* 81, no. 3 (1963): part 2, 181.

148. Havens to Heustis, Sept. 29, 1966, GWRGP, box 151, "Air Pollution" folder. See also Illinois Manufacturers' Association, For Release "As Desired," Oct. 14, 1966, p. 2n, IMAR, box 108, folder 2; Introductory Remarks by David Ferguson at Clean Air Committee Panel Discussion on Air Pollution Legislation, Oct. 23, 1967, p. 3, NIPCR, box 60, folder 6; C. William Hardell, "Air Quality and Public Understanding," Nov. 13, 1968, p. 7, HML, acc. 1411, ser. I, box 32, "AIP Point Clear—Speakers' Material" folder; Decker to Terry, June 27, 1963, NA, RG 90 A 1, entry 11, box 54, "OCC—Manufacturing Chemists Association, Inc." folder.

149. Nichols to Richards, Jan. 26, 1960, UCLA, coll. 1108, box 40, folder 2.

150. Burroughs to Schueneman, Jan. 6, 1964, p. 5, NA, RG 90 A 1, entry 11, box 40, "OCC—American Petroleum Institute Part 1" folder; National Association of Manufacturers, *Cinderella City*, 7.

151. Great Northern Oil Company, *What Is Minnesota's Largest Refinery Doing to Keep the Air Clean and the Water Pure* (n.l., n.d. [ca. 1965]), 13; Wright to Hanson, Dec. 15, 1965, NA, RG 90 A 1, entry 11, box 26, "IF-8—Publications (outside) Part 3" folder.

152. "Minutes of the Meeting of the IMA Clean Air and Water Committee," Nov. 13, 1967, p. 6, IMAR, box 108, folder 3.

153. Kerryn King, "Crisis of Concern. Air and Water Pollution," *Public Relations Journal* 23, no. 7 (1967): 12.

154. *Interlake Reporter* 2, no. 8 (1965): 2, IMAR, box 107, folder 4. See also Statement by Jules Brunner in Behalf of the Illinois Manufacturers' Association before the Illinois Air Pollution Control Board, Sept. 23, 1965, p. 1, IMAR, box 107, folder 4; Statement of Orville V. Bergren before the Committee on Health, Chicago City Council, Feb. 16, 1968, p. 5, IMAR, box 109, folder 1; Wakesberg to Hope, Feb. 3, 1959, NA, RG 90 A 1, entry 36, box 10, "Association—National" folder; Lenher to Wood, Oct. 4, 1966, HML, acc. 1801, box 10, "Environmental Quality" folder; Davey to Merritt, June 26, 1967, p. 1, GDAH, 26-21-143, box 1; Kuebler, "Air Contamination," 11; King, "Crisis," 13; James E. Kussman, "What Is Industry's Role in Pollution Abatement?" *Paper Trade Journal* 151, no. 52 (Dec. 25, 1967): 30; Walter J. Campbell, "New Goals for Industry," *Steel* 159, no. 16 (1966): 38; *Oil, Paint and Drug Reporter* 175, no. 27 (June 29, 1959): 7, 39; *Nation's Business* 56, no. 9 (Sept. 1968): 60.

155. Ludwig to MacKenzie, Apr. 4, 1962, NA, RG 90 A 1, entry 11, box 40, "OCC—American Petroleum Institute Part 1" folder.

156. Decker to "All Interested in Industrial Attention to Air Pollution Control in Arizona, Colorado, Idaho, Montana, Nevada, New Mexico, Utah, and Wyoming," Jan. 28, 1966, NA, RG 90 A 1, entry 11, box 54, "OCC—Manufacturing Chemists Association, Inc." folder; "Statement of Orville V. Bergren before the Joint Air Pollution Study Committee," Chicago, Dec. 16, 1968, p. 6, IMAR, box 109, folder 1.

157. Attachment to Stern to Chief, Bureau of State Services, Jan. 10, 1964, p. 2, NA, RG 90 A 1, entry 11, box 54, "OCC—National Advisory Committee on Air Pollution 1963 Part 1" folder; "Minutes of Meeting, New York State Action for Clean Air Committee," May 23, 1963, and Dec. 19, 1963, both HRCSP, cont. 38, "New York State Clean Air Committee" folder; National Air Conservation Commission, "Good News—A Report of Air Conservation Activity by TB-RD Associations Around the Country," Nov. 1967, p. 3, MCACR, box 2, "Air Pollution—NACC" folder; attachment to Dyksterhouse to Chief, Air Pollution Engineering, Apr. 17, 1959, NA, RG 90 A 1, entry 36, box 12, "Committee—Ohio Valley Air Pollution Control Council" folder; "Pollution," 407; New York Business Council for Clean Air, "News Release," Feb. 5, 1967, p. 2, HML, acc. 1411, ser. I, box 32, "Position Paper" folder.

158. King, "Crisis," 13; Meredith, "Industrial Planning for Air Pollution Control," p. 231.

159. "Minutes of Meeting, Allegheny County Air Pollution Control Advisory Committee," June 22, 1959, p. 3, AIS, 92:3, add. 1997, box 1, FF 1.

160. "Ohio Pure Air Association," July 27, 1964, WRHS, mss. 4074, folder 4; Cleveland Division of Air and Stream Pollution, *Annual Report 1965* (Cleveland, 1966).

161. Frank Graham Jr., *Since Silent Spring* (Boston, 1970), 48–68.

162. Michigan Department of Public Health, "Air Pollution," 10.

163. "Report on Meeting with Great Lakes Steel, Reference to Program for Control of Air Pollution, Cadillac Square Building," Aug. 22, 1966, p. 6, GWRGP, box 151, "Air Pollution" folder.

164. Havens to Heustis, Sept. 29, Dec. 15, 1966, both GWRGP, box 151, "Air Pollution" folder; "Statement by E. Edgerton Hart before Chicago City Council Committee on Air Pollution," May 24, 1963, IMAR, box 107, folder 2; Illinois Manufacturers' Association, Air and Water Pollution Committee, "Comments on Emission Limits to Prevent Air Pollution," Jan. 28, 1966, p. 5n, IMAR, box 108, folder 1.

165. "Proceedings of Illinois Air Pollution Control Board at Its Meeting of Thursday, September 23, 1965 Regarding Air Quality Standards and Emission Limits," p. 4, IMAR, box 107, folder 4.

166. "Remarks of Thomas W. Hunter at the September 8 Meeting of the Allegheny Conference on Community Development," p. 3, and G. A. Webb, "Comments on the Proposed Allegheny County Rules and Regulations," Sept. 8, 1969, p. 8, both HSWPA, mss. 285, "Smoke Control—General" file.

167. Morton Corn, "Comments on Proposed Allegheny County Health Department, Rules and Regulations, Article XVII, Smoke and Air Pollution Control, Draft for Public Hearing," Sept. 17, 1969, HSWPA, mss. 285, "Smoke Control—General" file.

168. James J. Hanks and Harold D. Kube, "Industry Action to Combat Pollution," *Harvard Business Review* 44, no. 5 (1966): 54.

169. U.S. Senate, *Steps*, 25; Stockton, "Air and Water Pollution," 11.

170. National Coal Policy Conference, "A Plea for Reason" (ca. 1969), UCLA, coll. 1276, box 137.

171. Holmes to Griswold, May 4, 1966, p. 1, NA, RG 90 A 1, entry 11, box 34, "LL-2-1—Bills and Acts Part 2" folder; Corn, "Comments," 5.

172. Michigan Department of Public Health, "Air Pollution," 15n, 16 (quotation).

173. Kuebler, "Air Contamination," 10; Hanks and Kube, "Industry Action," 49; Abed, *Air Pollution Control Policy*, 23; *Steel* 158, no. 20 (1966): 49; *Nation's Business* 56, no. 9 (1968): 65; Cleveland Division of Air Pollution Control, "Annual Report 1967," p. 8, WRHS, mss. 4370, box 74, folder 1424.

174. Cleveland Division of Air Pollution Control, "Annual Report 1967," 10.
175. *Nation's Business* 55, no. 3 (1967): 67.
176. *Let's Clear the Air* 1, no. 10 (1968): 4, GPMR, box 12, "Air Pollution" folder.
177. *Steel* 158, no. 20 (1966): 49.
178. King, "Crisis," 13.
179. Meredith, "Industrial Planning for Air Pollution Control," 231; Standard Oil Company of California, "Clearing the Air . . . Clearing the Waters," p. 1, OSA, ACC 70A-095, box 39, "Pollution—General" folder.
180. "Statement of Orville V. Bergren before the Joint Air Pollution Study Committee," Chicago, Dec. 16, 1968, p. 7, IMAR, box 109, folder 1.
181. Minnesota Emergency Conservation Committee, "Immediate Release," Mar. 11, 1968, SJGP, box 2, "Kaupanger, Olin L." folder.
182. Lenher to Heads of Departments, Plant Managers, Air and Water Resources Committee, Mar. 8, 1967, p. 2, HML, acc. 1801, box 10, "Environmental Quality" folder.
183. Campbell, "New Goals," 38.
184. Robert T. Derr, "Writing Practical Air Quality Ordinances, Presented to Area Industrial Problems Committee, National Association of Manufacturers, Point Clear, Alabama," Nov. 13, 1968, p. 2, HML, acc. 1411, ser. I, box 32, "AIP Point Clear—Speakers' Material" folder.
185. Bergren to Klassen, Mar. 18, 1966, IMAR, box 108, folder 1.
186. Havens to Romney, Sept. 29, 1966, GWRGP, box 151, "Air Pollution" folder.
187. "Memorandum to Members of the IMA Clean Air and Water Committee," Sept. 2, 1966, p. 3, IMAR, box 108, folder 2. The other members came from state universities.
188. Julie Freedman, Recorder, "Granite City Air Pollution Control Board," Jan. 24, 1968, p. 2, BML, FC 12, Helen T. Graham Papers, ser. 7, 12/3/1/12.
189. Vietor, *Environmental Politics*, 149.
190. Dyck, "Evolution," 222, 253.
191. Hazel Erskine, "The Polls: Pollution and Its Costs," *Public Opinion Quarterly* 36 (1972): 121.
192. Marshall Bridgewater, "Memorandum on the September 9, 1969 Hearing on Proposed Air Pollution Control Standards," Sept. 12, 1969, p. 1, HSWPA, mss. 285, "Smoke Control—General" file.
193. "Transcript of Proceedings at the Hearing on Chapter 270-5-24, Air Quality Control," Dec. 18, 1970, pp. 28, 42, GDAH, 26-21-181, box 10, "Georgia Ambient Air Standards Dec. 8, 1970 Hearing" folder.
194. "Statement of Honorable Abner J. Mikva of Illinois before the Air Pollution Control Board on Proposed Ambient Air Quality Standards," Aug. 5, 1969, p. 1, LMDP, box 87, folder 1.
195. "PCA Hearing—Statement on SO and Particulates," Mar. 25, 1970, p. 1, MCACR, box 1, "General" folder, no. 1.
196. "Statement Made on August 12, 1969 at Southern Illinois University, Edwardsville, by Mrs. Ralph P. Bieber at the Hearing of the Illinois Air Pollution Control Board on the Proposed Ambient Air Quality Standards for the St. Louis Metropolitan Air Quality Control Region," p. 1, BML, FC 12, Helen T. Graham Papers, ser. 7, 12/3/1/12.
197. "Gazette—20 Sept. 69," UVA, call no. 9883, box 1, "Air Quality Committee of C. C. Va." folder.
198. Elster to Sierra Club, Oct. 20, 1969, UCB, BANC MSS 71/103 c, cont. 122, folder 3.
199. *Clearing the Air* (May 1970): 1.
200. Kaupanger to Bethlehem Steel Corporation, Apr. 22, 1970, SJGP, box 1, "Horn, Charles Lilley" folder.
201. "Statement of Honorable Abner J. Mikva before the Air Pollution Control Board on Proposed Ambient Air Quality Standards," Aug. 5, 1969, p. 3, LMDP, box 87, folder 1.

202. Gordon to Stokes, July 26, 1970, WRHS, mss. 4370, box 21, folder 367.

203. "U.S. Steel's Clairton Coke Works—The Trail of Bad Faith Winds On," p. 1, HSWPA, mss. 43, box 2, folder 4. On the Group Against Smog and Pollution, see James Lewis Longhurst, "'Don't Hold Your Breath, Fight for It!' Women's Activism and Citizen Standing in Pittsburgh and the United States, 1965–1975" (Ph.D. diss., Carnegie Mellon University, 2004).

204. *Cleveland Press*, Dec. 5, 1970.

205. John T. Middleton, "Air Conservation—Whose Responsibility: Presented at the New England Conference on Air Pollution, Waterville, Maine," Dec. 2, 1968, p. 7, UMM, mss. 43, box 4, folder 10.

206. Conservation Foundation, "Clean Air Project," Feb. 14, 1969, UCB, BANC MSS 71/103 c, cont. 131, folder 1; "September State Board Report 1970, Environmental Quality," p. 21, SHSW, Milwaukee, mss. 80, box 11, folder 6; Bonham to Howe, Sept. 27, 1969, CHS, mss. 602, box 9, folder 5; *National Air Conservation Commission Newsletter*, no. 15 (Dec. 1969): 3, UMM, mss. 43, box 2, folder 3; U.S. Department of Health, Education, and Welfare, "For Release in A.M. Papers," Dec. 20, 1968, UMM, mss. 43, box 4, folder 9; Taylor to Willingham, Aug. 13, 1970, and Wolfe and Radin to Williams, Sept. 22, 1970, both OHS, mss. 2386, box 19, folder 10; Becque and Bergen to "Constituent and Affiliate Associations," Dec. 15, 1970, OHS, mss. 2386, box 22, folder 21; "Air Quality in the Atlanta Metropolitan Region: A Workshop for Citizens," fact sheet (n.d.), OHS, mss. 1900, box 17, folder 15; Head to Dennis, June 13, 1969, and *The Potomac View on Tuberculosis and Other Respiratory Disease* 6, no. 9 (1970): 1n, both UVA, call no. 10030, box 1, "Air Quality—Virginia (folder # 1)"; "Minutes of the Metro Clean Air Committee General Meeting," Nov. 25, 1969, p. 3, MCACR, box 1, "General" folder, no. 1; Irvine to State Board, Aug. 11, 1970, LWVMR, box 7, "Environmental Quality—Air" folder.

207. "Gazette—20 Sept. 69."

208. Terry H. Anderson, *The Movement and the Sixties: Protest in America from Greensboro to Wounded Knee* (New York and Oxford, 1995), 193.

209. Byron Kennard, "The Conservation Movement and the Fight for Clean Air," Aug. 6, 1969, pp. 7–9, UCB, BANC MSS 71/103 c, cont. 131, folder 2.

210. Bridgewater, "Memorandum on the September 9, 1969 Hearing"; "Written Statement Filed for the Record of Public Hearings on Ambient Air Standards for Georgia Business & Industry Association, Atlanta, Georgia, Submitted Dec. 22, 1970," GDAH, 26-21-181, box 10, "Georgia Ambient Air Standards Dec. 8, 1970 Hearing" folder.

211. R. A. Wilson, "Radicals Kidnap Pollution Baby," *Iron Age* 205, no. 5 (1970): 42.

212. "Hearing before the Department of Natural Resources on Proposed Air Pollution Control Rules," Dec. 15, 1969, p. 43, SHSW, ser. 2550, box 130, folder 20.

213. Bergren to "The I.M.A. Clean Air and Water Committee and Other Interested Individuals," Oct. 23, 1969, p. 3, IMAR, box 109, folder 2.

214. Edmund K. Faltermayer, "We Can Afford Clean Air," *Fortune* 72, no. 5 (1965): 158–63, 218, 223n; Hunter to Faltermayer, Dec. 3, 1965, NA, RG 90 A 1, entry 11, box 26, "IF-8—Publications (outside) Part 3" folder.

215. *IMA Executive Memo* 11, no. 18 (1970), and note, n.d., from Ray Hoewing, Director of Public Affairs, Quaker Oats Company, both IMAR, box 109, folder 3.

216. Citizens for Clean Air, New York City, "Position Paper—Con Ed's Proposed Astoria Expansion" (ca. 1970), UCB, BANC MSS 71/289 c, cont. 81, folder 9.

217. Marilyn Templeton, "Public Hearing," May 21, 1970, p. 3, UMM, mss. 43, box 3, folder 7; "Transcript of Proceedings at the Hearing on Chapter 270-5-24, Air Quality Control," Dec. 18, 1970, p. 76, GDAH, 26-21-181, box 10, "Georgia Ambient Air Standards Dec. 8, 1970 Hearing" folder.

218. Garrett De Bell, ed., *The Environmental Handbook: Prepared for the First National Environmental Teach-In* (New York, 1970), 309.

219. F. C. Olds, "Air Pollution Control: Good Intentions in Search of Directions," *Power Engineering* 73, no. 9 (1969): 35.

220. "Statement of the Cleaner Air Committee of Hyde Park and Kenwood before the Illinois Legislative Air Pollution Study Committee," Jan. 5, 1970, p. 3, CACR, folder 3.

221. James R. Wolf, "Comments on Proposed Air Pollution Control Regulations for Allegheny County," Aug. 26, 1969, p. 7, HSWPA, mss. 285, "Smoke Control—General" file.

222. Jones, *Clean Air*, 123–32; Krier and Ursin, *Pollution*, 190n; Vietor, *Environmental Politics*, 155n; Esposito and Silverman, *Vanishing Air*, 290; Bailey, *Congress*, 136; Tobin, *Social Gamble*, 71; Alfred Marcus, "Environmental Protection Agency," in James Q. Wilson, ed., *The Politics of Regulation* (New York, 1980), 272.

223. Andrews, *Managing*, 209n; Gottlieb, *Forcing*, 109, 127; Philip Shabecoff, *A Fierce Green Fire: The American Environmental Movement* (New York, 1993), 112, 203; Fiorino, *Making*, 152; Vietor, "Evolution," 133; Bailey, *Congress*, 78, 143; Lundqvist, *Hare*, 5; Nelson, "Air Quality Act," 355; *New York Times*, Mar. 22, 1970. On Muskie, see Paul Charles Milazzo, *Unlikely Environmentalists: Congress and Clean Water, 1945–1972* (Lawrence, 2006), 61–86.

224. Jones, *Clean Air*, 182; Hays, *Beauty*, 444; "Minutes of Board Meeting of the Area Industrial Problems Committee," May 12–13, 1970, HML, acc. 1411, ser. I, box 103, "Environmental Quality Control" folder.

225. *New York Times*, Dec. 8, 1970.

226. William D. Ruckelshaus, "The Beginning of the New American Revolution," *Annals of the American Academy of Political and Social Science* 396 (1971): 13–24.

227. Richard J. Lazarus, *The Making of Environmental Law* (Chicago and London, 2004), 69.

228. Coalition to Tax Pollution, "The Sulfur Tax," May 25, 1972, intro., UCLA, coll. 1199, box 85, folder 4.

229. D. M. Fort et al., *Proposal for a "Smog Tax"* (Santa Monica, 1959). The Coalition to Tax Pollution is one of the great blank spots in the history of environmentalism, having escaped even the attention of scholars who specifically looked at the history of emissions trading (Hugh S. Gorman and Barry D. Solomon, "The Origins and Practice of Emissions Trading," *Journal of Policy History* 14 [2002]: 293–320).

230. "Statement of Michael McCloskey on Economic Incentives to Control Environmental Pollution before the Subcommittee on Priorities and Economy in Government of the Joint Economic Committee," July 12, 1971, p. 1, UCLA, coll. 1199, box 85, folder 4.

231. Coalition to Tax Pollution: Cooperating Organizations, attachment to Fletcher to "Cooperating Organizations," Jan. 19, 1972, OHS, mss. 2386, box 19, folder 11.

232. Coalition to Tax Pollution, "Sulfur Tax," 2, 5n, 8.

233. Williams to Ullman, May 15, 1973, OHS, mss. 2386, box 19, folder 11.

234. Jones, *Clean Air*, 180. See also John F. Burby, "White House, Activists Debate Form of Sulfur Tax," *National Journal* 4 (1972): 1643–50, 1663–71.

235. Easterbrook, *Moment*, 188.

236. Joel A. Mintz, *Enforcement at the EPA: High Stakes and Hard Choices* (Austin, 1995), 23.

237. *Vorwärts*, May 3, 1961.

238. Peter Merseburger, *Willy Brandt 1913–1992. Visionär und Realist* (Stuttgart and Munich, 2002), 392.

239. STBK to Bundesminister für Verkehr, July 27, 1960, BArch, 136/5364.

240. Uekötter, "Kommunikation," 3–8; *Congressional Record* 40 (1905–06): 102.

241. "Protokoll der 8. Sitzung des Wirtschafts- und Verkehrsausschusses," Nov. 11, 1965, p. 42, LHAK, Bestand 930, no. 10319.

242. Holger Bonus and Ivo Bayer, "Symbolische Umweltpolitik aus der Sicht der Neuen Institutionenökonomik," in Bernd Hansjürgens and Gertrude Lübbe-Wolff, eds., *Symbolische Umweltpolitik* (Frankfurt, 2000), 291.

243. Volker von Prittwitz, "Symbolische Politik—Erscheinungsformen und Funktionen am Beispiel der Umweltpolitik," in Hansjürgens and Lübbe-Wolff, *Symbolische Umweltpolitik*, 274.

244. Bernd Hansjürgens and Gertrude Lübbe-Wolff, "Symbolische Umweltpolitik—Einführung und Überblick," in Hansjürgens and Lübbe-Wolff, *Symbolische Umweltpolitik*, 14n.

245. An enduring myth has it that people were laughing about Brandt's phrase because they favored jobs over environmental protection, and some environmental historians have unfortunately promoted this legend (Miranda A. Schreurs, *Environmental Politics in Japan, Germany, and the United States* [Cambridge and New York, 2002], 51). However, the reason for the widespread mockery did not lie in a lack of environmental awareness, and the choice between pollution and jobs was not yet prominent in the minds of people who were talking about pollution. The reception had nothing to do with environmental policy and everything with the broader context of Brandt's election campaign, which many criticized for its exceeding vagueness and abstinence from controversial political topics. "Promising blue skies" is a German proverb denoting exaggerated and unrealistic promises, and the implied critique was thus that Brandt was offering rhetoric rather than substance (Günter Struve, *Kampf um die Mehrheit* [Cologne, 1971], 53–67).

246. Regierender Bürgermeister von Berlin to Ziebill, Sept. 14, 1960, and notes of Jan. 19, 1961, Feb. 1, 1961, all LAB, Rep. 142/9, Acc. 5/24–30, vol. 7.

247. *Vorwärts*, May 3, 1961.

248. "Niederschrift über die 4. Sitzung des Beratenden Ausschusses nach § 16 Abs. 3 der Gewerbeordnung," Nov. 23, 1962, p. 5, LAB, Rep. 142/9, Acc. 5/24–31.

249. *Bundesgesetzblatt*, part 1 (1965), 413–15.

250. Deutscher Bundestag, *4. Wahlperiode, Drucksache 2097*, 1.

251. "Vermerk der Gruppe III B," Oct. 9, 1964, p. 1 (emphasis original), HStAD, NW 354, no. 51.

252. Bundesminister des Innern to Bundesverband der Deutschen Industrie, Nov. 2, 1970, BArch, B 106/38404.

253. "Auszug aus dem Kurzprotokoll über die 311. Koordinierungssitzung für Bundesratsangelegenheiten in der Bayerischen Staatskanzlei," Apr. 5, 1965, Punkt 1, BayHStA, StK-GuV 10245.

254. See J. Brooks Flippen, *Nixon and the Environment* (Albuquerque, 2000), 29.

255. "Resolution zur Luftverunreinigung des Zentralverbandes der Deutschen Haus- und Grundbesitzer e.V. und des Verbandes Deutscher Bürgervereine," Sept. 1957, NieHStA, Nds. 600, Acc. 153/92, no. 315; "Niederschrift über die Sitzung der Arbeitsgruppe 'Forschung,'" Oct. 23, 1957, p. 20, HStAD, NW 354, no. 35.

256. HStAD, NW 66, no. 352, p. 28.

257. Walther Liese, "Umweltschutz—staatliches und gesellschaftliches Ordnungsprinzip," *Gesundheits-Ingenieur* 93 (1972): 67.

258. "Aktion Lebensschutz," pamphlet, n.d., StAM, Polizeidirektion München 16075.

259. *Jahresbericht der Gewerbeaufsicht des Landes Hessen für das Jahr 1967*, 41.

260. Hans Willi Thoenes, "Energieerzeugung und Immissionsschutz," *Technische Überwachung* 15 (1974): 415.

261. Arbeits- und Sozialminister des Landes Nordrhein-Westfalen to Minister für Ernährung, Landwirtschaft und Forsten, Apr. 19, 1962, p. 1, HStAD, NW 354, no. 45. This was also the motivating force for local "round tables" in Cologne, Duisburg, and Essen (HStAD, NW 85, no. 163, p. 30n; "Niederschrift über die konstituierende Sitzung der Arbeitsgemeinschaft von Stadt und Handelskammer für Maßnahmen zur Verminderung der Verunreinigung der Luft," Dec. 30, 1959, StADu, 101/568; *Verwaltungsbericht der Stadt Essen für die Rechnungsjahre 1957, 1958 und 1959 [1.3.1957—31.3.1960]* [Essen, n.d.], 133).

262. HStAD, NW 50, no. 1226, pp. 23–27, 34 (quotation); invitation to a weekend conference on June 12–13, 1971, and Heinz Heinrich, "Forderungen des Bürgers an den Umweltschutz," both LASH, Abt. 761, no. 20317.

263. Bernd-A. Rusinek, "Wyhl," in Etienne François and Hagen Schulze (eds.), *Deutsche Erinnerungsorte II* (Munich, 2001), 654.

264. Deutscher Bundestag, *3. Wahlperiode, Drucksache 2855*, 2; Verhandlungen des Deutschen Bundestages, *3. Wahlperiode, Stenographische Berichte 49* (Bonn, 1961), 9619; Deutscher Bundestag, *4. Wahlperiode, Drucksache 1847*; note of Feb. 28, 1962, p. 1, BArch, B 136/5343; "Protokoll der 6. Sitzung des Ausschusses für Gesundheitswesen," Mar. 10, 1966, p. 5n, BArch, B 106/38369; note, Apr. 19, 1966, p. 2, and note, May 17, 1966, pp. 1–5, both BArch, B 106/38322.

265. Landtag Rheinland-Pfalz, *4. Wahlperiode, Drucksachen Abteilung III no. 433*; "Protokoll der 25. Sitzung des Rechtsausschusses," June 27, 1966, p. 4, LHAK, Bestand 930, no. 10320; "Begründung zum Gesetzentwurf," Apr. 1964, p. 3, and "Niederschrift über die Ministerratssitzung," Dec. 8, 1964, p. 5, both LHAK, Betand 860, no. 8581; Landtag des Saarlandes, *4. Wahlperiode, Drucksache no. 900*, 3.

266. Note, Nov. 27, 1963, and Schmidt to Schwarzhaupt, Jan. 7, 1964, both BArch, B 106/30403; "Niederschrift über die Besprechung mit den Ländern," Oct. 15, 1964, p. 10n, BayHStA, MArb 2600; "Auszug aus der Niederschrift über die 1. Sitzung des Ausschusses für Gesundheitswesen des Bundesrates," Jan. 28, 1965, pp. 28–30, BayHStA, StK-GuV 10245.

267. "Ergebnisniederschrift über eine Länderbesprechung zum Entwurf eines Bundesimmissionsschutzgesetzes im Arbeits- und Sozialministerium Düsseldorf," Sept. 19, 1968, p. 2, BayHStA, MWi 28368.

268. Bundesrat, *Drucksache 157/69*.

269. Deutscher Industrie- und Handelstag to Bundesminister für Gesundheitswesen, für Wirtschaft und für Arbeit und Sozialordnung, Sept. 24, 1968, p. 3, Wagner to Strobel, Sept. 17, 1968, p. 2, and "Kurzvermerk über die Sitzung des Vorstandes des Arbeitskreises 'Industrielle Immissionsfragen' des BDI in Cologne," Dec. 6, 1968, p. 1, all BArch, B 106/38324.

270. Note by Regierungsdirektor Feldhaus, July 26, 1968, and Bundesminister für Wirtschaft to Bundesminister für Gesundheitswesen, Nov. 29, 1968, p. 1, both BArch, B 106/38324; note, Feb. 27, 1969, p. 1, BArch, B 106/38323; "Kurzprotokoll über die Ressortbesprechung zum Entwurf eines Gesetzes zum Schutz vor schädlichen Umwelteinwirkungen durch Luftverunreinigungen, Geräusche, Erschütterungen und ähnliche Vorgänge (Bundes-Immissionsschutzgesetz)," June 16, 1971, p. 2n, BArch, B 136/5344.

271. BArch, B 106/38322, B 106/38323.

272. Verhandlungen des Deutschen Bundestages, *5. Wahlperiode, Stenographische Berichte 69* (Bonn, 1969), 12284.

273. "Kurzprotokoll über die 4. Sitzung des Arbeitskreises 'Bundes-Immissionsschutzgesetz' des Länderausschusses für Immissionsschutz," Nov. 26–27, 1969, p. 1, BayHStA, MWi 28369; Der Bundesminister des Innern to für den Immissionsschutz zuständige Landesbehörden, Jan. 30, 1970, MArb, 2598/I.

274. Verhandlungen des Deutschen Bundestages, *7. Wahlperiode, Stenographische Berichte 86* (Bonn, 1974), 4691.

275. Verhandlungen des Deutschen Bundestages, *6. Wahlperiode, Stenographische Berichte 78* (Bonn, 1971–72), 8914.

276. Kloepfer, *Geschichte*, 96n.

277. Kloepfer, *Geschichte*, 98.

278. Edda Müller, *Innenwelt der Umweltpolitik: Sozial-liberale Umweltpolitik— (Ohn)macht durch Organisation?* (Opladen, 1986), 45, 51.

279. Kai F. Hünemörder, *Die Frühgeschichte der globalen Umweltkrise und die Formierung der deutschen Umweltpolitik (1950–1973)* (Stuttgart, 2004).

280. Hans-Dietrich Genscher, *Planungen und Vorhaben des Bundesministeriums des Innern: Bericht vor dem Innenausschuss des Deutschen Bundestages am 22. Januar 1970 in Berlin* (Bonn, 1970), 3.

281. Figgen to Genscher, Apr. 28, 1970, BayHStA, MWi 28370.

282. Genscher to Figgen, June 3, 1970, BayHStA, MWi 28370.

283. *Bundesgesetzblatt*, part 1 (1974), 721–43; Feldhaus and Hansel, *Umweltschutz*; Hans Wiethaup, *Schutz vor Luftverunreinigungen, Geräuschen und Erschütterungen: Immissionsschutzgesetz Nordrhein-Westfalen* (Herne and Berlin, 1963).

284. *Bundesgesetzblatt*, part 1 (1974), 733–76.

285. *Bundesgesetzblatt*, part 1 (1974), 724, 726n.

286. Robert E. Swain, "Smoke and Fume Investigations: A Historical Review," *Industrial and Engineering Chemistry* 41 (1949): 2384.

287. Julius von Schroeder and Carl Reuss, *Die Beschädigung der Vegetation durch Rauch und die Oberharzer Hüttenrauchschäden* (Berlin, 1883, rpt., Hildesheim et al, 1986), 12.

288. *Wasser, Luft und Betrieb* 2 (1958): 180; *Gesundheits-Ingenieur* 84 (1963): 28.

289. Feldhaus and Hansel, *Umweltschutz*, 40n.

290. K. Husmann and G. Hänig, "Das Grillo-AGS-Verfahren zur Entschwefelung von Abgasen," *Brennstoff—Wärme—Kraft* 23 (1971): 85; Werner Brocke, "Stand der Technik der Rauchgas- und Brennstoffentschwefelung—Pro und Kontra," *Luftverunreinigung* (1972): 13–22.

291. Thoenes, "Energieerzeugung," 417.

292. Norbert Haug, "Entschwefelung bei Kohlekraftwerken," *Umweltmagazin* 11, no. 7 (1982): 41; Ekkehard Richter, "Der technische Stand der Rauchgasentschwefelung in der Bundesrepublik Deutschland," *Umwelt* 14 (1984): 192.

293. Jacob Jobst, "Regenschirme für unsere Bäume?" *Umweltmagazin* 11, no. 6 (1982): 20.

294. Kenneth Anders and Frank Uekötter, "Viel Lärm ums stille Sterben: Die Debatte über das Waldsterben in Deutschland," in Frank Uekötter and Jens Hohensee, eds., *Wird Kassandra heiser? Die Geschichte falscher Ökoalarme* (Stuttgart, 2004), 112–38.

295. Bernd Schärer and Norbert Haug, "Teurer Strom durch Umweltschutz?" *Umweltmagazin* 18, no. 8 (1989): 33.

296. Kiesinger to Filbinger, May 5, 1966, p. 2, HStAS, EA 8/301, Büschel 532.

297. Werner Weber, "Umweltschutz im Verfassungs- und Verwaltungsrecht: Stand und Tendenzen der Gesetzgebung," *Deutsches Verwaltungsblatt* 86 (1971): 811.

298. Werner Filmer and Heribert Schwan, *Hans-Dietrich Genscher* (Rastatt, 1993).

299. Georges Fülgraff, "Das Dilemma der Umweltpolitik—Eine Bilanz," in Eberhard Schmidt and Sabine Spelthahn, eds., *Umweltpolitik in der Defensive: Umweltschutz trotz Wirtschaftskrise* (Frankfurt, 1994), 17.

Chapter 8

1. Information on air-quality trends from Environmental Protection Agency, Office of Air Quality Planning and Standards, *Latest Findings on National Air Quality: 2000 Status and Trends* (Research Triangle Park, NC).

2. These remarks were inspired by the essays in Ute Frevert, ed., *Vertrauen: Historische Annäherungen* (Göttingen, 2003).

3. Uekoetter, "Strange Career."

4. LaFollette to Barlament, Mar. 13, 1971, SHSW, Green Bay, mss. 90, box 10, folder 4.

INDEX

Aachen, 62, 90, 102, 181
Abbott, System of Professions, 156
Academy for German Law, 104–6, 108
Acid Rain Program, 260
Adenauer, Konrad, 139, 179, 195, 247
Advocacy Coalition, 213–14
African Americans, 32, 82, 226
age of territoriality, 5–6
Air and Waste Management Association, 313n46
Air Pollution and Smoke Prevention Association of America, 86, 116, 156, 262. *See also* Air Pollution Control Association; Smoke Prevention Association
Air Pollution Control Act (1955), 216–17
Air Pollution Control Association, 129, 156–57, 163, 219, 262, 313n46. *See also* Smoke Prevention Association
Air Pollution Control Conference, 156
Air Pollution Control District. *See* Los Angeles County Air Pollution Control District
Air Pollution Control League of Greater Cincinnati, 9, 225
Air Pollution Control League of Rhode Island, 227
air pollution control cooperatives (Luftreinhaltegenossenschaften), 16, 137, 142, 168–73
Air Pollution Foundation, 201, 205, 210
Air Quality Act (1967), 14, 154, 220, 238, 241, 243
Air Repair, 129
air resource management approach, 218–20, 229
Air Resources Board, 159
Aktion Lebensschutz, 250
Alabama, 42
Alaska, 224
Alexander, I. Hope, 84–85
Allegheny Conference on Community Development, 294n124

Allegheny County, Pennsylvania, 22, 24, 37, 85, 126, 131, 151, 156, 158, 160, 165–66, 203, 206, 224, 233, 235
Allegheny County Citizens against Air Pollution, 233
Allegheny County Medical Society, 37
Allensbach Demoscopic Institute, 136
aluminum smelters, 111
Ambient Air Quality Standards, 220, 238, 244, 260
American Association of University Women, 239
American Civic Association, 11–12
American Industrial Hygiene Association, 129–30, 162
American Municipal Association, 125
American Petroleum Institute, 219, 231–32
American Public Health Association, 130
American Society of Mechanical Engineers, 24, 53, 72, 78, 124, 126, 130, 161–62, 175–77, 219
Anderson, Dennis LeRoy, 105
anthracite, 2, 42, 187
anti-industrial attitudes, 28, 119, 225, 228, 240
Anti-Smoke League (Baltimore), 21–24, 26, 28
Anti-Smoke League (Chicago), 22
APCA Abstracts, 129
Arendt, Hannah, 111
Arkansas, 81
Arnsberg, 138
arsenic, 2
ash content of coal: regulation of, 77–78
Associated Industries of New York State, 232
Association for the Promotion of Air Hygiene (Verein zur Förderung der Lufthygiene), 195
Association for the Promotion of Commerce and Industry (Verein zur Beförderung des Gewerbfleißes), 45, 49, 53, 94
Association of Commerce (Chicago), 40
Association of Commerce and Industry (Chicago), 128

341

Index

Association of German Cement Plants (Verein Deutscher Zementwerke), 144–45, 177
Association of German Chambers of Industry and Commerce (Deutscher Industrie- und Handelstag), 142
Association of German Engineers (Verein Deutscher Ingenieure, VDI), 17, 51–53, 100–103, 144, 146–47, 174–76, 265. *See also* Clean Air Commission
Association of German Industry (Bundesverband der Deutschen Industrie), 133, 142–44, 147–48, 170, 176, 180, 193
Association of High-Performance Steam Boiler Owners (Vereinigung der Großkesselbesitzer), 137, 142, 144, 147, 177, 184
associations for steam boiler surveillance (Dampfkesselüberwachungsvereine, DÜV), 48, 52, 55, 102–3
Associations for Technological Inspections (Technische Überwachungsvereine, TÜV), 135, 147, 169, 250
asthma, 198, 225, 239
Atlanta, 78, 239, 241–42
atomic power, 225, 251, 264
attorney general: of Pennsylvania, 161; U.S., 211–12
Augsburg, 133, 146
Auschwitz, 111
Austria, 44
automobile exhausts, 6, 15, 86, 139, 164, 180, 202–4, 206–17, 220, 244, 246, 251, 257–58
automobile industry, 11, 207–16, 220, 258, 327n41, 328n46. *See also* Ford Motor Company; General Motors
Automobile Manufacturers Association, 210, 213–14
Azbe, Victor J., 53, 158, 313n57

Bach, Carl Julius von, 64
Bach, Franz Josef, 179
Bachem, 194
bacteriology, 4
Baden, 89
Baden-Württemberg, 110, 186, 257
bakeries, 48–49, 62
Baltimore, Maryland, 20–24, 26–29, 34, 36, 39, 42, 62, 71, 74
Baltimore & Ohio Railroad, 126
banality of evil, 111
BASF, 185
Baur, John, 198
Bavaria, 93, 137, 182, 186, 249–50, 252
Bayer Corporation, 185
Becher, August, 44–45
Beckum, 133
Belgium, 118
Benjamin, Charles, 24–25, 53
Benline, Arthur, 163
Benner, Raymond 34, 128
Berge, Helmut 147, 176

Berlin, 45, 48–49, 53, 56, 87, 96, 100–1, 105, 161, 186, 194, 247–48
Bethlehem Steel, 239
Beverly Hills, California, 205
Bielefeld, 59, 63, 89, 96
Biersdorf, 176
Bird, Paul, 24, 33, 35, 37, 53, 69, 267
Birkenau, 111
Bismarck, Otto von, 247
Bitterfeld, 102
Black Survival Inc., 226
Black Tuesday (St. Louis, November 1939), 79
Blohm, Hermann, 64
"blue sky above the Ruhr," 246–47
Bochum, 89, 188, 190, 251
Bonn, 147, 194, 258
Boston, Massachusetts, 20, 24, 26, 34–35, 62, 70, 150, 242, 286n252
Boston Area Ecology Action, 242
Bottrop, 184, 192, 197
Boy Scouts, 75
boycott, call for against automobile and petroleum industries, 215
Brackwede, 96
Bramwell, Anna, 104
Brandt, Willy, 246–48, 253, 338n245
Braunschweig, 47–49, 55, 58, 62, 87
Breckenridge, Lester, 17, 31, 53, 267
Bremen, 44, 48, 53, 137, 186
Breslau, 60, 62
broadening of air pollution agenda, 86, 115, 118, 155–57, 159
Brockovich, Erin, 10, 267
Brokdorf, 251
Brower, David, 242
Brown, Freddie Mae, 226
brown coal, 138, 194–95
Brühl, 194
Bryce, James, 6
Buchka, Karl von, 172
Buffalo, New York, 20
Buhrer Air Pollution and Civic Association (Cleveland), 227–28
Bundesrat (German State Chamber), 249, 252
Bundestag (German Parliament), 139, 172, 196, 249, 252
Bureau of Conservation (U.S.), 67
Bureau of Mines (U.S.), 34, 51, 68, 216
Burhenne, Wolfgang, 174, 265, 267
Bush, George W., 13–14
Business and Professional Women's Club (Atlanta), 78
Business Men's Club (Cincinnati), 34
Business Men's League (St. Louis), 34
Busse, Fred, 32
Butler, James Gay, 27

Cahuenga Property Owners Association, 200
California, 96, 151, 153, 158–59, 198–207, 209–16, 229, 231, 239. *See also* California Manufacturers Association

Index

California Institute of Technology, 198, 201
California Manufacturers Association, 231
Canada, 39, 216
cancer, 135, 210, 239
carbon monoxide, 6, 135, 260
Carr, Donald, 216
Carson, Rachel, 225, 227–28, 234
Carter, James Earl, 246
Carter, James Hinman, 80, 128
case-by-case approach (Germany), 55–56, 59–61, 88–93, 99, 110, 138, 173
catalytic converters, 211, 215
Catholicism, 171
cement industry, 143–45, 147
Chamber of Commerce: Boston, 24, 35; Bremen, 48; Cleveland, 24; Duisburg, 143; Hamburg, 63–64; Pennsylvania, 241; Pittsburgh, 20, 22, 28, 36–37, 225; Rochester, 24, 38; San Francisco, 121; St. Louis, 72, 81; Syracuse, 118
Chambers, Frank, 75, 128, 158
chemical industry, 6, 81, 116, 120, 128, 184–85, 189, 228, 234
Chemnitz, 45, 59, 61–62
Chicago, Illinois, 8, 15, 17, 20, 22–24, 26–27, 29–42, 51, 58, 62, 68–69, 72–73, 75, 86, 117, 125–26, 128, 130, 151, 153, 155, 158, 164, 166, 223–24, 226, 230, 232, 239, 242–43, 246
Chicago Record-Herald, 28, 33
Chicago Woman's Club, 22, 24, 37, 39
Christian Democratic Party (Christlich-Demokratische Union, CDU), 171–72, 247–48
Christy, William, 70, 73–74, 86, 158
Chrysler Corporation, 212
Cincinnati, Ohio, 9, 21–22, 25–30, 34, 36, 39, 41, 62, 65, 68, 72, 78, 85, 132, 152, 158, 163–64, 225
Citizens Against Air Pollution (San Francisco), 229
Citizens Anti Smog Action Committee (Los Angeles), 199, 205
Citizens for Clean Air (Atlanta), 239
Citizens for Clean Air (New York City), 242
Citizens Union (New York City), 132
Citizens' Association (Chicago), 22, 25, 27
Citizens' Smoke Abatement Association (St. Louis), 23, 25
Citizens' Smoke Abatement League (St. Louis), 70, 72, 76, 80, 313n57
City Club (Chicago), 17, 22, 31–33, 35, 37–38, 276n28
Civic Club of Allegheny County (Pittsburgh), 22, 36, 224–25
civic groups (Germany), 16, 21, 43–45, 64, 133–35, 179, 194–97, 249–51, 254–55, 264, 308n118
Civic Improvement League (St. Louis). *See* Civic League
Civic League (St. Louis), 23–24, 30, 32, 34–38
Civil Code (Bürgerliches Gesetzbuch), 105, 107, 121, 142

civil rights struggle (U.S.), 224, 240
class, 3–4, 21, 81–83, 118–19, 190, 225–26
Clean Air Act: of 1963, 14, 153, 217–18; of 1970, 5, 114, 209, 216, 220, 243–45, 260
Clean Air Commission (VDI-Kommission Reinhaltung der Luft), 17, 139, 146, 172, 175–80, 182, 185, 193, 195, 219, 235, 248, 250, 264–67, 318n178
Clean Air Council (California), 239
Clean Air Week (Chicago 1919), 68
Cleaner Air Committee of Hyde Park-Kenwood (Chicago), 117, 223, 226, 230, 243
Cleaner Air Week (U.S.), 163–65, 265
cleanliness and air pollution, 1, 3, 6, 20, 44, 116, 133–34, 223, 263
Clear Air Clear Water Unlimited, 223, 227–28, 230
Cleveland, Ohio, 3, 24, 26, 38, 62, 69–73, 75, 86, 116–17, 119–20, 122–23, 130–31, 151, 155, 157–58, 166, 223, 226–28, 233, 236, 239–40
coal: prices of, 48, 62, 80, 82, 94; scarcity during World War I, (Germany) 95; types of, 2, 32, 42, 48, 68, 76–78, 81. *See also* coal industry; powdered coal
coal industry, 77–82, 85, 94–95, 126, 235, 238, 257, 314n80; crisis during World War I (U.S.), 67–68; regulation of, 68, 76–85, 197. *See also* coal; coke
Coal Producers Committee for Smoke Abatement, 85
Coalition for the Environment (St. Louis) 223
Coalition to Tax Pollution 245
coke, 48, 62–63; ovens for, 117, 184, 189, 191–92
Colby College, 240
collective goods. *See* Olson, Mancur
Cologne, 59–60, 88, 147, 185, 195, 338n261
Columbian Exposition, 27
Columbus, Ohio, 155
Committee for Clean Air Now (Palo Alto), 229
Committee for Nuclear Information (St. Louis), 221
compensation payments, 107–9, 111, 121, 180, 190–91
Conference of German Physicians (Deutscher Ärztetag), 134, 136
Congress of Women's Clubs (Pittsburgh), 37
conservation, 50, 52–53, 104, 218
Conservation Foundation, 224, 229, 240–41
Consolidated Edison, 122
conspiracy on automobile exhaust control, 211–15, 328n46
constitutional questions (Germany), 252
contagionism, 4
Cooke, Morris, 53
cooperation, 8–12, 16–17, 38–39, 60–61, 73, 92–93, 100–1, 113–115, 126–28, 130–32, 142–46, 157, 160, 164, 168, 193, 201–2, 220, 225, 228–32, 241, 243–44, 250–51, 263–67
cooperatives. *See* air pollution cooperatives
Corn, Morton, 132, 235

corporatism, 63, 94, 100–103, 131, 192–93. *See also* pseudocorporatism
corruption, 69, 162–63, 177
Cottrell, Frederick, 96
Council of Europe, 180, 253
Council of Experts for Environmental Questions (Sachverständigenrat für Umweltfragen), 183, 253
Council of State Governments, 153
county agencies, 13, 150–51, 200, 311n9
court proceedings, 11–12, 26–30, 34–35, 37, 96, 121
Crenson, Matthew, 275n10

Darré, Richard Walther, 107–8
Dayton, Ohio, 132, 165
Delaware Valley Citizens' Council for Clean Air (Philadelphia), 227–28
Dellwig, 196
Denver, Colorado, 151, 166, 232
Department of Health, Education, and Welfare (U.S.), 213, 216–17, 328n59
deregulation, 256, 258
desulphurization of flue gas, 257
Detmold, 181
Detroit, Michigan, 153, 214, 216, 223, 231, 234, 237
Dewey, Scott H., 304n44, 328n57
Dickmann, Bernard F., 77
disasters, 118, 120, 302n13
domestic furnaces, 40–42, 51, 63, 72, 74–85, 144, 161, 164, 197
Dominick, Raymond, 88
Donnelly, Thomas, 37–38, 128
Donora, Pennsylvania, 118, 120, 216, 328n57
Dortmund, 90, 182, 189, 196
Dresden, 44–45, 53, 58, 61–65, 87, 98, 107
Du Pont Company, 123, 162, 184, 228, 237
Duisburg, 133, 135, 143, 181, 189–90, 194, 196–97, 338n261
Duquesne Light, 126, 162
Düren, 194–95
Düsseldorf, 48, 56, 58, 64, 89, 147, 181, 188, 192, 195, 197
Dyktor, Herbert, 86, 155, 166

Earth Day, 7, 239, 241, 254
East Liverpool, Ohio, 233
Eastman Kodak, 162
Ecorse, Michigan, 234
Eiser, Ernst, 105–6, 108
electrification of railroads, 41
electrostatic precipitators, 96, 99–100, 161, 166
elite background of anti-smoke movement, 21
Ely, Sumner B., 128
Emory University, 239
emphysema, 239
engineers, 6, 8, 13, 16, 24, 31–32, 37–38, 45, 50–53, 56–57, 63, 66, 82, 96–98, 102–3, 156–57, 265
Engineers' Club of St. Louis, 23, 72, 77–78, 82

Engineers' Society of Western Pennsylvania, 22, 24
Environmental Action, 245
Environmental Protection Agency, 5, 15, 114, 244, 246
epidemiology, 4
Eschweiler, 92
Essen, 98, 111, 135, 147, 169, 190–93, 197, 250, 338n261
esthetic arguments, 21
Eugene, Oregon, 116
European Union, 5, 15, 258–59
Expert Committee on Dust Technology (Fachausschuß für Staubtechnik), 100–103, 147–48, 169, 174–76, 178, 265–66
eye irritation: from photochemical smog, 198–99

factory inspection (Germany), 47, 49, 53, 55–56, 59, 89–91, 94, 100, 138, 140, 144–46, 177, 180–82, 188, 190–92, 250
Fairfax, Virginia, 239–40
fallout. *See* nuclear fallout
farmers, 107–9, 143, 170, 176, 190, 199, 309n159
Federal Agency for Nature Conservation (Bundesamt für Naturschutz), 253
federal clean air act (Bundesimmissionsschutzgesetz), 186, 251–58
Federal Environmental Agency (Umweltbundesamt), 253
federal government (Germany), 136, 138–140, 180, 193, 248–49
federal government (U.S.), 12–14, 67–68, 129, 150, 154, 158–59, 208, 216–21, 240, 243–44, 246, 328n57, 328n59, 329n74. *See also* Public Health Service
Federal Health Agency (Bundesgesundheitsamt), 147, 175
Federal Ministry of Agriculture (Bundesministerium für Ernährung, Landwirtschaft und Forsten), 147
Federal Ministry of Health (Bundesministerium für Gesundheitswesen), 139, 178, 182, 184, 252
Federal Ministry of Labor (Bundesministerium für Arbeit und Sozialordnung), 139, 173–75, 178, 182, 257
Federal Ministry of the Interior (Bundesministerium des Innern), 251–58
Federation for Heimat Protection (Deutscher Heimatbund), 134, 136
Federation of American Scientists, 245
Federation of Women's Clubs of Missouri, 227
Fermi, Laura, 117, 126
Feynman, Richard, 198
Fields, Marshall, 25
Figgen, Werner, 255
Filbinger, Hans, 258
film and television reports on smoke, 70, 100, 223, 254
filter technology, 95–102

Index

fines, 8, 17, 26, 28, 30, 32, 35, 47, 122, 219, 229, 243
Finland, 64
firemen, 29, 32, 48, 63, 72–73, 119
Flanagan, Maureen, 276n28
Fleming, Arthur, 213
Florida, 122, 125, 152
fly ash, 95–102, 135, 138, 160–61, 189, 195
Ford Motor Company, 212
Ford, James, 79, 85–86
forestry, 147, 170, 178, 185
forests, damage to, 111, 192, 257–58, 264
foundries, 89, 91, 110, 236
Frankfurt, 44–45, 55, 133, 249
Franklin, Benjamin, 75
Franklin Institute, 25
Frechen, 194
Freiburg, 62
Friedman, Milton, 245
Friends of the Earth, 242, 245
Fuel Administration (U.S.), 67
fuel efficiency, 32, 35, 50–51, 67, 71, 94–95, 102
Fürth, 145

Gals Against Smog and Pollution (Missoula), 228, 242
Gary, Indiana, 226
Gelsenkirchen, 188, 193
gender, 3, 23–25, 45, 66, 71, 134, 165–66, 200, 223, 227, 315n102
General Motors, 212
Genscher, Hans-Dietrich, 253–58, 264–65
Georgia, 239, 241
Georgia Business and Industry Association, 241
Georgia Tuberculosis and Respiratory Diseases Association, 239
German Association of Farmers, 143
German Association of Industrialists (Bund der Industriellen), 47
German Association of Towns and Cities (Deutscher Städtetag), 133, 182, 248
German Democratic Republic, 274n26
German Society for Technical Physics (Deutsche Gesellschaft für Technische Physik), 97
German Trade Union Federation (Deutscher Gewerkschaftsbund), 134
Girl Scouts, 239–40
global dimensions of pollution, 113, 223–25, 253, 263
global warming, 7, 258
globalization, 5
Goklany, Indur, 12–14
Goldenberg power plant, 99–100
good neighbor policy, 121, 123, 164, 228
Goslar, 148
Goss, William, 53
Granite City, Illinois, 238
Granite City Steel, 231
Grasselli Chemicals, 116

Great Depression, 69–70, 72, 76, 93, 118, 192, 195–96
Great Lakes Steel Corporation, 234
Greater Detroit Board of Commerce, 231, 237
Green Party (Germany), 258
Griebling, Robert, 116, 166
Grinder, Robert Dale, 279n83
Griswold, S. Smith, 201, 203, 206, 210–11, 327n41
Grohnde, 251
Group Against Smog and Pollution (Pittsburgh), 240
Gruber, Charles, 78, 152, 156–57
Grundmann, Konrad ,136
Grünewald, Heinrich, 174, 176, 265
Gutehoffnungshütte, 107–8
Gymnich Castle, 258

Haagen-Smit, Arie Jan, 157, 201, 209, 213
Hagen, 138
Hahn, Kenneth 200, 209–10, 212, 214–15
Hamburg, 15, 34, 43, 58, 61, 63–64, 87, 133, 186, 308n118
Hamerschlag, Arthur, 35
Hamm, 188
Hanover, 44–45, 55, 60, 66, 100, 133
Harz, 111
Hauser, Karl, 57, 60, 64
Hawaii, 224
Hays, Samuel P., 15
health aspects, 4, 6, 44, 134, 142, 199–200, 213, 223, 226, 239, 242–44, 263
health departments, 26–27, 29–30, 84, 151, 153, 234, 236
Heidelberg, 55, 62
Heiligenhaus, 147
Heller, Arnold 140, 171
Helsinki, 64
Henderson, J. W. 37, 68
Herbede, 95
Herdecke, 98
Herne, 96, 196
Hesse, 186, 250
Hitchcock, Lauren, 201
Hochfeld, Germany,196
Hoever, Walter, 190
horror scenarios, 224
hospitals, 107
House of Representatives (U.S.), 154
Hudson County, New Jersey, 70, 74, 150, 158
Hudson River Conservation Society, 126, 153, 227
Humble Oil & Refining Company ,233
Hünemörder, Kai, 253
Hurley, Andrew, 226
Hyde Park-Kenwood Community Conference (Chicago), 224

Illinois, 77, 153, 232, 234, 237–38
Illinois Manufacturers' Association, 28, 206, 231–32, 234, 237, 241–42

implementation problems, 182–83, 259
incinerators, 143, 185
Industrial Code (Gewerbeordnung), 54, 109, 121, 138, 142–43, 172–73, 177–78, 248, 252, 255
Industrial Hygiene Foundation, 162
Industrial Smoke Abatement Association (Hudson County), 74
Inglehart, Ronald, 115–17
insider perspective, 162–68, 208, 230, 232–34, 249
Insull, Samuel, 68
interdisciplinary work, 101, 130, 178
Interlake Steel Corporation, 232
International Association for the Prevention of Smoke, 25, 28, 39 see also Smoke Prevention Association
Interparliamentary Working Group (Interparlamentarische Arbeitsgemeinschaft für naturgemäße Wirtschaft), 139, 169, 174–75, 178–79, 265
Interstate Commerce Commission, 81
interstate pollution problems, 154
Interstate Sanitation Commission, 154
inversion layers, 2, 118, 204
investments in control equipment, 12–13, 29, 35, 122, 185, 236, 257, 262–63
Isserman, Maurice, 224
Itzehoe, 133, 180
Izaak Walton League, 126

Japan, 65, 257
Joint Smoke Inspection Bureau of the Railroads of Chicago, 41
Jones & Laughlin, 120, 228
journals on air pollution, 43, 129
jurists and engineers, 50, 56–57

Kansas City, 78
Karlsruhe, 62
Kazin, Michael, 224
Kegel, Sturm, 137, 142–43, 146, 148, 168–74, 197
Kehoe, Robert, 225, 331n118
Kennedy, John F., 217
Kerschbaum, Emil, 47
Kettner, Helmut, 184
Kiel, 194, 251
Kiesinger, Kurt Georg, 257–58
King, Kerryn, 231, 233, 236
King, Martin Luther, Jr., 240
Kirchlengern, Germany,134
Klausing, Friedrich, 105–7
Kleinblittersdorf, 95–96, 135
Klingenberg power plant (Berlin-Rummelsburg), 96, 161
Knapsack, 99
Kohl, Helmut, 247
Krefeld, 110, 135
Krupp Industry, 190
Kuhn, David M., 128
Kuss, Robert, 31, 38

LaFollette, Douglas 267
Lang, Gustav, 66
Larson, Gordon, 201, 204–5
Law on Preventive Measures on Air Pollution Control (Gesetz über Vorsorgemaßnahmen zur Luftreinhaltung, 1965), 249
Law on the Change of the Industrial Code and Supplementation of the Civil Code (Gesetz zur Änderung der Gewerbeordnung und Ergänzung des Bürgerlichen Gesetzbuchs, 1959), 172, 179, 248
Lazarus, Richard, 244
lead, 6, 260. See also tetraethyl lead
League of Women Voters (Eugene), 116
League of Women Voters (Los Angeles), 216
League of Women Voters (Salem), 227
Leipzig, 44, 58, 91, 110
Lent, Heinrich, 176–77, 179, 185, 193, 195, 250–51, 257
Leontief, Wassily, 245
Leverkusen, 185
Lewis, Howard, 216, 228
liberal party (Freie Demokratische Partei, FDP), 253–54
libertarians, 12–14
licensing procedure (Germany), 62, 89, 109–12, 138, 178–79, 182, 252
Liesegang, Wilhelm, 92–93, 118, 173–74
lignite, 138, 194–95
Löbner, Alfred, 169
logic of collective action, 4, 21, 119, 230, 238
London smog disaster, (1952), 302n13
Los Angeles, California, 15, 132, 151, 158, 187–88, 198–207, 209–16, 222
Los Angeles Coat and Suit Manufacturers Association, 199
Los Angeles County Air Pollution Control District, 158, 200–207, 209–10, 214
Los Angeles County Medical Association, 199
Los Angeles Times, 202–3, 205
Lowell, Massachusetts, 36
Lower Saxony, 93, 186, 193
Lüdenscheid, 111
Ludwigshafen, 182, 185
Lufthansa German airlines, 183
Lukens, John, 158
Lund, Cornelia Sorenson, 72, 76
Lünen, 133
Lurgi Company, 98

MacKenzie, Vernon, 166, 218–21
Magdeburg, 48
Maier, Charles, 5
Maine, 240
Mannheim, 133, 145, 182
Manufacturers' and Citizens' Protective Association (Chicago), 29
Manufacturing Chemists' Association, 122–24, 130, 161, 232
market-based regulation, 245–46
Massachusetts, 125

measurements of air quality, 13, 39, 75, 91, 129, 131–32, 180, 184–85, 249, 260, 266
Meldau, Robert, 101, 133, 148, 169, 174–75
Meller, Harry B., 73, 150, 158
Mellon Institute of Industrial Research (Pittsburgh), 25, 33, 39, 73–74, 128, 150
Memphis, Tennessee, 40, 75
Metro Clean Air Committee (Minneapolis), 227–29, 239
Meuse Valley disaster, 118
Meyers, Franz, 172
miasma theory, 4
Michigan, 153, 206, 223, 226, 234, 236–37
Middleton, John, 240
Million Population Club (St. Louis), 23
Milwaukee, Wisconsin, 26, 62, 122, 126, 158
mining bureaus, 144, 189, 191–92
Ministries of Labor (Germany), 93–94
Ministry for Trade and Commerce (Preußisches Ministerium für Handel und Gewerbe), 49, 52, 57–58, 60, 87, 194
Ministry of Labor (North Rhine-Westphalia), 135–38, 144, 146–47, 169, 172, 175, 178, 181–82, 194, 197, 254–55
Minneapolis, Minnesota, 23–24, 62, 223, 227–30
Minnesota, 151, 223, 228, 231, 239
Minnesota Association of Commerce and Industry, 228, 239
Minnesota Emergency Conservation Committee, 239
Miss Cleaner Air, 165
Missoula, Montana, 228
Missouri, 227
Mitchell, John, 212
Möller, Erwin, 96
Mommer, Karl, 174, 317n146
Monnett, Osborn, 69–70, 75, 80, 85
Mont Cenis mine (Sodingen), 96, 98, 189
Montana, 219, 221, 228, 242
Montana Clean Air Association, 221
Motor Vehicle Pollution Control Act (California), 214
Motor Vehicle Pollution Control Board, 215
Mülheim an der Ruhr, 135, 170
Müller-Armack, Alfred, 134
Munich, 44–47, 55–57, 60, 62–65, 87, 134, 140, 145, 250
municipal enterprises, pollution from, 55, 64–65, 170
municipal role in air pollution control, 13, 62–65, 124–25, 127–28, 150–54, 157–58, 167–68
Muskie, Edmund, 243, 249
Mussolini, Benito, 71

Nader, Ralph, 216, 221
Nashville, Tennessee, 85
National Air Conservation Commission, 227. *See also* National Tuberculosis Association

National Air Pollution Control Administration, 240, 328n59
National Association of Manufacturers 194, 231, 237
National Audubon Society, 245
National Center for Air Pollution Control, 328n59
National Coal Policy Conference, 235
National Conference on Air Pollution (1966), 222
National League of Cities, 183
National Parks Association, 224
National Science Foundation, 157
National Smoke Abatement Conference, 53
national styles of regulation, 14–19, 259, 264
National Tuberculosis Association, 220, 227–28, 240. *See also* Georgia Tuberculosis and Respiratory Diseases Association; New York State Tuberculosis and Respiratory Disease Association
natural gas, 83, 198
Nazi Party, 110; regime of, 103–12, 195
Netherlands, 201
New Deal, 129
New Jersey, 121, 153–54, 220, 229, 235
New Orleans, Louisiana, 42, 74
New York Business Council for Clean Air, 233
New York City, New York, 2, 27, 30, 41–42, 53, 58, 122, 132, 154, 187, 233, 242
New York, 126, 152–54, 163, 232–33
New York State Action for Clean Air Committee, 232
New York State Tuberculosis and Respiratory Disease Association, 233
New York Times, 163, 244
Nickeson, Jean, 165
NIMBY ("Not in My Backyard"), 88
nitrogen oxides, 185, 260
Nixon, Richard, 243
North Rhine-Westphalia, 135–38, 140, 144, 146–47, 169–73, 175, 177–79, 181–62, 184–85, 188, 193–94, 197, 248–49, 251–52, 254–55, 262, 265
nuclear power 225, 251, 264; testing of, 223
nuisance law, 2, 12, 54, 121
Nuremberg, 59, 62, 145

O'Connor, John, 33
Oakland, California, 75
Oberhausen, 107, 188, 196
Obermeyer, Henry, 77, 150
Oberste-Brink, Karl, 143, 193
occupational medicine, 129–30, 156, 160, 162, 173, 194
odors, 90, 116–17, 155, 160, 185, 226
Oels, Franz, 137, 139, 182, 197, 267
Oels, Heinrich, 139, 182, 267
Oer-Erkenschwick, 134
Ohio, 153, 233
Ohio Pure Air Association, 223, 227–28, 233

Ohio River, 233
Olson, Mancur, 4, 21, 119, 230, 238
Opinion Research Corporation, 238
Orange County, California, 158
Oregon, 116, 127, 194, 227, 245
Oregon Environmental Council, 245
Osnabrück, 193
Ottobrunn, 134
Outdoor Art League (Chicago), 22, 24, 33
ozone, 260

Palo Alto, California, 229
panaceas for smoke problem, 57, 203–4
paper industry, 194, 228
Parkersburg, West Virginia, 233
Pasadena, California, 198–200
Pasinski, V., 48, 51, 56, 58
Pennsauken, New Jersey, 229
Pennsylvania, 161, 225, 241
Pennsylvania Railroad, 41
pesticides, 225, 253
Peters, Theodor, 52
Petersen, Wolfgang, 254
Petoskey, Michigan, 226
petroleum: favored over coal, 42, 78, 198
petroleum industry, 202, 212, 215, 219, 231–32
Philadelphia, Pennsylvania, 25–26, 151, 158, 227–28
photochemical smog, 146, 187, 198–207, 213
Pittsburgh, Pennsylvania, 3, 15, 18, 20, 22, 24, 26, 28, 33–37, 39–40, 58, 62, 67–69, 72–75, 83–86, 114, 118–19, 122, 124–26, 128, 131–32, 150–51, 155, 158–59, 161–62, 164–66, 188, 194, 197, 200, 203–4, 206, 222, 224–25, 231, 235, 239–41, 243, 246, 262
Platt, Harold, 275n10
Plittersdorf, 147
Poethke, Charles, 158
policy brokers, 53, 103, 259, 267
polluting businesses: closure of, 61, 73, 92, 96, 106, 200, 295n156
Portland, Oregon, 127
Posen, 57
postmaterial values, 115–17
powdered coal, 95–96, 98, 189
power plants, 65, 68, 95–102, 135, 138, 182, 189, 257
precautionary principle, 180
presidents of the United States. *See* Bush, George W.; Carter, James Earl; Kennedy, John F.; Nixon, Richard; Reagan, Ronald; Roosevelt, Theodore
Princeton, New Jersey, 238
Prittwitz, Volker von, 248
Procter and Gamble, 121
property, pollution and, 3, 6, 8, 20–21, 44, 199, 263
Providence, Rhode Island, 62, 286n252
Prussia, 6, 48–49, 52, 54, 56–58, 60, 62, 87, 90, 92, 102–3, 191, 194

Prussian Supreme Administrative Court (Preußisches Oberverwaltungsgericht), 54, 90
pseudocorporatism, 115, 131–32, 162, 167–68, 206, 208–9, 218–19, 226, 233–35, 243–44, 264–65
Public Health Service (U.S.), 122–24, 126–27, 129, 153–54, 157, 160–61, 166, 177, 203, 210, 213, 215–22, 229, 232, 240, 242, 328n59
public relations, 120–21, 231, 233, 235

Quaker Oats Company, 242

Rabeneick, Fritz, 171
race, 3, 32, 81–82, 117
railroads, 40–41, 126, 151. *See also* Baltimore & Ohio Railroad; electrification of railroads; Pennsylvania Railroad
Rand Corporation, 245
Reagan, Ronald, 203, 225, 246, 258
real estate interests, 3, 21, 23, 81
Ream, Jay, 294n124
Recklinghausen, 144
recruitment problems, 113, 157–59, 167
refineries, 6, 11, 160, 200, 202, 231
Reich Food Estate (Reichsnährstand), 107–9
Reich Health Office (Reichsgesundheitsamt), 90, 93
Reich Ministry of Trade and Commerce (Reichswirtschaftsministerium), 109, 111–12
Reichstein, Willy, 196
religious institutions: Protestant Seminary of Westphalia, 251; Twelve Apostles of the Mormon Church, 76
Renan, Ernest, 18
Republic Steel, 228
Retail Coal Merchants Association (Pittsburgh), 84
retrofitting, 33, 99, 122
Rhenish brown coal region, 194–95
Rhenish-Westphalian Protective Association against Smoke and Mining Damage (Rheinisch-Westfälischer Schutzverband gegen Rauch- und Bergschäden), 190
Rhineland-Palatinate, 182, 186, 247
Rhode Island, 227
Ringelmann Scale, 159, 266
Rio de Janeiro, Brazil, 253
Roberts, E. P., 38
Rochester, New York, 24, 26, 29, 38, 62
Rome, Adam, 4, 315n102, 331n109
Roosevelt, Theodore, 247
Rose-Ackerman, Susan, 318n178
Rosen, Christine Meisner, 54
Rosenheim, 143
Rowe, Gordon, 158
Royce, Stephen W., 199
Ruckelshaus, William, 15, 244
Ruhr area, 15, 133, 135, 169–71, 178, 184, 187–97, 246–47, 262

Ruhr Area Federation for Regional Planning (Siedlungsverband Ruhrkohlenbezirk), 133, 137, 168–69, 191–93

Saarbrücken, 9
Saarland, 96, 141, 186
Sabatier, Paul, 213–14
salaries, 157–58
Salem, Oregon, 227
Salt Lake City, Utah, 68, 72–76
Salzgitter, 133
Samuelson, Paul, 245
San Diego, California, 206
San Francisco, California, 121, 206, 213, 229
Sax, Joseph, 166
Saxony, 60, 102
Schäff, Karl, 178
Schenck, Paul, 217
Schiffer, Heinz, 105–6
Schleswig-Holstein, 133, 186, 251
Schmeck, Clemens, 196
Schmidt, Helmut, 258
Schueneman, Jean, 127
Schwarz, Karl, 169
scientific research, 91–92, 96–98, 100–101, 128–30, 146–48, 156–57, 180, 201, 213, 256
Scranton, Pennsylvania, 42
Searle, J. M., 24, 37
Seattle, Washington, 198
Seebohm, Hans-Christoph, 139
Seiler, Heinz, 146
Selby Smelter Commission, 216
Senate (U.S.), 152, 243, 246, 249
Sharon Civic Association (Pennsauken, New Jersey), 229
Siegerland, 176
Sierra Club, 245, 332n132
Silbert, Harold, 199
Silesia, 99
small towns: pollution in, 125, 153–54, 168
smelters, 2, 99, 111, 180, 189–90, 196, 216, 256
smog: alerts 202; disaster in 1952 London, 302n13; episode in 1962 Ruhr area, 178; tax, 245. *See also* photochemical smog
Smoke Abatement League (Cincinnati), 9, 21–22, 25, 27–30, 34, 36, 39, 65, 68–69, 72, 163, 225
Smoke and Dust Abatement League (Pittsburgh), 3, 22, 34–35, 37, 69, 86, 225
smoke clause, 54
smoke ordinances: Germany, 54, 58, 62; U.S., 26, 32–33, 62, 73–79, 125, 159, 161, 198
smoke-preventing devices, 57
Smoke Prevention Association, 25, 28, 39–40, 70, 73–74, 78, 125, 129, 156–58, 262
social costs of pollution, 20–21, 199–200
Social Democratic Party (Sozialdemokratische Partei Deutschlands, SPD), 139, 174, 196–97, 246–48, 253
Society for Fuel Economy and Smoke Abatement (Hamburger Verein für Feuerungsbetrieb und Rauchbekämpfung), 15, 34, 43, 63–64, 87
Society for the Prevention of Smoke (Chicago), 22, 27
Sodingen, 96
Soest, 138, 145
Southeast Air Pollution Committee (Cleveland), 119, 151, 228
Southeast Community Council (Cleveland), 117
Southwest Kiwanis Club (Los Angeles), 209
Speer, Albert, 110
SS (Schutzstaffel, Protective Detachment of the Nazi movement), 111
St. Louis, Missouri, 15, 18, 20, 23–27, 29–32, 34–39, 53, 62, 67, 70–72, 75–83, 85–86, 114, 118, 124, 128, 150, 158, 161, 198, 204, 221, 223, 239, 246, 262
St. Louis Medical Society, 81
staff of control agencies, 69, 113, 125, 157–59, 161, 167, 181, 201
Stamp Out Smog (Los Angeles), 200, 205, 215
Standard Oil, 231, 236
standards, development of, 159–62, 174–79, 218, 220, 234–35, 238, 244, 246
Stanford University, 256
State Agency for Air Pollution Control and Soil Protection (Landesanstalt für Immissions- und Bodennutzungsschutz), 193
State Agency for the Sanitation of Water, Ground, and Air (Landesanstalt für Wasser-, Boden- und Lufthygiene), 92–93, 99, 102, 112, 140, 146–47, 169, 184, 196
State Clean Air Act for North Rhine-Westphalia (Landesimmissionsschutzgesetz), 172, 179, 181, 185, 251, 255
State Clean Air Acts for other German states (Landesimmissionsschutzgesetze), 186, 255
state governments: Germany, 136–37, 139, 248; U.S., 12–14, 127, 150–54, 200, 246, 286n252, 305n73
States' Committee for Pollution Control (Länderausschuß für Immissionsschutz), 177, 193
Steinberg, Theodore, 21
steel industry, 9, 110, 120, 126, 141, 153–54, 164–66, 185, 187, 189, 226, 228, 231–32, 234–36, 239–40. *See also individual company names*
Stern, Arthur, 154, 161–62, 177, 203, 219
Stewart, David, 21, 26
Stockholm, 253
Stokes, Carl, 226
Stolberg, Michael, 283n187
Stradling, David, 278n74, 293n104
Strobel, Käte, 178, 252
Struthers, 153–54
students, 241
Stülpnagel, Adalbert von, 49
Stuttgart, 44, 46–47, 52, 55, 58–59, 62, 64–65
Styrum, 189
subsidiarity principle, 171

suburbanization, 118, 150–52, 167
sulfur dioxide, 2, 6–7, 77–78, 176–77, 185, 201, 205, 235, 245, 256–57, 260
sulfur tax, 245
Sunbelt, 116
symbolic policy, 205, 247–48, 252–58
Syracuse, New York, 118

Tarr, Joel, ix, 83, 95
Taylor, Frederick Winslow, 53
technology of smoke abatement, 25, 32
tetraethyl lead, 211, 225, 253, 331n118
Texaco, 231, 233
Thuringia, 102
Thyssen, 189
Tobin, James, 245
Toledo, Ohio, 157
tourism: in Dresden, 198–99
transatlantic exchange, 58, 65, 117, 146, 183, 197, 253–54, 256–58
Triberg, 89
Tschorn, 45, 53
tuberculosis, 198, 227 *See also* National Tuberculosis Association
Tucker, Raymond, 18, 67, 77–82, 150, 161, 198, 266–67, 314n80
Tufts University, 236
Tunnicliff, Sarah, 23, 36

unions, 119, 126, 134, 136, 226. *See also individual unions by name*
United Automobile, Aerospace, and Agricultural Implement Workers of America, 226
United Smoke Council (Pittsburgh), 119, 131, 225
United States Steel, 126, 164–65, 235, 240
United States Technical Conference on Air Pollution, 129, 146, 216
United Steelworkers of America, 226
University of Cincinnati, 225
University of Illinois, 17, 31
Unna, 94
Upper Silesia, 99

vegetation, damage to, 96, 134, 188, 191, 199, 256–57
Vietnam War, 224, 240
Virginia, 239
viscose, 109
Vogel, David, 16
Völklingen, 141

WaBoLu. *See* State Agency for the Sanitation of Water, Ground, and Air
Waldshut, 89
Waldsterben, 257–58, 264
war economies, 67–69, 109–12, 199–200
war on poverty, 224, 226
Washington, D.C., 85, 129, 247
Washington University, 77
Wayne County, Michigan, 153
Wedelstaedt, Karl von, 188, 193
Wednesday Club (St. Louis), 23–25
Wegener, Hermann, 53
Wellmann, Fritz, 99
Wernigerode, 89
West Virginia, 233
Western Oil and Gas Association, 202
Wetter an der Ruhr, 135
Weyl, Theodor, 57
Wilderness Society, 245
Wilhelmshaven, 257
Windsor, Ontario, 216
Wisconsin, 241, 267
Wolk, Abraham, 84, 161
Woman's City Club (Chicago), 22–24, 36, 39, 75, 86, 276n28
Woman's Civic Forum (Nashville), 85
Woman's Club (Minneapolis), 23–24
Women's Chamber of Commerce (Salt Lake City), 76
Women's City Club (Cleveland), 70–73, 75, 158
Women's Civic League (Baltimore), 22, 24, 36, 71
Women's Club (Cleveland), 3
Women's Health Protective Association of Allegheny County (Pittsburgh), 22, 24
Women's Organization for Smoke Abatement (St. Louis), 23, 27, 34, 36–37
Women's Smoke Abatement League (St. Louis), 72
Wood, Richard, 79
World Bank, 4
World War I, 67–69, 87
World War II, 80–81, 110–12, 138, 173, 199
Würselen, 90
Wurts, Thomas, 85
Württemberg, 45, 52, 102
Wüstenberg, Joachim, 193

Zafonte, Matthew, 213–14
Zero Population Growth, 245